大数据分析：R语言实现（影印版）
Big Data Analytics With R

Simon Walkowiak 著

南京　东南大学出版社

图书在版编目(CIP)数据

大数据分析:R 语言实现:英文/(英)西蒙·沃克威克(Simon Walkowiak)著. —影印本. —南京:东南大学出版社,2017.10(2018.10 重印)

书名原文:Big Data Analytics With R

ISBN 978-7-5641-7361-6

Ⅰ.①大… Ⅱ.①西… Ⅲ.①程序语言-程序设计-英文 Ⅳ.①TP312

中国版本图书馆 CIP 数据核字(2017)第 192628 号

图字:10-2017-115 号

© 2016 by PACKT Publishing Ltd

Reprint of the English Edition, jointly published by PACKT Publishing Ltd and Southeast University Press, 2017. Authorized reprint of the original English edition, 2017 PACKT Publishing Ltd, the owner of all rights to publish and sell the same.

All rights reserved including the rights of reproduction in whole or in part in any form.

英文原版由 PACKT Publishing Ltd 出版 2016。

英文影印版由东南大学出版社出版 2017。此影印版的出版和销售得到出版权和销售权的所有者—— PACKT Publishing Ltd 的许可。

版权所有,未得书面许可,本书的任何部分和全部不得以任何形式重制。

大数据分析:R 语言实现(影印版)

出版发行:东南大学出版社
地　　址:南京四牌楼 2 号　邮编:210096
出 版 人:江建中
网　　址:http://www.seupress.com
电子邮件:press@seupress.com
印　　刷:江苏凤凰数码印务有限公司
开　　本:787 毫米×980 毫米　16 开本
印　　张:31.5
字　　数:617 千字
版　　次:2017 年 10 月第 1 版
印　　次:2018 年 10 月第 2 次印刷
书　　号:ISBN 978-7-5641-7361-6
定　　价:94.00 元

本社图书若有印装质量问题,请直接与营销部联系。电话(传真):025-83791830

Credits

Authors

Simon Walkowiak

Reviewer

Zacharias Voulgaris

Dipanjan Sarkar

Commissioning Editor

Akram Hussain

Acquisition Editor

Sonali Vernekar

Content Development Editor

Onkar Wani

Technical Editor

Sushant S Nadkar

Copy Editor

Safis Editing

Project Coordinator

Ulhas Kambali

Proofreader

Safis Editing

Indexer

Tejal Daruwale Soni

Graphics

Kirk D'Penha

Production Coordinator

Arvindkumar Gupta

About the Author

Simon Walkowiak is a cognitive neuroscientist and a managing director of Mind Project Ltd – a Big Data and Predictive Analytics consultancy based in London, United Kingdom. As a former data curator at the UK Data Service (UKDS, University of Essex) – European largest socio-economic data repository, Simon has an extensive experience in processing and managing large-scale datasets such as censuses, sensor and smart meter data, telecommunication data and well-known governmental and social surveys such as the British Social Attitudes survey, Labour Force surveys, Understanding Society, National Travel survey, and many other socio-economic datasets collected and deposited by Eurostat, World Bank, Office for National Statistics, Department of Transport, NatCen and International Energy Agency, to mention just a few. Simon has delivered numerous data science and R training courses at public institutions and international companies. He has also taught a course in *Big Data Methods in R* at major UK universities and at the prestigious Big Data and Analytics Summer School organized by the Institute of Analytics and Data Science (IADS).

Acknowledgement

The inspiration for writing this book came directly from the brilliant work and dedication of many R developers and users, whom I would like to thank first for creating a vibrant and highly-supportive community that nourishes the progress of publicly accessible data analytics and development of R language. However, this book would never be completed if I wasn't surrounded with love and unconditional support from my partner Ignacio, who always knew how to encourage and motivate me, particularly in moments of my weakness and when I lacked creativity.

I would also like to thank other members of my family, especially my father Peter, who despite not sharing my excitement of data science, always listens patiently to my stories about emerging Big Data technologies and their use cases.

Also, I dedicate this book to my friends and former colleagues from UK Data Service at the University of Essex, where I had an opportunity to work with amazing individuals and experience the best practices in robust data management and processing.

Finally, I highly appreciate the hard work, expertise and feedback offered by many people involved in the creation of this book at Packt Publishing – especially my content development editor Onkar Wani, publishers, and the reviewers, who kindly shared their knowledge with me in order to create a quality and well-received publication.

About the Reviewers

Dr. Zacharias Voulgaris was born in Athens, Greece. He studied Production Engineering and Management at the Technical University of Crete, shifted to Computer Science through a Masters in Information Systems & Technology (City University, London), and then to Data Science through a PhD on Machine Learning (University of London). He has worked at Georgia Tech as a Research Fellow, at an e-marketing startup in Cyprus as an SEO manager, and as a Data Scientist in both Elavon (GA) and G2 (WA). He also was a Program Manager at Microsoft, on a data analytics pipeline for Bing.

Zacharias has authored two books and several scientific articles on Machine Learning and as well as a couple of articles on AI topics. His first book, Data Scientist - The Definitive Guide to Becoming a Data Scientist (Technics Publications), has been translated into Korean and Chinese, while his latest one, Julia for Data Science (Technics Publications) is coming out this September. He has also reviewed a number of data science books (mainly on Python and R) and has a passion for new technologies, literature, and music.

I'd like to thank the people at Packt for inviting me to review this book and for promoting Data Science and particularly Julia through their books. Also, a big thanks to all the great authors out there who choose to publish their work through the lesser-known publishers, keeping the whole process of sharing knowledge a democratic endeavor.

Dipanjan Sarkar is a Data Scientist at Intel, the world's largest silicon company which is on a mission to make the world more connected and productive. He primarily works on analytics, business intelligence, application development and building large scale intelligent systems. He received his Master's degree in Information Technology from the International Institute of Information Technology, Bangalore. His area of specialization includes software engineering, data science, machine learning and text analytics.

Dipanjan's interests include learning about new technology, disruptive start-ups, data science and more recently deep learning. In his spare time he loves reading, writing, gaming and watching popular sitcoms. He has authored a book on Machine Learning titled *R Machine Learning by Example, Packt Publishing* and also acted as a technical reviewer for several books on Machine Learning and Data Science from Packt Publishing.

www.PacktPub.com

eBooks, discount offers, and more

Did you know that Packt offers eBook versions of every book published, with PDF and ePub files available? You can upgrade to the eBook version at `www.PacktPub.com` and as a print book customer, you are entitled to a discount on the eBook copy. Get in touch with us at `customercare@packtpub.com` for more details.

At `www.PacktPub.com`, you can also read a collection of free technical articles, sign up for a range of free newsletters and receive exclusive discounts and offers on Packt books and eBooks.

https://www2.packtpub.com/books/subscription/packtlib

Do you need instant solutions to your IT questions? PacktLib is Packt's online digital book library. Here, you can search, access, and read Packt's entire library of books.

Why subscribe?

- Fully searchable across every book published by Packt
- Copy and paste, print, and bookmark content
- On demand and accessible via a web browser

Table of Contents

Preface	1
Chapter 1: The Era of Big Data	7
Big Data – The monster re-defined	7
Big Data toolbox – dealing with the giant	11
Hadoop – the elephant in the room	12
Databases	15
Hadoop Spark-ed up	16
R – The unsung Big Data hero	17
Summary	24
Chapter 2: Introduction to R Programming Language and Statistical Environment	25
Learning R	25
Revisiting R basics	28
Getting R and RStudio ready	28
Setting the URLs to R repositories	30
R data structures	32
Vectors	32
Scalars	35
Matrices	35
Arrays	37
Data frames	38
Lists	41
Exporting R data objects	42
Applied data science with R	47
Importing data from different formats	48
Exploratory Data Analysis	50
Data aggregations and contingency tables	53
Hypothesis testing and statistical inference	56
Tests of differences	57
Independent t-test example (with power and effect size estimates)	57
ANOVA example	60
Tests of relationships	63
An example of Pearson's r correlations	63
Multiple regression example	65
Data visualization packages	70

Summary	71
Chapter 3: Unleashing the Power of R from Within	**73**
Traditional limitations of R	**74**
Out-of-memory data	74
Processing speed	75
To the memory limits and beyond	**76**
Data transformations and aggregations with the ff and ffbase packages	76
Generalized linear models with the ff and ffbase packages	87
Logistic regression example with ffbase and biglm	89
Expanding memory with the bigmemory package	97
Parallel R	**106**
From bigmemory to faster computations	107
An apply() example with the big.matrix object	108
A for() loop example with the ffdf object	108
Using apply() and for() loop examples on a data.frame	109
A parallel package example	110
A foreach package example	113
The future of parallel processing in R	115
Utilizing Graphics Processing Units with R	115
Multi-threading with Microsoft R Open distribution	117
Parallel machine learning with H2O and R	118
Boosting R performance with the data.table package and other tools	**118**
Fast data import and manipulation with the data.table package	118
Data import with data.table	119
Lightning-fast subsets and aggregations on data.table	120
Chaining, more complex aggregations, and pivot tables with data.table	123
Writing better R code	126
Summary	**127**
Chapter 4: Hadoop and MapReduce Framework for R	**129**
Hadoop architecture	**130**
Hadoop Distributed File System	130
MapReduce framework	131
A simple MapReduce word count example	132
Other Hadoop native tools	134
Learning Hadoop	136
A single-node Hadoop in Cloud	**137**
Deploying Hortonworks Sandbox on Azure	138
A word count example in Hadoop using Java	159
A word count example in Hadoop using the R language	169
RStudio Server on a Linux RedHat/CentOS virtual machine	169

Installing and configuring RHadoop packages	177
HDFS management and MapReduce in R – a word count example	179

HDInsight – a multi-node Hadoop cluster on Azure — 194
Creating your first HDInsight cluster — 194
- Creating a new Resource Group — 195
- Deploying a Virtual Network — 197
- Creating a Network Security Group — 200
- Setting up and configuring an HDInsight cluster — 203
- Starting the cluster and exploring Ambari — 211
- Connecting to the HDInsight cluster and installing RStudio Server — 215
- Adding a new inbound security rule for port 8787 — 218
- Editing the Virtual Network's public IP address for the head node — 221

Smart energy meter readings analysis example – using R on HDInsight cluster — 229
Summary — 241

Chapter 5: R with Relational Database Management Systems (RDBMSs) — 243

Relational Database Management Systems (RDBMSs) — 244
- A short overview of used RDBMSs — 244
- Structured Query Language (SQL) — 245

SQLite with R — 247
- Preparing and importing data into a local SQLite database — 248
- Connecting to SQLite from RStudio — 250

MariaDB with R on a Amazon EC2 instance — 255
- Preparing the EC2 instance and RStudio Server for use — 255
- Preparing MariaDB and data for use — 257
- Working with MariaDB from RStudio — 266

PostgreSQL with R on Amazon RDS — 281
- Launching an Amazon RDS database instance — 281
- Preparing and uploading data to Amazon RDS — 290
- Remotely querying PostgreSQL on Amazon RDS from RStudio — 304

Summary — 314

Chapter 6: R with Non-Relational (NoSQL) Databases — 315

Introduction to NoSQL databases — 315
- Review of leading non-relational databases — 316

MongoDB with R — 319
- Introduction to MongoDB — 319
  - MongoDB data models — 319
- Installing MongoDB with R on Amazon EC2 — 322

Processing Big Data using MongoDB with R	325
Importing data into MongoDB and basic MongoDB commands	326
MongoDB with R using the rmongodb package	333
MongoDB with R using the RMongo package	346
MongoDB with R using the mongolite package	350
HBase with R	**355**
Azure HDInsight with HBase and RStudio Server	355
Importing the data to HDFS and HBase	363
Reading and querying HBase using the rhbase package	367
Summary	**372**
Chapter 7: Faster than Hadoop - Spark with R	**373**
Spark for Big Data analytics	**374**
Spark with R on a multi-node HDInsight cluster	**375**
Launching HDInsight with Spark and R/RStudio	375
Reading the data into HDFS and Hive	383
Getting the data into HDFS	385
Importing data from HDFS to Hive	386
Bay Area Bike Share analysis using SparkR	393
Summary	**411**
Chapter 8: Machine Learning Methods for Big Data in R	**413**
What is machine learning?	**414**
Supervised and unsupervised machine learning methods	415
Classification and clustering algorithms	416
Machine learning methods with R	417
Big Data machine learning tools	418
GLM example with Spark and R on the HDInsight cluster	**419**
Preparing the Spark cluster and reading the data from HDFS	419
Logistic regression in Spark with R	425
Naive Bayes with H2O on Hadoop with R	**437**
Running an H2O instance on Hadoop with R	437
Reading and exploring the data in H2O	441
Naive Bayes on H2O with R	446
Neural Networks with H2O on Hadoop with R	**458**
How do Neural Networks work?	458
Running Deep Learning models on H2O	461
Summary	**469**
Chapter 9: The Future of R - Big, Fast, and Smart Data	**471**
The current state of Big Data analytics with R	**471**

Out-of-memory data on a single machine	471
Faster data processing with R	473
Hadoop with R	475
Spark with R	476
R with databases	477
Machine learning with R	478
The future of R	**478**
Big Data	479
Fast data	480
Smart data	481
Where to go next	**482**
Summary	**482**
Index	**483**

Preface

We live in times of Internet of Things—a large, world-wide network of interconnected devices, sensors, applications, environments, and interfaces. They generate, exchange, and consume massive amounts of data on a daily basis, and the ability to harness these huge quantities of information can provide us with novel understanding of physical and social phenomena.

The recent rapid growth of various open source and proprietary big data technologies allows deep exploration of these vast amounts of data. However, many of them are limited in terms of their statistical and data analytics capabilities. Some others implement techniques and programming languages that many classically educated statisticians and data analysts are simply unfamiliar with and find them difficult to apply in real-world scenarios.

R programming language—an open source, free, extremely versatile statistical environment, has a potential to fill this gap by providing users with a large variety of highly optimized data processing methods, aggregations, statistical tests, and machine learning algorithms with a relatively user-friendly and easily customizable syntax.

This book challenges traditional preconceptions about R as a programming language that does not support big data processing and analytics. Throughout the chapters of this book, you will be exposed to a variety of core R functions and a large array of actively maintained third-party packages that enable R users to benefit from most recent cutting-edge big data technologies and frameworks, such as Hadoop, Spark, H2O, traditional SQL-based databases, such as SQLite, MariaDB, and PostgreSQL, and more flexible NoSQL databases, such as MongoDB or HBase, to mention just a few. By following the exercises and tutorials contained within this book, you will experience firsthand how all these tools can be integrated with R throughout all the stages of the Big Data Product Cycle, from data import and data management to advanced analytics and predictive modeling.

What this book covers

Chapter 1, *The Era of "Big Data"*, gently introduces the concept of Big Data, the growing landscape of large-scale analytics tools, and the origins of R programming language and the statistical environment.

Chapter 2, *Introduction to R Programming Language and Statistical Environment*, explains the most essential data management and processing functions available to R users. This chapter also guides you through various methods of Exploratory Data Analysis and hypothesis testing in R, for instance, correlations, tests of differences, ANOVAs, and Generalized Linear Models.

Chapter 3, *Unleashing the Power of R From Within*, explores possibilities of using R language for large-scale analytics and out-of-memory data on a single machine. It presents a number of third-party packages and core R methods to address traditional limitations of Big Data processing in R.

Chapter 4, *Hadoop and MapReduce Framework for R*, explains how to create a cloud-hosted virtual machine with Hadoop and to integrate its HDFS and MapReduce frameworks with R programming language. In the second part of the chapter, you will be able to carry out a large-scale analysis of electricity meter data on a multinode Hadoop cluster directly from the R console.

Chapter 5, *R with Relational Database Management Systems (RDBMSs)*, guides you through the process of setting up and deploying traditional SQL databases, for example, SQLite, PostgreSQL and MariaDB/MySQL, which can be easily integrated with their current R-based data analytics workflows. The chapter also provides detailed information on how to build and benefit from a highly scalable Amazon Relational Database Service instance and query its records directly from R.

Chapter 6, *R with Non-Relational (NoSQL) Databases*, builds on the skills acquired in the previous chapters and allows you to connect R with two popular nonrelational databases a.) a fast and user-friendly MongoDB installed on a Linux-run virtual machine, and b.) HBase database operated on a Hadoop cluster run as part of the Azure HDInsight service.

Chapter 7, *Faster than Hadoop: Spark with R*, presents a practical example and a detailed explanation of R integration with the Apache Spark framework for faster Big Data manipulation and analysis. Additionally, the chapter shows how to use Hive database as a data source for Spark on a multinode cluster with Hadoop and Spark installed.

Chapter 8, *Machine Learning Methods for Big Data in R*, takes you on a journey through the most cutting-edge predictive analytics available in R. Firstly, you will perform fast and highly optimized Generalized Linear Models using Spark MLlib library on a multinode Spark HDInsight cluster. In the second part of the chapter, you will implement Naïve Bayes and multilayered Neural Network algorithms using R's connectivity with H2O-an award-winning, open source, big data distributed machine learning platform.

Chapter 9, *The Future of R: Big, Fast and Smart Data*, wraps up the contents of the earlier chapters by discussing potential areas of development for R language and its opportunities in the landscape of emerging Big Data tools.

Online Chapter, Pushing R Further, available at https://www.packtpub.com/sites/default/files/downloads/5396_6457OS_PushingRFurther.pdf, enables you to configure and deploy their own scaled-up and Cloud-based virtual machine with fully operational R and RStudio Server installed and ready to use.

What you need for this book

All the code snippets presented in the book have been tested on a Mac OS X (Yosemite) running on a personal computer equipped with 2.3 GHz Intel Core i5 processor, 1 TB Solid State hard drive, and 16 GB of RAM. It is recommended that readers run the scripts on a Mac OS X or Windows machine with at least 4 GB of RAM. In order to benefit from the instructions presented throughout the book, it is advisable that readers install most recent R and RStudio on their machines as well as at least one of the popular web browsers: Mozilla Firefox, Chrome, Safari, or Internet Explorer.

Who this book is for

This book is intended for middle level data analysts, data engineers, statisticians, researchers, and data scientists, who consider and plan to integrate their current or future big data analytics workflows with R programming language.

It is also assumed that readers will have some previous experience in data analysis and the understanding of data management and algorithmic processing of large quantities of data. However, they may lack specific R skills related to particular open source big data tools.

Conventions

In this book, you will find a number of text styles that distinguish between different kinds of information. Here are some examples of these styles and an explanation of their meaning.

Code words in text, database table names, folder names, filenames, file extensions, pathnames, dummy URLs, user input, and Twitter handles are shown as follows: "The -getmerge option allows to merge all data files from a specified directory on HDFS."

Any command-line input or output is written as follows:

```
$ sudo -u hdfs hadoop fs -ls /user
```

New terms and important words are shown in bold. Words that you see on the screen, for example, in menus or dialog boxes, appear in the text like this: "Clicking the **Next** button moves you to the next screen."

 Warnings or important notes appear in a box like this.

 Tips and tricks appear like this.

Reader feedback

Feedback from our readers is always welcome. Let us know what you think about this book—what you liked or disliked. Reader feedback is important for us as it helps us develop titles that you will really get the most out of.

To send us general feedback, simply e-mail `feedback@packtpub.com`, and mention the book's title in the subject of your message.

If there is a topic that you have expertise in and you are interested in either writing or contributing to a book, see our author guide at `www.packtpub.com/authors`.

Customer support

Now that you are the proud owner of a Packt book, we have a number of things to help you to get the most from your purchase.

Downloading the example code

You can download the example code files for this book from your account at `http://www.packtpub.com`. If you purchased this book elsewhere, you can visit `http://www.packtpub.com/support` and register to have the files e-mailed directly to you.

You can download the code files by following these steps:

1. Log in or register to our website using your e-mail address and password.
2. Hover the mouse pointer on the **SUPPORT** tab at the top.
3. Click on **Code Downloads & Errata**.
4. Enter the name of the book in the **Search** box.
5. Select the book for which you're looking to download the code files.
6. Choose from the drop-down menu where you purchased this book from.
7. Click on **Code Download**.

Once the file is downloaded, please make sure that you unzip or extract the folder using the latest version of:

- WinRAR / 7-Zip for Windows
- Zipeg / iZip / UnRarX for Mac
- 7-Zip / PeaZip for Linux

The code bundle for the book is also hosted on GitHub at https://github.com/PacktPublishing/Big-Data-Analytics-with-R. We also have other code bundles from our rich catalog of books and videos available at https://github.com/PacktPublishing/. Check them out!

Errata

Although we have taken every care to ensure the accuracy of our content, mistakes do happen. If you find a mistake in one of our books-maybe a mistake in the text or the code-we would be grateful if you could report this to us. By doing so, you can save other readers from frustration and help us improve subsequent versions of this book. If you find any errata, please report them by visiting http://www.packtpub.com/submit-errata, selecting your book, clicking on the **Errata Submission Form** link, and entering the details of your errata. Once your errata are verified, your submission will be accepted and the errata will be uploaded to our website or added to any list of existing errata under the Errata section of that title.

To view the previously submitted errata, go to https://www.packtpub.com/books/content/support and enter the name of the book in the search field. The required information will appear under the **Errata** section.

Piracy

Piracy of copyrighted material on the Internet is an ongoing problem across all media. At Packt, we take the protection of our copyright and licenses very seriously. If you come across any illegal copies of our works in any form on the Internet, please provide us with the location address or website name immediately so that we can pursue a remedy.

Please contact us at `copyright@packtpub.com` with a link to the suspected pirated material.

We appreciate your help in protecting our authors and our ability to bring you valuable content.

Questions

If you have a problem with any aspect of this book, you can contact us at `questions@packtpub.com`, and we will do our best to address the problem.

1
The Era of Big Data

Big Data – The monster re-defined

Every time Leo Messi scores at Camp Nou in Barcelona, almost one hundred thousand Barca fans cheer in support of their most prolific striker. Social media services such as Twitter, Instagram, and Facebook are instantaneously flooded with comments, views, opinions, analyses, photographs, and videos of yet another wonder goal from the Argentinian goalscorer. One such goal, scored in the semifinal of the UEFA Champions League, against Bayern Munich in May 2015, generated more than 25,000 tweets per minute in the United Kingdom alone, making it the most tweeted sports moment of 2015 in this country. A goal like this creates a widespread excitement, not only among football fans and sports journalists. It is also a powerful driver for the marketing departments of numerous sportswear stores around the globe, who try to predict, with a military precision, day-to-day, in-store, and online sales of Messi's shirts, and other FC Barcelona related memorabilia. At the same time, major TV stations attempt to outbid each other in order to show forthcoming Barca games, and attract multi-million revenues from advertisement slots during the half-time breaks. For a number of industries, this one goal is potentially worth much more than Messi's 20 million Euro annual salary. This one moment also creates an abundance of information, which needs to be somehow collected, stored, transformed, analyzed, and redelivered in the form of yet another product, for example, sports news with a slow-motion replay of Messi's killing strike, additional shirts dispatched to sportswear stores, or a sales spreadsheet and a marketing briefing outlining Barca's TV revenue figures.

Such moments, like memorable Messi's goals against Bayern Munich, happen on a daily basis. Actually, they are probably happening right now, while you are holding this book in front of your eyes. If you want to check what currently makes the world buzz, go to the Twitter web page and click on the **Moments** tab to see the most trending hashtags and topics at this very moment. Each of these less, or more, important events generates vast amounts of data in many different formats, from social media status updates to YouTube videos and blog posts to mention just a few. These data may also be easily linked with other sources of the event-related information to create complex unstructured deposits of data that attempt to explain one specific topic from various perspectives and using different research methods. But here is the first problem: the simplicity of data mining in the era of the World Wide Web means that we can very quickly fill up all the available storage on our hard drives, or run out of processing power and memory resources to crunch the collected data. If you end up having such issues when managing your data, you are probably dealing with something that has been vaguely denoted as **Big Data**.

Big Data is possibly the scariest, deadliest and the most frustrating phrase which can ever be heard by a traditionally trained statistician or a researcher. The initial problem lies in how the concept of Big Data is defined. If you ask ten, randomly selected, students what they understand by the term *Big Data* they will probably give you ten, very different, answers. By default, most will immediately conclude that Big Data has something to do with the size of a data set, the number of rows and columns; depending on their fields they will use similar wording. Indeed they will be somewhat correct, but it's when we inquire about when exactly *normal* data becomes *Big* that the argument kicks off. Some (maybe psychologists?) will try to convince you that even 100 MB is quite a big file or big enough to be scary. Some others (social scientists?) will probably say that 1 GB heavy data would definitely make them anxious. Trainee actuaries, on the other hand, will suggest that 5 GB would be problematic, as even Excel suddenly slows down or doesn't want to open the file. In fact, in many areas of medical science (such as human genome studies) file sizes easily exceed 100 GB each, and most industry data centers deal with data in the region of 2 TB to 10 TB at a time. Leading organizations and multi-billion dollar companies such as Google, Facebook, or YouTube manage petabytes of information on a daily basis. What is then the threshold to qualify data as Big?

The answer is not very straightforward, and the exact number is not set in stone. To give an approximate estimate we first need to differentiate between simply storing the data, and processing or analyzing the data. If your goal was to preserve 1,000 YouTube videos on a hard drive, it most likely wouldn't be a very demanding task. Data storage is relatively inexpensive nowadays, and new rapidly emerging technologies bring its prices down almost as you read this book. It is amazing just to think that only 20 years ago, $300 would merely buy you a 2GB hard drive for your personal computer, but 10 years later the same amount would suffice to purchase a hard drive with a 200 times greater capacity. As of December 2015, having a budget of $300 can easily afford you a 1TB SATA III internal solid-state drive: a fast and reliable hard drive, one of the best of its type currently available to personal users. Obviously, you can go for cheaper and more traditional hard disks in order to store your 1,000 YouTube videos; there is a large selection of available products to suit every budget. It would be a slightly different story, however, if you were tasked to process all those 1,000 videos, for example by creating shorter versions of each or adding subtitles. Even worse if you had to analyze the actual footage of each movie, and quantify, for example, how many seconds per video red colored objects of the size of at least 20×20 pixels are shown. Such tasks do not only require considerable storage capacities, but also, and primarily, the processing power of the computing facilities at your disposal. You could possibly still process and analyze each video, one by one, using a top-of-the-range personal computer, but 1,000 video files would definitely exceed its capabilities and most likely your limits of patience too. In order to speed up the processing of such tasks, you would need to quickly find some extra cash to invest into further hardware upgrades, but then again this would not solve the issue. Currently, personal computers are only **vertically scalable** to a very limited extent. As long as your task does not involve heavy data processing, and is simply restricted to file storage, an individual machine may suffice. However, at this point, apart from large enough hard drives, we would need to make sure we have a sufficient amount of **Random Access Memory (RAM)**, and fast, heavy-duty processors on compatible motherboards installed in our units. Upgrades of individual components, in a single machine, may be costly, short-lived due to rapidly advancing new technologies, and unlikely to bring a real change to complex data crunching tasks. Strictly speaking, this is not the most efficient and flexible approach for Big Data analytics to say the least. A couple of sentence back, I used the plural *units* intentionally, as we would most probably have to process the data on a cluster of machines working in parallel. Without going into details at this stage, the task would require our system to be **horizontally scalable,** meaning that we would be capable of easily increasing (or decreasing) the number of units (nodes) connected in our cluster as we wish. A clear advantage of horizontal scalability over vertical scalability is that we would simply be able to use as many nodes working in parallel as required by our task, and we would not be bothered too much with the individual configuration of each and every machine in our cluster.

Let's go back now for a moment to our students and the question of when normal data becomes *Big*? Amongst the many definitions of Big Data, one is particularly neat and generally applicable to a very wide range of scenarios. One byte more than you are comfortable with is a well-known phrase used by Big Data conference speakers, but I can't deny that it encapsulates the meaning of Big Data very precisely, and yet it is non-specific enough it leaves the freedom to make a subjective decision to each one of us as to what and when to qualify data as Big. In fact, all our students, whether they said Big Data was as little as 100MB or as much as 10 petabytes, were more or less correct in their responses. As long as an individual (and his/her equipment) is not comfortable with a certain size of data, we should assume that this is Big Data for them. The size of data is not, however, the only factor that makes the data Big. Although the simplified definition of Big Data, previously presented, explicitly refers to the *one byte* as a measurement of size, we should dissect the second part of the statement, in a few sentences, to have a greater understanding of what Big Data actually means. Data do not just come to us and *sit* in a file. Nowadays, most data change, sometimes very rapidly. Near real-time analytics of Big Data currently gives huge headaches to in-house data science departments, even at international large financial institutions or energy companies. In fact stock-market data, or sensor data, are pretty good, but still quite extreme examples of high-dimensional data that are stored and analyzed at milliseconds intervals. Several seconds of delay in producing data analyses, on near real-time information, may cost investors quite substantial amounts, and result in losses in their portfolio value, so the speed of processing fast-moving data is definitely a considerable issue at the moment. Moreover, data are now more complex than ever before. Information may be scrapped off the websites as unstructured text, JSON format, HTML files, through service APIs, and so on. Excel spreadsheets and traditional file formats such as **Comma-Separated Values** (CSV) or tab-delimited files that represent structured data are not in the majority any more. It is also very limiting to think of data as of only numeric or textual types. There is an enormous variety of available formats that store, for instance, audio and visual information, graphics, sensors, and signals, 3D rendering and imaging files, or data collected and compiled using highly specialized scientific programs or analytical software packages such as **Stata** or **Statistical Package for the Social Sciences** (**SPSS**) to name just a few (a large list of most available formats is accessible through Wikipedia at `https://en.wikipedia.org/wiki/List_of_file_formats`).

The size of data, the speed of their inputs/outputs and the differing formats and types of data were in fact the original three *Vs: Volume, Velocity*, and *Variety*, described in the article titled *3D Data Management: Controlling Data Volume, Velocity, and Variety* published by Doug Laney back in 2001, as major conditions to treat any data as Big Data. Doug's famous three Vs were further extended by other data scientists to include more specific and sometimes more qualitative factors such as data *variability* (for data with periodic peaks of data flow), *complexity* (for multiple sources of related data), *veracity* (coined by IBM and denoting trustworthiness of data consistency), or *value* (for examples of insight and interpretation). No matter how many Vs or Cs we use to describe Big Data, it generally revolves around the limitations of the available IT infrastructure, the skills of the people dealing with large data sets and the methods applied to collect, store, and process these data. As we have previously concluded that Big Data may be defined differently by different entities (for example individual users, academic departments, governments, large financial companies, or technology leaders), we can now rephrase the previously referenced definition in the following general statement:

> Big Data any data that cause significant processing, management, analytical, and interpretational problems.

Also, for the purpose of this book, we will assume that such problematic data will generally start from around 4 GB to 8 GB in size, the standard capacity of RAM installed in most commercial personal computers available to individual users in the years 2014 and 2015. This arbitrary threshold will make more sense when we explain traditional limitations of the R language later on in this chapter, and methods of Big Data in-memory processing across several chapters in this book.

Big Data toolbox – dealing with the giant

Just like doctors cannot treat all medical symptoms with generic paracetamol and ibuprofen, data scientists need to use more potent methods to store and manage vast amounts of data. Knowing already how Big Data can be defined, and what requirements have to be met in order to qualify data as *Big*, we can now take a step forward and introduce a number of tools that are specialized in dealing with these enormous data sets. Although traditional techniques may still be valid in certain circumstances, Big Data comes with its own ecosystem of scalable frameworks and applications that facilitate the processing and management of unusually large or fast data. In this chapter, we will briefly present several most common Big Data tools, which will be further explored in greater detail later on in the book.

Hadoop – the elephant in the room

If you have been in the Big Data industry for as little as one day, you surely must have heard the unfamiliar sounding word *Hadoop*, at least every third sentence during frequent tea break discussions with your work colleagues or fellow students. Named after Doug Cutting's child's favorite toy, a yellow stuffed elephant, Hadoop has been with us for nearly 11 years. Its origins began around the year 2002 when Doug Cutting was commissioned to lead the **Apache Nutch** project-a scalable open source search engine. Several months into the project, Cutting and his colleague Mike Cafarella (then a graduate student at University of Washington) ran into serious problems with the scaling up and robustness of their Nutch framework owing to growing storage and processing needs. The solution came from none other than Google, and more precisely from a paper titled *The Google File System* authored by Ghemawat, Gobioff, and Leung, and published in the proceedings of the *19th ACM Symposium on Operating Systems Principles*. The article revisited the original idea of **Big Files** invented by Larry Page and Sergey Brin, and proposed a revolutionary new method of storing large files partitioned into fixed-size 64 MB chunks across many nodes of the cluster built from cheap commodity hardware. In order to prevent failures and improve efficiency of this setup, the file system creates copies of chunks of data, and distributes them across a number of nodes, which were in turn mapped and managed by a master server. Several months later, Google surprised Cutting and Cafarella with another groundbreaking research article known as *MapReduce: Simplified Data Processing on Large Clusters*, written by Dean and Ghemawat, and published in the Proceedings of the *6th Conference on Symposium on Operating Systems Design and Implementation*.

The MapReduce framework became a kind of mortar between bricks, in the form of data distributed across numerous nodes in the file system, and the outputs of data transformations and processing tasks.

The MapReduce model contains three essential stages. The first phase is the **Mapping** procedure, which includes indexing and sorting data into the desired structure based on the specified key-value pairs of the mapper (that is, a script doing the mapping). The Shuffle stage is responsible for the redistribution of the mapper's outputs across nodes, depending on the key; that is, the outputs for one specific key are stored on the same node. The Reduce stage results in producing a kind of summary output of the previously mapped and shuffled data, for example, a descriptive statistic such as the arithmetic mean for a continuous measurement by each key (for example a categorical variable). A simplified data processing workflow, using the MapReduce framework in Google and **Distributed File System,** is presented in the following figure:

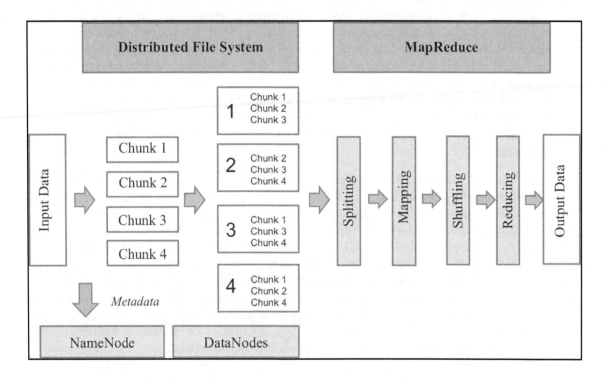

A diagram depicting a simplified Distributed File System architecture and stages of the MapReduce framework

The ideas of the **Google File System** model, and the MapReduce paradigm, resonated very well with Cutting and Cafarella's plans, and they introduced both frameworks into their own research on Nutch. For the first time their web crawler algorithm could be run in parallel on several commodity machines with minimal supervision from a human engineer.

In 2006, Cutting moved to Yahoo! and in 2008, Hadoop became a separate Nutch independent Apache project. Since then, it's been on a never-ending journey towards greater reliability and scalability to allow bigger and faster workloads of data to be effectively crunched by gradually increasing node numbers. In the meantime, Hadoop has also become available as an add-on service on leading cloud computing platforms such as Microsoft Azure, Amazon **Elastic Cloud Computing** (**EC2**), or Google Cloud Platform. This new, unrestricted, and flexible way of accessing shared and affordable commodity hardware, enabled numerous companies as well as individual data scientists and developers to dramatically cut their production costs and process larger than ever data in a more efficient and robust manner. A few Hadoop record-breaking milestones are worth mentioning at this point. In the well-known real-world example at the end of 2007, the New York Times was able to convert more than 4TB of images, within less than 24 hours, using a cluster built of 100 nodes on Amazon EC2, for as little as $200. A job that would have potentially taken weeks of hard labor, and a considerable amount of working man-hours, could now be achieved at a fraction of the original cost, in a significantly shorter scope of time. A year later 1TB of data was already sorted in 209 seconds and in 2009, Yahoo! set a new record by sorting the same amount of data in just 62 seconds. In 2013, Hadoop reached its best **Gray Sort Benchmark** (`http://sortbenchmark.org/`) score so far. Using 2,100 nodes, Thomas Graves from Yahoo! was able to sort 102.5TB in 4,328 seconds, so roughly 1.42TB per minute.

In recent years, the Hadoop and MapReduce frameworks have been extensively used by the largest technology businesses such as Facebook, Google, Yahoo!, major financial and insurance companies, research institutes, and Academia, as well as Big Data individual enthusiasts. A number of enterprises offering commercial distributions of Hadoop such as **Cloudera** (headed by Tom Reilly with Doug Cutting as Chief Architect) or **Hortonworks** (currently led by Rob Bearden-a former senior officer at Oracle and CEO of numerous successful open source projects such as SpringSource and JBoss, and previously run by Eric Baldeschwieler-a former Yahoo! engineer working with Cutting) have spun off from the original Hadoop project, and evolved into separate entities, providing additional Big Data proprietary tools, and extending the applications and usability of Hadoop ecosystem. Although MapReduce and Hadoop revolutionized the way we process vast quantities of data, and hugely propagated Big Data analytics throughout the business world and individual users alike, they also received some criticism for their still present performance limitations, reliability issues, and certain programming difficulties. We will explore these limitations and explain other Hadoop and MapReduce features in much greater detail using practical examples in `Chapter 4`, *Hadoop and MapReduce Framework for R*.

Databases

There are many excellent online and offline resources and publications available to readers, on both SQL-based **Relational Database Management System (RDBMS),** and more modern non-relational and **Not Only SQL (NoSQL)** databases. This book will not attempt to describe each and every one of them with a high level of detail, but in turn it will provide you with several practical examples on how to store large amounts of information in such systems, carry out essential data crunching and processing of the data using known and tested R packages, and extract the outputs of these Big Data transformations from databases directly into your R session.

As mentioned earlier, in `Chapter 5`, *R with Relational Database Management System (RDBMSs)*, we will begin with a very gentle introduction to standard and more traditional databases built on the relational model developed in 1970s by Oxford-educated English computer scientist Edgar Codd, while working at IBM's laboratories in San Jose. Don't worry if you have not got too much experience with databases yet. At this stage, it is only important for you to know that, in the RDBMS, the data is stored in a structured way in the form of tables with fields and records. Depending on the specific industry that you come from, fields can be understood as either variables or columns, and records may alternatively be referred to as observations or rows of data. In other words, fields are the smallest units of information and records are collections of these fields. Fields, like variables, come with certain attributes assigned to them, for example, they may contain only numeric or string values, double or long, and so on. These attributes can be set when inputting the data into the database. The RDBMS have proved extremely popular and most (if not all) currently functioning enterprises that collect and store large quantities of data have some sort of relational databases in place. The RDBMS can be easily queried using the **Structured Query Language (SQL)**-an accessible and quite natural database management programming language, firstly created at IBM by Donald Chamberlin and Raymond Boyce, and later commercialized and further developed by Oracle. Since the original birth of the first RDBMS, they have evolved into fully supported commercial products with Oracle, IBM, and Microsoft being in control of almost 90% of the total RDBMS market share. In our examples of R's connectivity with RDBMS (`Chapter 5`, *R with Relational Database Management System (RDBMSs)*) we will employ a number of the most popular relational and open source databases available to users, including MySQL, PostgreSQL, SQLite, and MariaDB.

However, this is not where we are going to end our journey through the exciting landscape of databases and their Big Data applications. Although RDBMS perform very well in the cases of heavy transactional load, and their ability to process quite complex SQL queries is one of their greatest advantages, they are not so good with (near) real-time and streaming data. Also they generally do not support unstructured and hierarchical data, and they are not easily horizontally scalable. In response to these needs, a new type of database has recently evolved, or to be more precise, it was revived from a long state of inactivity as non-relational databases were known and in use, parallel with RDBMS, even forty years ago, but they never became as popular. NoSQL and non-relational databases, unlike SQL-based RDBMS, come with no predefined schema, thus allowing the users a much needed flexibility without altering the data. They generally scale horizontally very well and are praised for the speed of processing, making them ideal storage solutions for (near) real-time analytics in such industries as retail, marketing, and financial services. They also come with their own flavors of SQL like queries; some of them, for example, the MongoDB NoSQL language, are very expressive and allow users to carry out most data transformations and manipulations as well as complex data aggregation pipelines or even database-specific MapReduce implementations. The rapid growth of interactive web-based services, social media, and streaming data products resulted in a large array of such purpose-specific NoSQL databases being developed. In `Chapter 6`, *R with Non-Relational and (NoSQL) Databases*, we will present several examples of Big Data analytics with R using data stored in three leading open source non-relational databases: a popular document-based NoSQL, MongoDB, and a distributed Apache Hadoop-complaint **HBase database**.

Hadoop Spark-ed up

In the section *Hadoop – the elephant in the room*, we introduced you to the essential basics of Hadoop, its **Hadoop Distributed File System** (**HDFS**) and the MapReduce paradigm. Despite the huge popularity of Hadoop in both academic and industrial settings, many users complain that Hadoop is generally quite slow and some computationally demanding data processing operations can take hours to complete. **Spark**, which makes use of, and is deployed on top of, the existing HDFS, has been designed to excel in iterative calculations in order to fine-tune Hadoop's in-memory performance by up to 100 times faster, and about 10 times faster when run on disk.

Spark comes with its own small but growing ecosystem of additional tools and applications that support large-scale processing of structured data by implementing SQL queries into Spark programs (through **Spark SQL**), enabling fault-tolerant operations on streaming data (**Spark Streaming**), allowing users to perform sophisticated machine-learning models (**MLlib**), and carrying out *out-of-the-box* parallel community detection algorithms such as PageRank, label propagation, and many others on graphs and collections through the GraphX module. Owing to the open source, community-run Apache status of the project, Spark has already attracted an enormous interest from people involved in Big Data analysis and machine learning. As of the end of July 2016, there are over 240 third-party packages developed by independent Spark users available at the `http://spark-packages.org/` website. A large majority of them allow further integration of Spark with other more or less common Big Data tools on the market. Please feel free to visit the page and check which tools or programming languages, that are known to you, are already supported by the packages indexed in the directory.

In `Chapter 7`, *Faster than Hadoop: Spark and R* and `Chapter 8`, *Machine Learning for Big Data in R* we will discuss the methods of utilizing Apache Spark in our Big Data analytics workflows using the R programming language. However before we do so, we need to familiarize ourselves with the most integral part of this book-the R language itself.

R – The unsung Big Data hero

By now you have been introduced to the notion of Big Data, its features, and its characteristics, as well as the most widely used Big Data analytics tools and frameworks such as Hadoop, HDFS, MapReduce framework, relational and non-relational databases, and Apache Spark project. They will be explored more thoroughly in the next few chapters of this book, but now the time has finally come to present the true hero and the main subject of this book-the R language. Although R as a separate language on its own has been with us since the mid 90s, it is derived from a more archaic **S language** developed by John Chambers in the mid 1970s. During Chambers' days in Bell Laboratories, one of the goals of his team was to design a user-friendly, interactive, and quick-in-deployment interface for statistical analysis and data visualizations. As they frequently had to deal with non-standard data analyses, and different data formats, the flexibility of this new tool, and its ability to make the most of the previously used Fortran algorithms, were the highest-order priorities. Also, the project was developed to implement certain graphical capabilities in order to visualize the outputs of numeric computations. The first version was run on a Honeywell operated machine, which wasn't the ideal platform, owing to a number of limitations and impracticalities.

The continuous work on S language progressed quite quickly and by 1981 Chambers and his colleagues were able to release a Unix implementation of the S environment along with a first book-cum-manual titled *S: A Language and System for Data Analysis*. In fact S was a hybrid-an environment or an interface allowing access to Fortran-based statistical algorithms through a set of built-in functions, operating on data with the flexibility of custom programmable syntax, for those users who wished to go beyond default statistical methods and implement more sophisticated computations. This hybrid, quasi object-oriented and functional programming language-like, statistical computing tool created a certain amount of ambiguity and confusion, even amongst its original developers. It is a well-known story that John Chambers, Richard Becker, and other Bell Labs engineers working on S at that time had considerable problems in categorizing their software as either a programming language, a system, or an environment. More importantly, however, S found a fertile ground in Academia and statistical research, which resulted in a sharp increase in external contributions from a growing community of S users. Future versions of S would enhance the functional and object-based structure of the S environment by allowing users to store their data, subsets, outputs of the computations, and even functions and graphs as separate objects. Also, since the third release of S in 1986, the core of the environment has been written in the C language, and the interface with Fortran modules has become accessible dynamically through the evaluator directly from S functions, rather than through a preprocessor and compilation, which was the case in the earlier versions. Of course, the new releases of the modernized and polished S environment were followed by a number of books authored by Chambers, Becker, and now also Allan Wilks, in which they explained the structure of the S syntax and frequently used statistical functions. The S environment laid the foundations for the R programming language, which has been further redesigned in 1990s by Robert Gentleman and Ross Ihaka from the University of Auckland in New Zealand. Despite some differences (a comprehensive list of differences between the S and R languages is available at `https://cran.r-project.org/doc/FAQ/R-FAQ.html#R-and-S`), the code written in the S language runs almost unaltered in the R environment. In fact R and also the S-Plus language are the implementations of S-in its more evolved versions, and it is advisable to treat them as such. Probably the most striking difference between S and R, comes from the fact that R blended in an evaluation model based on the lexical scoping adopted from Steele and Sussman's Scheme programming language. The implementation of lexical scoping in R allowed you to assign free variables, depending on the environment in which a function referencing such variables was created, whereas in S free variables could have only been defined globally. The evaluation model used in R meant that the functions, in comparison with S, were more expressive and written in a more *natural* way giving the developers or analysts more flexibility and greater capabilities by defining variables specific to particular functions within their own environments. The link provided earlier, lists other minor, or very subtle, differences between both languages-many of which are sometimes not very apparent unless they are explicitly referred to in error messages when attempting to run a line of R code.

Currently, R comes in various shapes or forms, as there are several open source and commercial implementations of the language. The most popular, and the ones used throughout this book, are the free R GUI available to download from the **R Project CRAN** website at `https://cran.r-project.org/` and our preferred and also free-of-charge **RStudio IDE** available from `https://www.rstudio.com/`. Owing to their popularity and functionalities, both implementations deserve a few words of introduction.

The most generic R implementation is a free, open source project managed and distributed by the R Foundation for Statistical Computing headquartered in Vienna (Austria) at the Institute for Statistics and Mathematics. It is a not-for-profit organization, founded and set up by the members of the R Development Core Team, which includes a number of well-known academics in statistical research, R experts, and some of the most prolific R contributors over the past several years. Amongst its members, there are still the original *fathers* of the S and R languages such as John Chambers, Robert Gentleman, and Ross Ihaka, as well as the most influential R developers: Dirk Eddelbuettel, Peter Dalgaard, Bettina Grun, and Hadley Wickham to mention just a few. The team is responsible for the supervision and the management of the R core source code, approval of changes and alterations to the source code, and the implementation of community contributions. Through the web pages of the **Comprehensive R Archive Network (CRAN)** (`https://cran.r-project.org/`), the R Development Core Team releases new versions of the R base installation and publishes recent third-party packages created by independent R developers and R community members. They also release research-oriented, open access, refereed periodic R Journals, and organize extremely popular annual useR! conferences, which regularly gather hundreds of passionate R users from around the world.

The R core is the basis for enterprise-ready open source and commercial license products developed by RStudio operating from Boston, MA and led by the Chief Scientist Dr. Hadley Wickham-a creator of numerous crucial data analysis and visualization packages for R, for example, `ggplot2`, `rggobi`, `plyr`, `reshape`, and many more. Their IDE is probably the best and most user-friendly R interface currently available to R users, and if you still don't have it, we recommend that you install it on your personal computer as soon as possible. The RStudio IDE consists of an easy-to-navigate workspace view, with a console window, and an editor, equipped with code highlighting and a direct code execution facility, as well as additional views allowing user-friendly control over plots and visualizations, file management within and outside the working directory, core and third-party packages, history of functions and expressions, management of R objects, code debugging functionalities, and also a direct access to help and support files.

The Era of Big Data

As the R core is a multi-platform tool, RStudio is also available to users of the Windows, Mac, or Linux operating systems. We will be using RStudio desktop edition as well as the open source version of the RStudio Server throughout this book extensively, so please make sure you download and install the desktop free version on your machine in order to be able to follow some of the examples included in Chapter 2, *Introduction to R Programming Language and Statistical Environment* and Chapter 3, *Unleashing the Power of R from Within*. When we get to cloud computing (*Online Chapter, Pushing R Further,* https://www.packtpub.com/sites/default/files/downloads/5396_6457OS_PushingRFurther.pdf) we will explain how to install and use RStudio Server on a Linux run server.

The following are screenshots of graphical user interfaces (on Mac OS X) of both the R base installation, available from CRAN, and the RStudio IDE for desktop. Please note that for Windows users, GUIs may look a little different than the attached examples; however, the functionalities remain the same in most cases. RStudio Server has the same GUI as the desktop version; there are however some very minor differences in available options and settings:

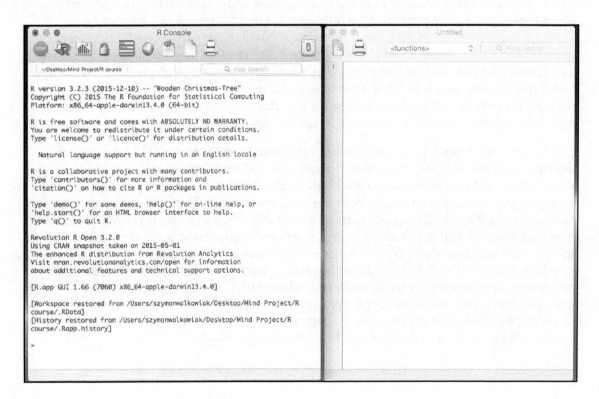

R core GUI for Mac OS X with the console panel (on the left) and the code editor (on the right)

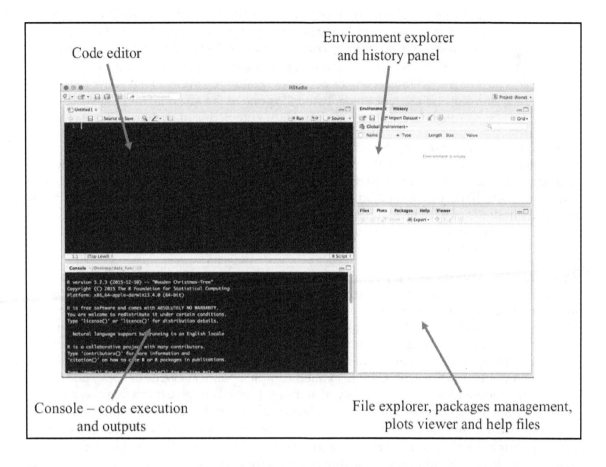

RStudio IDE for Mac OS X with execution console, code editor panels, and other default views. The Cobalt theme was used in the preceding example

After many years of R being used almost exclusively in Academia and research, recent years have witnessed an increased interest in the R language coming from business customers and the financial world. Companies such as Google and Microsoft have turned to R for its high flexibility and simplicity of use. In July 2015, Microsoft completed the acquisition of Revolution Analytics-a Big Data and predictive analytics company that gained its reputation for their own implementations of R with built-in support for Big Data processing and analysis.

The exponential growth in popularity of the R language made it one of the 20 most commonly used programming languages according to **TIOBE Programming Community Index** in years 2014 and 2015 (http://www.tiobe.com/index.php/content/paperinfo/tpci/index.html). Moreover **KDnuggets** (http://www.kdnuggets.com/)-a leading data mining and analytics blog, listed R as one of the essential must-have skills for a career in data science along with knowledge of Python and SQL.

The growth of R does not surprise. Thanks to its vibrant and talented community of enthusiastic users, it is currently the world's most widely-used statistical programming language with nearly 8,400 third-party packages available in the CRAN (as of June 2016) and many more libraries featured on BioConductor, Forge.net, and other repositories, not to mention hundreds of packages under development on GitHub, as well as many other unachieved and accessible only through the personal blogs and websites of individual researchers and organizations.

Apart from the obvious selling points of the R language, such as its open source license, a lack of any setup fees, unlimited access to a comprehensive set of ready-made statistical functions, and a highly active community of users, R is also widely praised for its data visualization capabilities and ability to work with data of many different formats. Popular and very influential newspapers and magazines such as the Washington Post, Guardian, the Economist, or the New York Times, use R on a daily basis to produce highly informative diagrams and info graphics, in order to explain complex political events or social and economical phenomena. The availability of static graphics through extremely powerful packages such as `ggplot2` or `lattice`, has lately been extended even further to offer interactive visualizations using the `shiny` or `ggobi` frameworks, and a number of external packages (for example `rCharts` created and maintained by Ramnath Vaidyanathan) that support JavaScript libraries for spatial analysis, for example `leaflet.js`, or interactive data-driven documents through `morris.js`, `D3.js`, and others. Moreover, R users can now benefit from Google Visualization API charts, such as the famous motion graph, directly through `googleVis` package developed by Markus Gesmann, Diego de Castillo, and Joe Cheng (the following screenshot shows that `googleVis` package in action):

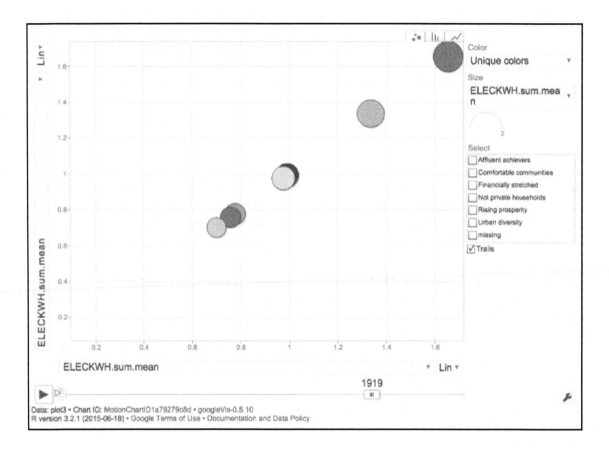

An example of a motion chart created by the Google Visualization package-an R interface to Google charts

As mentioned earlier, R works with data coming from a large array of different sources and formats. This is not just limited to *physical* file formats such as traditional comma-separated or tab-delimited formats, but it also includes less common files such as JSON (a widely used format in web applications and modern NoSQL databases and that we will explore in detail in Chapter 6, *R with Non-Relational and (NoSQL) Databases*) or images, other statistical packages and proprietary formats such as Excel workbooks, Stata, SAS, SPSS, and Minitab files, *scrapping* data from the Internet, or direct access to SQL or NoSQL databases as well as other Big Data containers (such as Amazon S3), or files stored in the HDFS. We will explore most of the data import capabilities of R in a number of real-world examples throughout this book. However if you would like to get a feel for a variety of data import methods in R right now, please visit the CRAN page at https://cran.r-project.org/manuals.html, which lists a set of the most essential manuals including the *R Data Import/Export* document outlining the import and export methods of data to and from R.

Finally, the code in the R language can easily be called from other programming platforms such as Python or **Julia**. Moreover, R itself is able to implement functions and statements from other programming languages such as the C family, Java, SQL, Python, and many others. This allows for greater veracity and helps to integrate the R language into existing data science workflows. We will explore many examples of such implementations throughout this book.

Summary

In the first chapter we explained the ambiguity of Big Data definitions and highlighted its major features. We also talked about a deluge of Big Data sources, and mentioned that even one event, such as Messi's goal, can lead to an avalanche of large amounts of data being created almost instantaneously.

You were then introduced to some most commonly used Big Data tools we will be working with later, such as Hadoop, its Distributed File System and the parallel MapReduce framework, traditional SQL and NoSQL databases, and the Apache Spark project, which allows faster (and in many cases easier) data processing than in Hadoop.

We ended the chapter by presenting the origins of the R programming language, its gradual evolution into the most widely-used statistical computing environment, and the current position of R amongst a spectrum of Big Data analytics tools.

In the next chapter you will finally have a chance to get your hands dirty and learn, or revise, a number of frequently used functions in R for data management, transformations, and analysis.

2
Introduction to R Programming Language and Statistical Environment

In Chapter 1, *The era of "Big Data"*, you have become familiar with the most useful Big Data terminology, and a small selection of typical tools applied to unusually large or complex data sets. You have also gained essential insights into how R was developed and how it became the leading statistical computing environment and programming language favored by technology giants and the best universities in the world. In this chapter you will have the opportunity to learn some most important R functions from base R installation and well-known third party packages used for data crunching, transformation, and analysis. More specifically in this chapter you will:

- Understand the landscape of available R data structures
- Be guided through a number of R operations allowing you to import data from standard and proprietary data formats
- Carry out essential data cleaning and processing activities such as subsetting, aggregating, creating contingency tables, and so on
- Inspect the data by implementing a selection of Exploratory Data Analysis techniques such as descriptive statistics
- Apply basic statistical methods to estimate correlation parameters between two (Pearson's r) or more variables (multiple regressions) or find the differences between means for two (t-tests) or more groups **Analysis of Variance (ANOVA)**
- Be introduced to more advanced data modeling tasks like logistic and Poisson regressions

Learning R

This book assumes that you have been previously exposed to R programming language, and this chapter will serve more as a revision, and an overview, of the most essential operations, rather than a very thorough handbook on R. The goal of this work is to present you with specific R applications related to Big Data and the way you can combine R with your existing Big Data analytics workflows instead of teaching you basics of data processing in R. There is a substantial number of great introductory and beginner-level books on R available at IT specialized bookstores or online, directly from Packt Publishing, and other respected publishers, as well as on the Amazon store. Some recommendations include the following:

- *R in Action: Data Analysis and Graphics with R* by Robert Kabacoff (2015), 2nd edition, Manning Publications
- *R Cookbook* by Paul Teetor (2011), O'Reilly
- *Discovering Statistics Using R* by Andy Field, Jeremy Miles, and Zoe Field (2012), SAGE Publications
- *R for Data Science* by Dan Toomey (2014), Packt Publishing

An alternative route to the acquisition of good practical R skills is through a large number of online resources, or more traditional tutor-led in-class training courses. The first option offers you an almost limitless choice of websites, blogs, and online guides. A good starting point is the main and previously mentioned Comprehensive R Archive Network (CRAN) page (`https://cran.r-project.org/`), which, apart from the R core software, contains several well-maintained manuals and **Task Views**-community run indexes of R packages dealing with specific statistical or data management issues. R-bloggers on the other hand (`http://www.r-bloggers.com/`) deliver regular news on R in the form of R-related blog posts or tutorials prepared by R enthusiasts and data scientists. Other interesting online sources, which you will probably find yourself using quite often, are as follows:

- `http://www.inside-r.org/`-news and information from and by the R community
- `http://www.rdocumentation.org/`-a useful search engine of R packages and functions
- `http://blog.rstudio.org/`-a blog run and edited by RStudio engineers
- `http://www.statmethods.net/`-a very informative tutorial-laden website based on the popular R book *R in Action* by Rob Kabacoff

However, it is very likely that after some initial reading, and several months of playing with R, your most frequent destinations for further R-related information and help on more complex use cases for specific functions will become StackOverflow(http://stackoverflow.com/) and, even better, StackExchange (http://stackexchange.com/). StackExchange is in fact a network of support and question-and-answer community-run websites, which address many problems related to statistical, mathematical, biological, and other methods or concepts, whereas StackOverflow, which is currently one of the sub-sites under the StackExchange label, focuses more on applied programming issues and provides users with coding hints and solutions in most (if not all) programming languages known to developers. Both tend to be very popular amongst R users, and as of late December 2015, there were almost 120,000 R-tagged questions asked on StackOverflow. The http://stackoverflow.com/tags/r/info page also contains numerous links and further references to free interactive R learning resources, online books and manuals, and many others.

Another good idea is to start your R adventure from user-friendly online training courses available through online-learning providers like Coursera (https://www.coursera.org), DataCamp (https://www.datacamp.com), edX (https://www.edx.org), or CodeSchool (https://www.codeschool.com). Of course, owing to the nature of such courses, a successful acquisition of R skills is somewhat subjective, however, in recent years, they have grown in popularity enormously, and they have also gained rather positive reviews from employers and recruiters alike. Online courses may then be very suitable, especially for those who, for various reasons, cannot attend a traditional university degree with R components, or just prefer to learn R at their own leisure or around their working hours.

Before we move on to the practical part, whichever strategy you are going to use to learn R, please do not be discouraged by initial difficulties. R, like any other programming language, or should I say, like any other language (including foreign languages), needs time, patience, long hours of practice, and a large number of varied exercises to let you explore many different dimensions and complexities of its syntax and rich libraries of functions. If you are still struggling with your R skills, however, I am sure the next section will get them off the ground.

Revisiting R basics

In the following section we will present a short revision of the most useful and frequently applied R functions and statements. We will start from a quick R and RStudio installation guide and then proceed to creating R data structures, data manipulation, and transformation techniques, and basic methods used in **Exploratory Data Analysis** (**EDA**). Although the R codes listed in this book have been tested extensively, as always in such cases, please make sure that your equipment is not faulty and note that you will be running all the following scripts at your own risk.

Getting R and RStudio ready

Depending on your operating system (whether Mac OS X, Windows, or Linux) you can download and install specific base R files directly from https://cran.r-project.org/. If you prefer to use RStudio IDE you still need to install the R core available from CRAN website first and then download and run installers of the most recent version of RStudio IDE specific for your platform from https://www.rstudio.com/products/rstudio/download/.

> Personally I prefer to use RStudio, owing to its practical add-ons such as code highlighting and more user-friendly GUI; however, there is no particular reason why you can't use just the simple R core installation if you want to. Having said that, in this book we will be using RStudio in most of the examples.

All code snippets have been executed and run on a MacBook Pro laptop with Mac OS X (Yosemite) operating system, 2.3 GHz Intel Core i5 processor, 1 TB solid-state hard drive and 16 GB of RAM memory, but you should also be fine with a much weaker configuration. In this chapter we won't be using large data, and even in the remaining parts of this book the data sets used are limited to approximately 100 MB to 130 MB in size each. You are also provided with links and references to full Big Data whenever possible.

> If you would like to follow the practical parts of this book you are advised to download and unzip the R code and data for each chapter from the web page created for this book by Packt Publishing. If you use this book in PDF format it is not advisable to copy the code and paste it into the R console. When printed, some characters (like quotation marks " ") may be encoded differently than in R and the execution of such commands may result in errors being returned by the R console.

Once you have downloaded both R core and RStudio installation files, follow the on-screen instructions for each installer. When you have finished installing them, open your RStudio software. Upon initialization of the RStudio you should see its GUI with a number of windows distributed on the screen. The largest one is the console in which you input and execute the code, line by line. You can also invoke the editor panel (it is recommended) by clicking on the white empty file icon in the top left corner of the RStudio software or alternatively by navigating to **File | New File | R Script**. If you have downloaded the R code from the book page of the Packt Publishing website, you may also just click on the **Open an existing file (Ctrl + O)** (a yellow open folder icon) and locate the downloaded R code on your computer's hard drive (or navigate to **File | Open File...**).

Now your RStudio session is open and we can adjust some most essential settings. First, you need to set your *working directory* to the location on your hard drive where your data files are. If you know the specific location you can just type the `setwd()` command with a full and exact path to the location of your data as follows:

```
> setwd("/Users/simonwalkowiak/Desktop/data")
```

Of course your actual path will differ from mine, shown in the preceding code, however please mind that if you copy the path from the Windows Explorer address bar you will need to change the backslashes \ to forward slashes / (or to double backslashes \\). Also, the path needs to be kept within the quotation marks "...". Alternatively you can set your working directory by navigating to **Session | Set Working Directory | Choose Directory...** to manually select the folder in which you store the data for this session.

Apart from the ones we have already described, there are other ways to set your working directory correctly. In fact most of the operations, and even more complex data analysis and processing activities, can be achieved in R in numerous ways. For obvious reasons, we won't be presenting all of them, but we will just focus on the frequently used methods and some tips and hints applicable to special or difficult scenarios.

You can check whether your working directory has been set correctly by invoking the following line:

```
> getwd()
[1] "/Users/simonwalkowiak/Desktop/data"
```

From what you can see, the `getwd()` function returned the correct destination for my previously defined working directory.

Setting the URLs to R repositories

It is always good practice to check whether your R repositories are set correctly. R repositories are servers located at various institutes and organizations around the world, which store recent updates and new versions of third-party R packages. It is recommended that you set the URL of your default repository to the CRAN server and choose a mirror that is located relatively close to you. To set the repositories you may use the following code:

```
> setRepositories(addURLs = c(CRAN = "https://cran.r-project.org/"))
```

You can check your current, or default, repository URLs by invoking the following function:

```
> getOption("repos")
```

The output will confirm your URL selection:

```
                    CRAN
"https://cran.r-project.org/"
```

You will be able to choose specific mirrors when you install a new package for the first time during the session, or you may navigate to **Tools | Global Options… | Packages**. In the **Package management** section of the window you can alter the default CRAN mirror location; click on the **Change…** button to adjust.

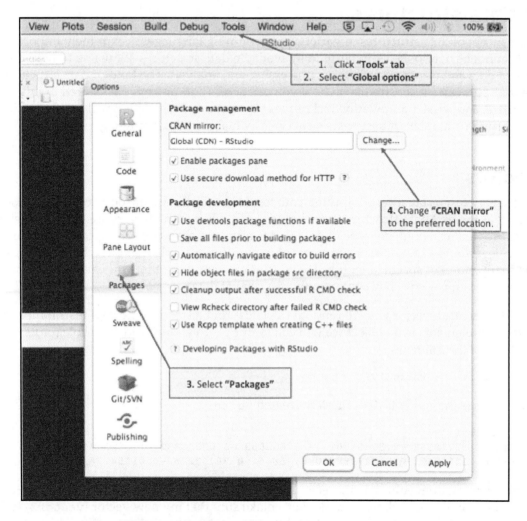

Once your repository URLs and working directory are set, you can go on to create data structures that are typical for R programming language.

R data structures

The concept of data structures in various programming languages is extremely important and cannot be overlooked. Similarly in R, available data structures allow you to hold any type of data and use them for further processing and analysis. The kind of data structure you use puts certain constraints on how you can access and process data stored in this structure, and what manipulation techniques you can use. This section will briefly guide you through a number of basic data structures available in the R language.

Vectors

Whenever I teach statistical computing courses, I always start by introducing R learners to vectors as the first data structure they should get familiar with. Vectors are one-dimensional structures that can hold any type of data that is numeric, character, or logical. In simple terms, a vector is a sequence of some sort of values (for example numeric, character, logical, and many more) of specified length. The most important thing that you need to remember is that an atomic vector may contain only one type of data.

Let's then create a vector with 10 random deviates from a standard normal distribution, and store all its elements in an object we will call `vector1`. In your RStudio console (or its editor) type the following:

```
> vector1 <- rnorm(10)
```

Let's now see the contents of our newly created `vector1`:

```
> vector1
 [1] -0.37758383 -2.30857701  2.97803059 -0.03848892  1.38250714
 [6]  0.13337065 -0.51647388 -0.81756661  0.75457226 -0.01954176
```

As we drew random values, your vector most likely contains different elements from the ones shown in the preceding example. Let's make sure that my new vector (`vector2`) is the same as yours. In order to do this we need to set a seed from which we will be drawing the values:

```
> set.seed(123)
> vector2 <- rnorm(10, mean=3, sd=2)
> vector2
 [1] 1.8790487 2.5396450 6.1174166 3.1410168 3.2585755 6.4301300
 [7] 3.9218324 0.4698775 1.6262943 2.1086761
```

In the preceding code we've set the seed to an arbitrary number (`123`) in order to allow you to replicate the values of elements stored in `vector2` and we've also used some optional parameters of the `rnorm()` function, which enabled us to specify two characteristics of our data, that is the arithmetic mean (set to 3) and standard deviation (set to 2). If you wish to inspect all available arguments of the `rnorm()` function, its default settings, and examples of how to use it in practice, type `?rnorm` to view help and information on that specific function.

However, probably the most common way in which you will be creating a vector of data is by using the `c()` function (c stands for concatenate) and then explicitly passing the values of each element of the vector:

```
> vector3 <- c(6, 8, 7, 1, 2, 3, 9, 6, 7, 6)
> vector3
[1] 6 8 7 1 2 3 9 6 7 6
```

In the preceding example we've created `vector3` with 10 numeric elements. You can use the `length()` function of any data structure to inspect the number of elements:

```
> length(vector3)
[1] 10
```

The `class()` and `mode()` functions allow you to determine how to handle the elements of `vector3` and how the data are stored in `vector3` respectively.

```
> class(vector3)
[1] "numeric"
> mode(vector3)
[1] "numeric"
```

The subtle difference between both functions becomes clearer if we create a vector that holds levels of categorical variables (known as a *factor* in R) with character values:

```
> vector4 <- c("poor", "good", "good", "average", "average", "good",
"poor", "good", "average", "good")
> vector4
 [1] "poor"    "good"    "good"    "average" "average" "good"    "poor"
 [8] "good"    "average" "good"
> class(vector4)
[1] "character"
> mode(vector4)
[1] "character"
> levels(vector4)
NULL
```

In the preceding example, both the `class()` and `mode()` outputs of our character vector are the same, as we still haven't set it to be treated as a categorical variable, and we haven't defined its levels (the content of the `levels()` function is empty-NULL). In the following code we will explicitly set the vector to be recognized as categorical with three levels:

```
> vector4 <- factor(vector4, levels = c("poor", "average", "good"))
> vector4
[1] poor    good    good    average average good    poor    good
[8] average good
Levels: poor average good
```

The sequence of levels doesn't imply that our vector is ordered. We can order the levels of factors in R using the `ordered()` command. For example, you may want to arrange the levels of `vector4` in reverse order, starting from `"good"`:

```
> vector4.ord <- ordered(vector4, levels = c("good", "average", "poor"))
> vector4.ord
[1] poor    good    good    average average good    poor    good
[8] average good
Levels: good < average < poor
```

You can see from the output that R has now properly recognized the order of our levels, which we had defined. We can now apply `class()` and `mode()` functions on the `vector4.ord` object:

```
> class(vector4.ord)
[1] "ordered" "factor"
> mode(vector4.ord)
[1] "numeric"
```

You may very likely be wondering why the `mode()` function returned the `"numeric"` type instead of `"character"`. The answer is simple. By setting the levels of our factor, R has assigned values 1, 2, and 3 to `"good"`, `"average"`, and `"poor"` respectively, exactly in the same order as we had defined them in the `ordered()` function. You can check this using the `levels()` and `str()` functions:

```
> levels(vector4.ord)
[1] "good"    "average" "poor"
> str(vector4.ord)
 Ord.factor w/ 3 levels "good"<"average"<..: 3 1 1 2 2 1 3 1 2 1
```

Just to finalize the subject of vectors, let's create a logical vector, that contains only TRUE and FALSE values:

```
> vector5 <- c(TRUE, FALSE, TRUE, FALSE, FALSE, FALSE, TRUE, FALSE, FALSE,
FALSE)
> vector5
 [1]  TRUE FALSE  TRUE FALSE FALSE FALSE  TRUE FALSE FALSE FALSE
```

Similarly, for all other vectors already presented, feel free to check their structure, class, mode, and length using appropriate functions shown in this section. What outputs did those commands return?

Scalars

The reason why I always start from vectors is that scalars just seem trivial when they follow vectors. To simplify things even more, think of scalars as one-element vectors that are traditionally used to hold some constant values, for example:

```
> a1 <- 5
> a1
[1] 5
```

Of course you may use scalars in computations and also assign any one-element outputs of mathematical or statistical operations to another, arbitrary named scalar for example:

```
> a2 <- 4
> a3 <- a1 + a2
> a3
[1] 9
```

In order to complete this short subsection on scalars, create two separate scalars that will hold a character and a logical value.

Matrices

A matrix is a two-dimensional R data structure in which each of its elements must be of the same type; that is numeric, character, or logical. As matrices consist of rows and columns, their *shape* resembles tables. In fact, when creating a matrix, you can specify how you want to distribute values across its rows and columns, for example:

```
> y <- matrix(1:20, nrow=5, ncol=4)
> y
     [,1] [,2] [,3] [,4]
[1,]    1    6   11   16
```

```
[2,]    2    7   12   17
[3,]    3    8   13   18
[4,]    4    9   14   19
[5,]    5   10   15   20
```

In the preceding example we have allocated a sequence of 20 values (from 1 to 20) into five rows and four columns, and by default they have been distributed by column. We may now create another matrix in which we will distribute the values by rows and give names to rows and columns using the dimnames argument (dimnames stands for names of dimensions) in the matrix() function:

```
> rows <- c("R1", "R2", "R3", "R4", "R5")
> columns <- c("C1", "C2", "C3", "C4")
> z <- matrix(1:20, nrow=5, ncol=4, byrow=TRUE, dimnames=list(rows,
columns))
> z
   C1 C2 C3 C4
R1  1  2  3  4
R2  5  6  7  8
R3  9 10 11 12
R4 13 14 15 16
R5 17 18 19 20
```

As we are talking about matrices it's hard not to mention anything about how to extract specific elements stored in a matrix. This skill will actually turn out to be very useful when we get to subsetting *real* data sets. Looking at the matrix y, for which we didn't define any names for its rows and columns, notice how R denotes them. The row numbers come in the format [r,], where r is a consecutive number of a row, whereas columns are identified by [,c], where c is a consecutive number of a column. If you then wished to extract a value stored in the fourth row of the second column of our matrix y, you could use the following code to do so:

```
> y[4,2]
[1] 9
```

In case you wanted to extract the whole column number three from our matrix y, you could type the following:

```
> y[,3]
[1] 11 12 13 14 15
```

As you can see, we don't even need to allow an empty space before the comma in order for this short script to work. Let's now imagine you would like to extract three values stored in the second, third, and fifth rows of the first column in our vector z with named rows and columns. In this case, though you may still want to use the previously shown notation, you do not need to refer explicitly to the names of dimensions of our matrix z. Additionally, notice that for several values to extract we have to specify their row locations as a vector; hence we will put their row coordinates inside the c() function that we had previously used to create vectors:

```
> z[c(2, 3, 5), 1]
R2 R3 R5
 5  9 17
```

Similar rules for extracting data will apply to other data structures in R such as arrays, lists, and data frames, which we are going to present next.

Arrays

Arrays are very similar to matrices with only one exception: they contain more dimensions. However, just like matrices or vectors, they may only hold one type of data. In the R language, arrays are created using the array() function:

```
> array1 <- array(1:20, dim=c(2,2,5))
> array1
, , 1
     [,1] [,2]
[1,]    1    3
[2,]    2    4
, , 2
     [,1] [,2]
[1,]    5    7
[2,]    6    8
, , 3
     [,1] [,2]
[1,]    9   11
[2,]   10   12
, , 4
     [,1] [,2]
[1,]   13   15
[2,]   14   16
, , 5
     [,1] [,2]
[1,]   17   19
[2,]   18   20
```

The `dim` argument, which was used within the `array()` function, specifies how many dimensions you want to distribute your data across. As we had 20 values (from 1 to 20) we had to make sure that our array can hold all 20 elements; therefore we decided to assign them into two rows, two columns, and five dimensions (2 x 2 x 5 = 20). You can check the dimensionality of your multi-dimensional R objects with the `dim()` command:

```
> dim(array1)
[1] 2 2 5
```

As with matrices, you can use standard rules for extracting specific elements from your arrays. The only difference is that now you have additional dimensions to take care of. Let's assume you want to extract a specific value located in the second row of the first column in the fourth dimension of our `array1`:

```
> array1[2, 1, 4]
[1] 14
```

Also, if you need to find a location of a specific value, for example 11, within the array, you can simply type the following line:

```
> which(array1==11, arr.ind=TRUE)
     dim1 dim2 dim3
[1,]    1    2    3
```

Here, the `which()` function returns indices of the array (`arr.ind=TRUE`), where the sought value equals `11` (hence `==`). As we had only one instance of value `11` in our array, there is only one row specifying its location in the output. If we had more instances of `11`, additional rows would be returned indicating indices for each element equal to `11`.

Data frames

The following two short subsections concern two of probably the most widely used R data structures. Data frames are very similar to matrices, but they may contain different types of data. Here you might have suddenly thought of a typical *rectangular* data set with rows and columns or observations and variables. In fact you are correct. Most data sets are indeed imported into R as data frames. You can also create a simple data frame manually with the `data.frame()` function, but as each column in the data frame may be of a different type, we must first create vectors that will hold data for specific columns:

```
> subjectID <- c(1:10)
> age <- c(37,23,42,25,22,25,48,19,22,38)
> gender <- c("male", "male", "male", "male", "male", "female", "female", "female", "female", "female")
> lifesat <- c(9,7,8,10,4,10,8,7,8,9)
```

```
> health <- c("good", "average", "average", "good", "poor", "average",
"good", "poor", "average", "good")
> paid <- c(T, F, F, T, T, T, F, F, F, T)
> dataset <- data.frame(subjectID, age, gender, lifesat, health, paid)
> dataset
   subjectID age gender lifesat health  paid
1          1  37   male       9    good  TRUE
2          2  23   male       7 average FALSE
3          3  42   male       8 average FALSE
4          4  25   male      10    good  TRUE
5          5  22   male       4    poor  TRUE
6          6  25 female      10 average  TRUE
7          7  48 female       8    good FALSE
8          8  19 female       7    poor FALSE
9          9  22 female       8 average FALSE
10        10  38 female       9    good  TRUE
```

The preceding example presents a simple data frame which contains some dummy data, possibly a sample from a basic psychological experiment, which measured the subjects' life satisfaction (`lifesat`) and their health status (`health`) and also collected other socio-demographic information such as `age` and `gender`, and whether the participant was a paid subject or a volunteer. As we deal with various types of data, the elements for each column had to be amalgamated into a single structure of a data frame using the `data.frame()` command, and specifying the names of objects (vectors) in which we stored all values. You can inspect the structure of this data frame with the previously mentioned `str()` function:

```
> str(dataset)
'data.frame':   10 obs. of  6 variables:
 $ subjectID: int  1 2 3 4 5 6 7 8 9 10
 $ age      : num  37 23 42 25 22 25 48 19 22 38
 $ gender   : Factor w/ 2 levels "female","male": 2 2 2 2 1 1 1 1 1
 $ lifesat  : num  9 7 8 10 4 10 8 7 8 9
 $ health   : Factor w/ 3 levels "average","good",..: 2 1 1 2 3 1 2 3 1 2
 $ paid     : logi  TRUE FALSE FALSE TRUE TRUE TRUE ...
```

The output of `str()` gives you some basic insights into the shape and format of your data in the `dataset` object, for example, the number of observations and variables, the names of variables, types of data they hold, and examples of values for each variable.

Introduction to R Programming Language and Statistical Environment

While discussing data frames, it may also be useful to introduce you to another way of creating subsets. As presented earlier, you may apply standard extraction rules to subset data of your interest. For example, suppose you want to print only those columns which contain age, gender, and life satisfaction information from our dataset data frame. You may use the following two alternatives (the output is not shown to save space, but feel free to run it):

```
> dataset[,2:4] #or
> dataset[, c("age", "gender", "lifesat")]
```

Both lines of code will produce exactly the same results. The subset() function however gives you the additional capabilities of defining conditional statements that will filter the data, based on the output of logical operators. You can replicate the preceding output using subset() in the following way:

```
> subset(dataset[c("age", "gender", "lifesat")])
```

Assume now that you want to create a subset with all subjects who are over 30 years old, and with a score of greater than or equal to eight on the life satisfaction scale (lifesat). The subset() function comes in very handy:

```
> subset(dataset, age > 30 & lifesat >= 8)
   subjectID age gender lifesat  health  paid
1          1  37   male       9    good  TRUE
3          3  42   male       8 average FALSE
7          7  48 female       8    good FALSE
10        10  38 female       9    good  TRUE
```

Or you want to produce an output with two socio-demographic variables of age and gender, relating only to those subjects who were paid to participate in this experiment:

```
> subset(dataset, paid==TRUE, select=c("age", "gender"))
   age gender
1   37   male
4   25   male
5   22   male
6   25 female
10  38 female
```

We will perform much more thorough and complex data transformations on real data frames in the second part of this chapter.

Lists

A list in R is a data structure, which is a collection of other objects. For example, in the list you can store vectors, scalars, matrices, arrays, data frames, and even other lists. In fact, lists in R are vectors, but they differ from atomic vectors, which we introduced earlier in this section, as lists can hold many different types of data. In the following example, we will construct a simple list (using the `list()` function) that will include a variety of other data structures:

```
> simple.vector1 <- c(1, 29, 21, 3, 4, 55)
> simple.matrix <- matrix(1:24, nrow=4, ncol=6, byrow=TRUE)
> simple.scalar1 <- 5
> simple.scalar2 <- "The List"
> simple.vector2 <- c("easy", "moderate", "difficult")
> simple.list <- list(name=simple.scalar2, matrix=simple.matrix,
vector=simple.vector1, scalar=simple.scalar1, difficulty=simple.vector2)
>simple.list
$name
[1] "The List"
$matrix
     [,1] [,2] [,3] [,4] [,5] [,6]
[1,]    1    2    3    4    5    6
[2,]    7    8    9   10   11   12
[3,]   13   14   15   16   17   18
[4,]   19   20   21   22   23   24
$vector
[1]  1 29 21  3  4 55
$scalar
[1] 5
$difficulty
[1] "easy"      "moderate"   "difficult"
> str(simple.list)
List of 5
 $ name      : chr "The List"
 $ matrix    : int [1:4, 1:6] 1 7 13 19 2 8 14 20 3 9 ...
 $ vector    : num [1:6] 1 29 21 3 4 55
 $ scalar    : num 5
 $ difficulty: chr [1:3] "easy" "moderate" "difficult"
```

Looking at the preceding output, you can see that we have assigned names to each component in our list and the `str()` function prints them as if they were variables of a standard rectangular data set.

In order to extract specific elements from a list, you first need to use a double square bracket notation [[x]] to identify a component x within the list. For example, assuming you want to print an element stored in its first row and the third column of the second component, you may use the following line in R:

```
> simple.list[[2]][1,3]
[1] 3
```

Owing to their flexibility, lists are commonly used as preferred data structures in the outputs of statistical functions. It is then important for you to know how you can deal with lists and what sort of methods you can apply to extract and process data stored in them.

Once you are familiar with the basic features of data structures available in R, you may wish to visit Hadley Wickham's online book at http://adv-r.had.co.nz/ in which he explains various more advanced concepts related to each native data structure in R language, and different techniques of subsetting data, depending on the way they are stored.

Exporting R data objects

In the previous section we created numerous objects, which you can inspect in the **Environment** tab window in RStudio. Alternatively, you may use the ls() function to list all objects stored in your global environment:

```
> ls()
```

If you've worked through the chapter, and run the script for this book line-by-line, the output of the ls() function should hopefully return 27 objects:

```
 [1] "a1"              "a2"               "a3"
 [4] "age"             "array1"           "columns"
 [7] "dataset"         "gender"           "health"
[10] "lifesat"         "paid"             "rows"
[13] "simple.list"     "simple.matrix"    "simple.scalar1"
[16] "simple.scalar2"  "simple.vector1"   "simple.vector2"
[19] "subjectID"       "vector1"          "vector2"
[22] "vector3"         "vector4"          "vector4.ord"
[25] "vector5"         "y"                "z"
```

In this section, we will present various methods of saving the created objects to your local drive and exporting their contents to a number of the most commonly used file formats.

Sometimes, for various reasons, it may happen that you need to leave your project and exit RStudio or shut your PC down. If you do not save your created objects, you will lose all of them, the moment you close RStudio. Remember that R stores created data objects in the RAM of your machine, and whenever these objects are not in use any longer, R frees them from the memory, which simply means that they get deleted. Of course this might turn out to be quite costly, especially if you had not saved your original R script, which would have enabled you to replicate all the steps of your data processing activities when you start a new session in R. In order to prevent the objects from being deleted, you can save all or selected ones as .RData files on your hard drive. In the first case, you may use the save.image() function, which saves your whole current workspace with all objects to your current working directory:

```
> save.image(file = "workspace.RData")
```

If you are dealing with large objects, first make sure you have enough storage space available on your drive (this is normally not a problem any longer), or alternatively you can reduce the size of the saved objects using one of the compression methods available. For example, the above workspace.RData file was 3,751 bytes in size without compression, but when xz compression was applied the size of the resulting file decreased to 3,568 bytes.

```
> save.image(file = "workspace2.RData", compress = "xz")
```

Of course, the difference in sizes in the presented example is minuscule, as we are dealing with very small objects; however it gets much more significant for bigger data structures. The trade-off with applying one of the compression methods is the time it takes for R to save and load .RData files.

If you prefer to save only chosen objects (for example the dataset data frame and simple.list list) you can achieve this with the save() function:

```
> save(dataset, simple.list, file = "two_objects.RData")
```

You may now test whether the above solutions worked by cleaning your global environment of all objects, and then loading one of the created files, for example:

```
> rm(list=ls())
> load("workspace2.RData")
```

Introduction to R Programming Language and Statistical Environment

As an additional exercise, feel free to explore other functions that allow you to write text representations of R objects, for example `dump()` or `dput()`. More specifically, run the following commands and compare the returned outputs:

```
> dump(ls(), file = "dump.R", append = FALSE)
> dput(dataset, file = "dput.txt")
```

The `save.image()` and `save()` functions only create images of your workspace or selected objects on the hard drive. It is an entirely different story if you want to export some of the objects to data files of specified formats, for example, comma-separated, tab-delimited, or proprietary formats like Microsoft Excel, SPSS, or Stata.

The easiest way to export R objects to generic file formats like CSV, TXT, or TAB is through the `cat()` function, but it only works on atomic vectors:

```
> cat(age, file="age.txt", sep=",", fill=TRUE, labels=NULL, append=TRUE)
> cat(age, file="age.csv", sep=",", fill=TRUE, labels=NULL, append=TRUE)
```

The preceding code creates two files, one as a text file and another one as a comma-separated format, both of which contain values from the `age` vector that we had previously created for the `dataset` data frame. The `sep` argument is a character vector of strings to append after each element, the `fill` option is a logical argument that controls whether the output is automatically broken into lines (if set to `TRUE`), the `labels` parameter allows you to add a character vector of labels for each printed line of data in the file, and the `append` logical argument enables you to append the output of the call to the already existing file with the same name.

In order to export vectors and matrices to TXT, CSV, or TAB formats, you can use the `write()` function, which writes out a matrix or a vector in a specified number of columns for example:

```
> write(age, file="agedata.csv", ncolumns=2, append=TRUE, sep=",")
> write(y, file="matrix_y.tab", ncolumns=2, append=FALSE, sep="\t")
```

Another method of exporting matrices provides the `MASS` package (make sure you install it with the `install.packages("MASS")` function) through the `write.matrix()` command:

```
> library(MASS)
> write.matrix(y, file="ymatrix.txt", sep=",")
```

For large matrices, the `write.matrix()` function allows users to specify the size of blocks in which the data are written through the `blocksize` argument.

Probably the most common R data structure that you are going to export to different file formats will be a data frame. The generic `write.table()` function gives you an option to save your processed data frame objects to standard data formats, for example TAB, TXT, or CSV:

```
> write.table(dataset, file="dataset1.txt", append=TRUE, sep=",", na="NA",
col.names=TRUE, row.names=FALSE, dec=".")
```

The `append` and `sep` arguments should already be clear to you as they were explained earlier. In the `na` option you may specify an arbitrary string to use for missing values in the data. The logical parameter `col.names` allows users to append the names of columns to the output file, and the `dec` parameter sets the string used for decimal points and must be a single character. In the example, we used `row.names` set to `FALSE`, as the names of the rows in the data are the same as the values of the `subjectID` column. However, it is very likely that in other data sets the ID variable may differ from the names (or numbers) of rows, so you may want to control it depending on the characteristics of your data.

Two similar functions `write.csv()` and `write.csv2()` are just convenience wrappers for saving CSV files, and they only differ from the generic `write.table()` function by default settings of some of their parameters, for example `sep` and `dec`. Feel free to explore these subtle differences at your leisure.

To complete this section of the chapter we need to present how to export your R data frames to third-party formats. Amongst several frequently used methods, at least four of them are worth mentioning here. First, if you wish to write a data frame to a proprietary Microsoft Excel format, such as XLS or XLSX, you should probably use the WriteXLS package (please use `install.packages("WriteXLS")` if you have not done it yet) and its `WriteXLS()` function:

```
> library(WriteXLS)
> WriteXLS("dataset", "dataset1.xlsx", SheetNames=NULL,  row.names=FALSE,
col.names=TRUE, AdjWidth=TRUE,   envir=parent.frame())
```

The `WriteXLS()` command offers users a number of interesting options; for instance you can set the names of the worksheets (`SheetNames` argument), adjust the widths of columns depending on the number of characters of the longest value (`AdjWidth`), or even freeze rows and columns just as you do in Excel (the `FreezeRow` and `FreezeCol` parameters).

Please note that, in order for the `WriteXLS` package to work, you need to have **Perl** installed on your machine. The package creates Excel files using Perl scripts called `WriteXLS.pl` for Excel 2003 (XLS) files, and `WriteXLSX.pl` for Excel 2007 and later version (XLSX) files. If Perl is not present on your system, please make sure to download and install it from https://www.perl.org/get.html. After the Perl installation, you may have to restart your R session and load the `WriteXLS` package again to apply the changes. For solutions to common Perl issues please visit the following websites: https://www.perl.org/docs.html, http://www.ahinea.com/en/tech/perl-unicode-struggle.html, and http://www.perl.com/pub/2012/04/perlunicook-standard-preamble.html or search StackOverflow and similar websites for R- and Perl-related specific problems.

Another very useful way of writing R objects to the XLSX format is provided by the `openxlsx` package through the `write.xlsx()` function, which, apart from data frames, also allows lists to be easily written to Excel spreadsheets. Please note that Windows users may need to install the `Rtools` package in order to use `openxlsx` functionalities. The `write.xlsx()` function gives you a large choice of possible options to set, including a custom style to apply to column names (through the `headerStyle` argument), the color of cell borders (`borderColour`), or even line style (`borderStyle`). The following example utilizes only the most common and minimal arguments required to write a list to the XLSX file, but be encouraged to explore other options offered by this very flexible function:

```
> write.xlsx(simple.list, file = "simple_list.xlsx")
```

A third-party package called `foreign` makes it possible to write data frames to other formats used by well-known statistical tools such as SPSS, Stata, or SAS. When creating files, the `write.foreign()` function requires users to specify the names of both the data and code files. Data files hold raw data, whereas code files contain scripts with the data structure and metadata (value and variable labels, variable formats, and so on) written in the proprietary syntax. In the following example, the code writes the `dataset` data frame to the SPSS format:

```
> library(foreign)
> write.foreign(dataset, "datafile.txt", "codefile.txt", package="SPSS")
```

Finally, another package called `rio` contains only three functions, allowing users to quickly `import()`, `export()`, and `convert()` data between a large array of file formats, (for example TSV, CSV, RDS, RData, JSON, DTA, SAV, and many more). The package, in fact, is dependent on a number of other R libraries, some of which, for example `foreign` and `openxlsx`, have already been presented in this chapter. The `rio` package does not introduce any new functionalities apart from the default arguments characteristic for underlying export functions, so you still need to be familiar with the original functions and their parameters if you require more advanced exporting capabilities. But, if you are only looking for a no-fuss general export function, the `rio` package is definitely a good shortcut to take:

```
> export(dataset, format = "stata")
> export(dataset, "dataset1.csv", col.names = TRUE, na = "NA")
```

In this part, we have provided you with quite a bit of theory, and hopefully a lot of practical examples of data structures available to R users. You've created several objects of different types, and you've become familiar with a variety of data and file formats on offer. We then showed you how to save R objects held in your R workspace to external files on your hard drive, or to export them to various standard and proprietary file formats.

In the second part of the chapter we will attempt to go through all essential stages of data management, processing, transformations, and analysis, from data input to data output in the form of statistical tests, and simple visualization techniques, to inform our findings. From now on, we will also be using real data sets to highlight the most common data science problems and their solutions.

Applied data science with R

Applied datascience covers all the activities and processes data analysts must typically undertake to deliver evidence-based results of their analyses. This includes data collection, preprocessing data that may contain some basic but frequently time-consuming data transformations, and manipulations, EDA to describe the data under investigation, research methods, and statistical models applicable to the data and related to the research questions, and finally, data visualizations and reporting the insights. Data science is an enormous field, covering a great number of specific disciplines, techniques, and tools, and there are hundreds of very good printed and online resources explaining the particulars of each method or application.

In this section, we will merely focus on a small fraction of selected topics in data science using the R language. From this moment on, we will also be using real data sets from socio-economic domains. These data sets, however, will not be large-at least not big enough to call them Big Data. As the aim of this part of the book is to revisit some introductory concepts of the R language and data analysis, there is no need for large data sets right now.

As we saw in the preceding section, if you wish to follow all the practical instructions in your RStudio session, download the data and R script zip bundles from Packt Publishing's website.

Importing data from different formats

In the same way you exported R objects to external files on your hard drive, you can just as easily import data into R from almost all available file formats. But this is not everything R has to offer in terms of data import capabilities. R also connects very well with traditional SQL and NoSQL databases, other statistical tools such as SPSS, SAS, Minitab, Stata, Systat, and many others, online data repositories, and distributed file systems (such as HDFS); and besides, it allows you to do web scraping, or import and filter textual data. All in all, R is a great tool for serious data mining, and we will spend some time in the next several chapters presenting different methods of accessing and processing data through R's connectivity with SQL and NoSQL databases or HDFS.

The CRAN website provides a good manual on data import in R at https://cran.r-project.org/doc/manuals/r-release/R-data.html. There are also a large number of other online resources elaborating on that subject, for example:

- http://blog.datacamp.com/r-data-import-tutorial/
- http://www.r-tutor.com/r-introduction/data-frame/data-import
- http://www.statmethods.net/input/importingdata.html

More detailed information about data import can be found in general R books, or in help files of packages we used in the previous section when exporting R objects. Make sure to familiarize yourself with some of the most commonly used functions for importing data.

The first set of data that we will be exploring in this section comes from Queen's University in Belfast (Northern Ireland) and is based on the **Northern Ireland Life and Times Survey (NILT)** 2012. For our instruction purposes, we will be using an open access NILT 2012 Teaching Dataset, which, apart from the standard socio-demographic NILT variables, contains several additional variables concerning the *Good relations* topic, which refers to social attitudes of people living in Northern Ireland towards ethnic minorities, migrants, and cultural diversity. The data (in the SPSS *.sav file), along with the accompanying documentation, are downloadable from the Access Research Knowledge (ARK) Northern Ireland website at http://www.ark.ac.uk/teaching/teachqm.html. Tab-delimited and Stata versions of the same data are available from UK Data Service at https://discover.ukdataservice.ac.uk/catalogue/?sn=7547.

Upon downloading the data to the current working directory, we are now ready to import it to the R session. As the dataset comes in the SPSS format, we will use the `foreign` package to upload it to the workspace:

```
> library(foreign)
> grel <- read.spss("NILT2012GR.sav", to.data.frame=TRUE)
```

We will then quickly check the structure of the newly created `grel` data frame by listing the top lines of the dataset and its structure (please note that we have restricted the output to several lines in order to save space in the book):

```
> head(grel, n=5)
  serial househld rage spage    rsex nadult nkids nelderly nfamily
1      1        1    4    44  52 Male      3     1        0       4
2      2        2    2    86  81 Male      2     0        2       2
3      3        3    3    26  26 Female    2     1        0       3
4      4        4    1    72  NA Female    1     0        1       1
5      5        5    2    40  38 Male      2     0        0       2
...
> str(grel)
'data.frame':   1204 obs. of  133 variables:
 $ serial   : num  1 2 3 4 5 6 7 8 9 10 ...
 $ househld : num  4 2 3 1 2 2 1 1 2 1 ...
 $ rage     : atomic  44 86 26 72 40 44 76 23 89 56 ...
  ..- attr(*, "value.labels")= Named num
  .. ..- attr(*, "names")= chr
...
```

From the output, you can see that we are dealing with 1,204 observations over 133 variables. However not all variables will be useful during the analysis, so for our convenience we can simply create a smaller subset and store it in a separate object:

```
> grel.data <- subset(grel[, c(1:3, 5, 7:8, 10, 12, 16:19, 38:39, 41, 47,
52, 55, 60, 64, 66, 76:77, 80:83, 105:112)])
> str(grel.data)
'data.frame':   1204 obs. of  35 variables:
 $ serial    : num  1 2 3 4 5 6 7 8 9 10 ...
...
```

Multiple R objects versus R memory limitations
Generally it's a good idea to store the outputs of consecutive stages of your data transformations to separate objects, in order to avoid data loss in cases when you are not satisfied with the results or you need to go back to earlier steps. But sometimes, especially when dealing with large datasets, this approach is not recommended. As you probably remember, R stores all objects in its memory, which substantially restricts the amount of free processing resources. Later in the book, we will show you how to perform data manipulations outside R (for example within data storage systems like databases or HDFS) and how to import only small outputs of these manipulations to R or other computations for final processing, analysis, and visualizations.

Exploratory Data Analysis

Once we have the data ready, we can carry out the EDA. It includes a variety of statistical techniques allowing analysts and researchers to explore and understand the main characteristics of a given dataset, and formulate research hypotheses for further testing. It was pioneered by John W Tukey, an American mathematician who worked at Bell Labs and Princeton University, and became famous for his **Fast Fourier Transform (FFT)** algorithm. But Tukey was also known for his strong statistical interests. In 1977 he published a book titled *Exploratory Data Analysis,* in which he introduced box plots-a graphical visualization of the classic *five number summary* which displays the measurements of median, first quartile, third quartile, minimum, and maximum values. Currently, graphical methods of EDA also include histograms, scatter plots, bar plots, q-q plots, and many more.

In R there are many ways of obtaining descriptive statistics about the data. The `psych` package is one of them:

```
> library(psych)
> describe(grel.data)
         vars    n    mean      sd  median  trimmed     mad  min
serial      1 1204  602.50  347.71   602.5   602.50  446.26    1
househld    2 1204    2.36    1.34     2.0     2.19    1.48    1
rage        3 1201   49.62   18.53    48.0    49.03   22.24   18
...
          max range  skew kurtosis     se
serial   1204  1203  0.00    -1.20  10.02
househld   10     9  1.11     1.32   0.04
rage       97    79  0.23    -0.87   0.53
...
```

Depending on your specific needs you may want to obtain only individual measures of either central tendency (for example arithmetic mean, median, and mode) or dispersion (for example variance, standard deviation, and range), for a chosen variable like the age of respondents (`rage`):

```
> mean(grel.data$rage, na.rm=TRUE)
[1] 49.61532
> median(grel.data$rage, na.rm=TRUE)
[1] 48
```

The mode (a statistic) in R is calculated using the `modeest` package and its `mlv()` function with the most frequent value (`mfv`) argument:

```
> library(modeest)
> mlv(grel.data$rage, method="mfv", na.rm=TRUE)
Mode (most likely value): 42
Bickel's modal skewness: 0.2389675
...
```

For the basic measures of dispersion or variability of the `rage` variable, you may use the following calls:

```
> var(grel.data$rage, na.rm=TRUE) #variance
[1] 343.3486
> sd(grel.data$rage, na.rm=TRUE) #standard deviation
[1] 18.52967
> range(grel.data$rage, na.rm=TRUE) #range
[1] 18 97
```

Instead of iterating the name of the data object and the chosen variable every time you compute a single measurement, you may want to put all selected statistics into one function as in the listing below:

```
> cent.tend <- function(data) {
+     library(modeest)
+     data <- as.numeric(data)
+     m <- mean(data, na.rm=TRUE)
+     me <- median(data, na.rm=TRUE)
+     mo <- mlv(data, method="mfv", na.rm=TRUE)[1]
+     stats <- data.frame(c(m, me, mo), row.names="Totals:")
+     names(stats)[1:3] <- c("Mean", "Median", "Mode")
+     return(stats)
+ }
> cent.tend(grel.data$rage)
            Mean Median Mode
Totals:  49.61532     48   42
```

The `summary()` function makes it possible to print some basic descriptive statistics (including the count of missing values) for not only one, but also multiple variables, for example:

```
> summary(grel.data[c("rage", "persinc2")])
      rage          persinc2
 Min.   :18.00   Min.   :  260
 1st Qu.:35.00   1st Qu.: 6760
 Median :48.00   Median :11960
 Mean   :49.62   Mean   :16395
 3rd Qu.:64.00   3rd Qu.:22100
 Max.   :97.00   Max.   :75000
 NA's   :3       NA's   :307
```

Of course, our examples do not include all the different descriptive statistics available to you in the R language; there are many more and you are encouraged to research and test those of your particular interest.

A topic of EDA visualizations deserves a separate publication, and the actual methods used for graphical EDA are largely dependent on the type of your data and what specific information you would like to visualize. The last section of this chapter provides a list of R packages designed for static and interactive data visualizations including box plots, histograms, scatter plots, line graphs, density plots, bar charts, and many more.

Data aggregations and contingency tables

It's very unlikely that your data analysis process will be satisfied by calculating simple descriptive statistics. Data aggregations and contingency tables provide a much deeper perspective on the distribution of the data and may inform you of possible patterns between variables.

In data aggregations, you are able to calculate a certain statistic as cross-tabulated between all levels of a categorical variable. For example, assume we want to calculate mean values of age (`rage`), personal income (`persinc2`) and the number of people in the household (`househld`) for two levels (male and female) of the factor `rsex` (sex of respondents):

```
> aggregate(grel.data[c("rage", "persinc2", "househld")],
by=list(rsex=grel.data$rsex), FUN=mean, na.rm=TRUE)
     rsex     rage persinc2 househld
1    Male 50.70467 19154.56 2.210428
2  Female 48.74024 14203.16 2.485757
```

Unfortunately, the `aggregate()` function can only perform a calculation of one statistic (the `FUN` argument) per call. In such cases, you can create your own function (named `stats` in the following listing) that will estimate the values of several statistics and then pass them to the `summaryBy()` function from the `doBy` package:

```
> stats <- function(x, na.omit=TRUE) {
+    if(na.omit)
+       x <- x[!is.na(x)]
+    n <- length(x)
+    m <- mean(x)
+    s <- sd(x)
+    r <- range(x)
+    return(c(n=n, mean=m, stdev=s, range=r))
+ }
> library(doBy)
> summaryBy(rage+persinc2+househld~rsex, data=grel.data, FUN=stats)
     rsex rage.n rage.mean rage.stdev rage.range1 rage.range2
1    Male    535  50.70467   18.30921          18          94
2  Female    666  48.74024   18.67257          18          97
  persinc2.n persinc2.mean persinc2.stdev persinc2.range1
1        397      19154.56       14792.73             260
2        500      14203.16       11877.14             260
  persinc2.range2 househld.n househld.mean househld.stdev
1           75000        537      2.210428       1.245306
2           75000        667      2.485757       1.402815
  househld.range1 househld.range2
1               1               7
2               1              10
```

Introduction to R Programming Language and Statistical Environment

The previously mentioned `psych` package also allows you to aggregate summary statistics by the levels of a factor through its `describeBy()` function. As an additional exercise, run the two lines of code listed below and check their outputs. How do they differ? What does the `mat` argument do? What estimates can you obtain using `describeBy()`?

```
> library(psych)
> describeBy(grel.data[c("rage", "persinc2", "househld")], grel.data$rsex)
> describeBy(grel.data[c("rage", "persinc2", "househld")], grel.data$rsex,
mat=TRUE)
```

Contingency tables let us inspect the frequency (the occurrence or the count) of specific values in a numeric variable, or a number of records for each level of a categorical variable. The easiest way to print a contingency table for a single variable is through the `table()` function available in the base R installation:

```
> table(grel.data$househld)
  1   2   3   4   5   6   7   8  10
369 406 188 144  72  16   6   2   1
> table(grel.data$uprejmeg)
  Very prejudiced     A little prejudiced   Not prejudiced at all
               31                     300                     845
Other (please specify)
                    7
```

In the preceding example we've obtained frequencies of one continuous variable (`househld`-the number of people in the household), and one factor (`uprejmeg`-the self-reported level of prejudice against people of minority ethnic communities), with four labeled levels.

You may also obtain frequency tables for all variables of your data in one go, using the `describe()` function from the `Hmisc` package. As you have probably noticed, we have already introduced one function called `describe()` (in the `psych` package), so if you would like now to use the `Hmisc` implementation, you need to explicitly *force* R to execute the `Hmisc` package's version of `describe()`:

```
> library(Hmisc)
> Hmisc::describe(grel.data)
... #output truncated
househld
         n  missing  unique     Info    Mean
      1204        0       9     0.93   2.363
              1    2    3    4    5   6  7  8 10
Frequency   369  406  188  144   72  16  6  2  1
%            31   34   16   12    6   1  0  0  0
... #output truncated
```

Hmisc provides not only a simple frequency table, but it also supplies several basic descriptive statistics such as the relative percentages of counts.

However, the most useful applications of contingency tables are when we want to count the occurrences of specific values between two or more categorical variables. The generic table() and xtabs() functions allow such multidimensional frequency tables:

```
> attach(grel.data)
> table(uprejmeg, househld)
                        househld
uprejmeg                   1   2   3   4   5   6   7   8  10
  Very prejuiced          14   8   4   3   1   0   1   0   0
  A little prejudiced     96 102  53  28  13   5   1   2   0
  Not prejudiced at all  248 288 125 112  56  11   4   0   1
  Other (please specify)   2   1   2   1   1   0   0   0   0
> detach(grel.data)
> xtabs(~uprejmeg+househld+rsex, data=grel.data)
, , rsex = Male
                        househld
uprejmeg                   1   2   3   4   5   6   7   8  10
  Very prejuiced           9   1   2   2   0   0   0   0   0
  A little prejudiced     52  46  22   8   7   2   1   0   0
  Not prejudiced at all  118 135  49  45  20   4   0   0   0
  Other (please specify)   1   0   0   0   1   0   0   0   0
, , rsex = Female
                        househld
uprejmeg                   1   2   3   4   5   6   7   8  10
  Very prejuiced           5   7   2   1   1   0   1   0   0
  A little prejudiced     44  56  31  20   6   3   0   2   0
  Not prejudiced at all  130 153  76  67  36   7   4   0   1
  Other (please specify)   1   1   2   1   0   0   0   0   0
```

If you do not find the arrangement of the xtabs() function output visually appealing, you can assign it to another R object and apply ftable() for more user-friendly cross-tabulation (the actual output is not shown in the listing below):

```
> xTab <- xtabs(~uprejmeg+househld+rsex, data=grel.data)
> ftable(xTab)
```

Data aggregations and contingency tables are just two of many data transformations and management activities you will probably see yourself performing in R. There are a number of extremely useful data *cleaning* packages which you should become familiar with if you wish to turn into a true R wizard. Some of them are:

- `dplyr`: Hadley Wickham's well known package for data manipulation
- `tidyr`: another product of Wickham's R genius; suitable for data tidying (splitting, joining variables, and so on) and it supports the `dplyr` package's data pipelines
- `reshape2`: authored by, you guessed it, Hadley Wickham; it provides easy aggregations and restructuring of data
- `data.table`: a product of a team led by Matt Dowle; allows fast aggregations and transformations of large data sets; we will explore it in `Chapter 3`, *Unleashing the Power of R from Within*

Data manipulations very often include operations on character variables, strings, and text or date and time information. Some packages specialize in addressing this sort of transformations:

- `stringi` and `stringr` two packages that support work with strings and allow, for example, pattern searching, string generation, date and time formatting, parsing, and many more.
- `lubridate`: a team effort that helps R users with date and time stamps, their extraction and advanced manipulation.

Once we've covered some essential data management issues we can now proceed to the essence of data analysis statistical tests.

Hypothesis testing and statistical inference

It's very unlikely that you've ever attempted any data analysis without asking yourself, *a priori*, what sort of results you were expecting to obtain. Although many of my data consultancy clients haven't got skills in statistics, they usually come to me with a good understanding of their data, and they are often capable of identifying certain patterns with a quite impressive level of accuracy. The EDA we covered in the previous sections allows us to learn about the characteristics of our data, but it is hypothesis testing, and inferential statistics, that in fact let us measure whether our expectations of relationships and patterns in the data are correct.

As with the preceding sections, this part of the chapter will only very briefly touch a number of the most popular statistical tests that could be applicable to our data. As this is not a pure statistics or research methods book, we will also abstain from any explanations underlying specific statistical computations. Depending on your field of specialization, and your individual level of mathematical and/or statistical knowledge, we are certain that you can find a suitable printed or online resource, that can be used either to learn new concepts, or to revise your existing skills. At the very beginning of this chapter, we have provided you with a list of websites and books, that may improve both your theoretical understanding of statistics and its implementations using the R language, but feel free to research other potentially more relevant sources.

Tests of differences

In R, t.test() is the most generic function that can be used to perform a variety of t-tests estimations. Depending on the type of t-test (whether *one-sample, matched-sample,* or *independent two-sample*), users can either specify the function's formula argument (for the independent t-test) or set the paired parameter to TRUE (for the paired/matched sample t-test). By default the function runs a standard one-sample test.

Independent t-test example (with power and effect size estimates)

Assume that you want to investigate whether there is a statistical difference between male and female respondents in their perceived happiness score as measured with the ruhappy variable (where 1 is *Very happy* and 4 is *Not at all happy*).

First of all, it might be advisable to convert the ruhappy factor into a numeric variable, check the frequencies of scores, set all scores outside the range of **Likert scale** 1-4 as **Not Applicable** (NA) (as per the documentation file), and then calculate the mean values of happiness for male and female respondents:

```
> grel.data$ruhappy <- as.numeric(grel.data$ruhappy)
> table(grel.data$ruhappy)
  1   2   3   4   5
404 656  95  12  28
> library(car)
> grel.data$ruhappy <- recode(grel.data$ruhappy, "5=NA")
> table(grel.data$ruhappy)
  1   2   3   4
404 656  95  12
> aggregate(grel.data$ruhappy, by=list(rsex=grel.data$rsex), FUN=mean,
```

```
  na.rm=TRUE)
     rsex        x
1    Male 1.799235
2  Female 1.720497
```

From the output, we can deduct that male respondents perceived themselves as generally less happy than females did (remember that ruhappy values were reverse coded); however, we still don't know whether this difference was significant. Let's run a t-test to find out:

```
> t.test(ruhappy ~ rsex, data=grel.data, alternative="two.sided")
    Welch Two Sample t-test
data:  ruhappy by rsex
t = 2.0728, df = 1066.9, p-value = 0.03843
alternative hypothesis: true difference in means is not equal to 0
95 percent confidence interval:
  0.004201224 0.153275350
sample estimates:
  mean in group Male mean in group Female
            1.799235              1.720497
```

The t.test() function calculated the **t-statistic**, 95% confidence intervals, and mean values of happiness for both groups in our sample. The **p-value** was below the conventional threshold of 0.05, which supports the alternative hypothesis that both means differed significantly.

At this stage you may wish to run some **ad hoc tests** like **power analysis** or **effect size** estimate. As both calculations make use of several descriptive statistics, we can compute them beforehand, and then apply them in power and effect size tests.

```
> library(doBy)
> stats <- function(x, na.omit=TRUE){
  if(na.omit)
    x <- x[!is.na(x)]
  m <- mean(x)
  n <- length(x)
  s <- sd(x)
  return(c(n=n, mean=m, stdev=s))
}
> sum.By <- summaryBy(ruhappy~rsex, data=grel.data, FUN=stats)
> sum.By
    rsex ruhappy.n ruhappy.mean ruhappy.stdev
1   Male       523     1.799235     0.6722748
2 Female       644     1.720497     0.6105468
```

As both our groups are not equal in sizes, we need to calculate the effective sample size based on the **harmonic mean** of n. We will take the values from the sum.By object:

```
> attach(sum.By)
> n.harm <- (2*ruhappy.n[1]*ruhappy.n[2])/(ruhappy.n[1]+ruhappy.n[2])
> detach(sum.By)
> n.harm
[1] 577.2271
```

We also have to estimate a pooled standard deviation. For that reason, we will use all ruhappy scores regardless of the sex of respondents:

```
> pooledsd.ruhappy <- sd(grel.data$ruhappy, na.rm=TRUE)
> pooledsd.ruhappy
[1] 0.6398692
```

From the sum.By object we already know the means for both groups, therefore we can now put all the stats together to calculate the power:

```
> power.t.test(n=round(n.harm, 0), delta=sum.By$ruhappy.mean[1]-
sum.By$ruhappy.mean[2], sd=pooledsd.ruhappy, sig.level=.03843,
type="two.sample", alternative="two.sided")
Two-sample t test power calculation
              n = 577
          delta = 0.07873829
             sd = 0.6398692
      sig.level = 0.03843
          power = 0.5071487
    alternative = two.sided
NOTE: n is number in *each* group
```

Based on the achieved p-value and characteristics of our data, the power was moderate, but it could, and probably even should, be better. By convention, researchers should aim to reach 0.6 as the acceptable level of power.

Last but not least, we will estimate **the effect size** by dividing the means difference by the pooled standard deviation:

```
> d <- (sum.By$ruhappy.mean[1]-sum.By$ruhappy.mean[2])/pooledsd.ruhappy
> d
[1] 0.1230537
```

The effect size was small.

In the next part we will carry out the **Analysis of Variance (ANOVA)**-a test of difference between multiple groups.

ANOVA example

In our ANOVA example we will investigate whether there is any significant difference between mean scores of happiness (the ruhappy variable) based on the respondents' place of living (placeliv), which contains the following five categories:

```
> levels(grel.data$placeliv)
[1] "...a big city"
[2] "the suburbs or outskirts of a big city"
[3] "a small city or town"
[4] "a country village"
[5] "a farm or home in the country"
```

We will not be covering in any depth the details of assessing whether the data meets the main assumptions for the ANOVA, such as normal distribution and homoscedasticity (homogeneity of variances); here, however, you may wish to run the qqPlot() from the car library and the generic bartlett.test() to test these two assumptions. It is worth saying that both tests return satisfactory outputs, and hence we can proceed to estimate the means of happiness scores for each place of living in question:

```
> ruhappy.means <- aggregate(grel.data$ruhappy,
by=list(grel.data$placeliv), FUN=mean, na.rm=TRUE)
> ruhappy.means
                                     Group.1          x
1                              ...a big city   1.951515
2   the suburbs or outskirts of a big city   1.807018
3                      a small city or town   1.718954
4                         a country village   1.627219
5             a farm or home in the country   1.722222
```

It is easy to notice that the differences between the mean scores are quite large in some cases. Let's run the actual ANOVA to check if the differences between means are in fact statistically significant, and whether we can claim that people living in certain places tend to perceive themselves as happier than those living elsewhere:

```
> fit <- aov(ruhappy~placeliv, data=grel.data)
> summary(fit)
              Df  Sum Sq  Mean Sq  F value    Pr(>F)
placeliv       4    10.5    2.624    6.529  0.0000339 ***
Residuals   1160   466.3    0.402
---
Signif. codes:  0 '***' 0.001 '**' 0.01 '*' 0.05 '.' 0.1 ' ' 1
39 observations deleted due to missingness
```

Based on the **F statistic,** and the p-value, it's clear that the differences are statistically significant overall in our data, but ANOVA doesn't reveal precisely which means differ significantly from one another. You can visualize all mean scores using the `plotmeans()` function from the `gplots` package:

```
> library(gplots)
> plotmeans(grel.data$ruhappy~grel.data$placeliv, xlab="Place of living",
  ylab="Happiness", main="Means Plot with 95% CI")
```

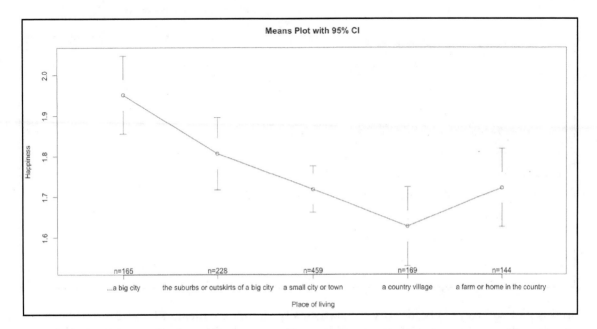

We will also apply the `TukeyHSD()` function from the core R installation to compare pairs of means:

```
> TukeyHSD(fit)
```

As the output is quite long we won't show it here. Upon quick inspection of the *p adj* section you will probably notice that there are a number of pairs of means that differ significantly. You may display these differences visually, using the base `plot()` applied on the `TukeyHSD(fit)` expression, but because the value labels of the `placeliv` variable are quite long, make sure you adjust the margin size to fit their labels. In the end your plot should look similar to the following one:

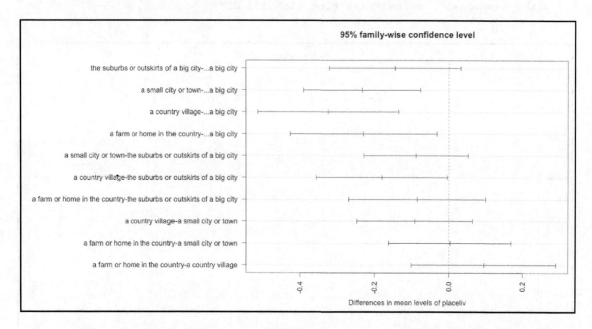

In this plot you can easily identify the statistically significant mean differences as they (along with their 95% confidence intervals) do not include the vertical line crossing the 0.0 point on the x-axis.

Following the ANOVA, you may also perform a number of post hoc model diagnostics such as **Bonferonni's test** for the presence of outliers through the `outlierTest()` function in the `car` package, or **the standard Eta Squared effect size** estimate for ANOVA using the `etaSquared()` function from the `lsr` package. Feel free to explore these additional tests at your leisure.

As we have given you a very brief review of two major tests of differences, which we can apply to our dataset, we will now present two simple ways of estimating relationships between variables through correlations and their extensions: multiple regressions.

Tests of relationships

We will begin this section with a quick revision of the most frequently used R functions for estimating covariance and Pearson's r correlation coefficients. We will then explore methods and data visualizations applied to multiple regressions.

An example of Pearson's r correlations

As we don't have too many continuous variables in our data, we will use one of the ordinal variables miecono to correlate with the perceived level of happiness (ruhappy). The miecono variable measures, on a scale from 0 to 10, to what extent the respondents agree with the statement that it is good for Northern Ireland's economy that migrants come there from other countries. The score of 0 means that it is *Extremely bad for the economy* and the score of 10 denotes that it is *Extremely good for the economy*. As the numeric scores were coded from 1 to 11, we need to clean up the variable a little bit so that the labeling is consistent with the actual data:

```
> grel.data$miecono <- as.numeric(grel.data$miecono)
> library(car)
> grel.data$miecono <- recode(grel.data$miecono,
"1=0;2=1;3=2;4=3;5=4;6=5;7=6;8=7;9=8;10=9;11=10")
> table(grel.data$miecono)

  0   1   2   3   4   5   6   7   8   9  10
 50  64  92  97  81 296  95 173 128  36  43
```

It is generally a good idea to test the data for normality assumptions by displaying q-q plots for both variables using qqnorm() and plotting density graphs with plot(density()). We will skip the steps in this section as we are only interested in presenting R methods applicable to measuring relationships between variables, and not in finding robust effects in our data. We may now calculate the **covariance** between the ruhappy and miecono variables:

```
> cov(grel.data$ruhappy, grel.data$miecono, method="pearson", use =
"complete.obs")
[1] -0.2338568
```

Owing to the several missing values, we have set the `use` argument to include only complete observations. The obtained negative covariance simply means that an increase in values of one variable is associated with a decrease in values of another (note that the `ruhappy` variable is reverse coded-a low score denotes a greater level of happiness, so, based on our example, the happier the respondents are, the more they agree with the statement that it is good for Northern Ireland's economy that migrants come from other countries). Also, as you probably remember from an introductory level statistics course, correlations do not imply causation. Finally, we must stress that judging from the value of covariance we do not know whether the degree of relationship is small or large, as the measurement of covariance is dependent on the value of the standard deviation. Therefore, let's estimate a less biased **Pearson's r correlation coefficient**. The simplest way to achieve it is by using the generic `cor()` function; however it doesn't return significance testing. The recommended method is through the `cor.test()` function:

```
> cor.test(grel.data$ruhappy, grel.data$miecono, alternative="two.sided",
method="pearson")
Pearson's product-moment correlation
data:  grel.data$ruhappy and grel.data$miecono
t = -5.0001, df = 1118, p-value = 0.0000006649
alternative hypothesis: true correlation is not equal to 0
95 percent confidence interval:
 -0.2046978 -0.0900985
sample estimates:
       cor
-0.1478945
```

The function returned not only the Pearson's r value (-0.1479), but also the **t statistic**, 95% confidence intervals, and p-value, which was found to be far below the conventional threshold of 0.05. We may then conclude that the negative small correlation of -0.1479 is statistically significant.

At this moment, you can also calculate the **Adjusted Correlation Coefficient (radj)**, which is an unbiased estimate of the population correlation. In order to do so, we first need to extract the number of complete observations for both variables and the value of the r correlation coefficient:

```
> N.compl <- sum(complete.cases(grel.data$ruhappy, grel.data$miecono))
#number of complete cases
> N.compl
[1] 1120
> cor.grel <- cor(grel.data$ruhappy, grel.data$miecono, method="pearson",
use = "complete.obs") #correlation coefficient
> cor.grel
[1] -0.1478945
```

Calculating the adjusted r:

```
> adjusted.cor <- sqrt(1 - (((1 - (cor.grel^2))*(N.compl - 1))/(N.compl - 2)))
> adjusted.cor
[1] 0.1449065
```

Owing to quite a large sample size, our adjusted r is not much different from the previously estimated Pearson's r. You may also want to carry out a post hoc power analysis using `pwr.r.test()` from the `pwr` package:

```
> library(pwr)
> pwr.cor <- pwr.r.test(n=N.compl, r=cor.grel, sig.level=0.0000006649, alternative="two.sided")
> pwr.cor

     approximate correlation power calculation (arctangh transformation)
              n = 1120
              r = 0.1478945
      sig.level = 0.0000006649
          power = 0.5008763
    alternative = two.sided
```

Based on the values of Pearson's r, it's p-value, and the number of valid observations in the sample, the power was found to be moderate.

Let's now take one step forward and move on to the multiple regression example.

Multiple regression example

A **multiple regression** is in fact an extension of the correlation-also known as a simple regression. There is, however, a subtle theoretical difference between linear regression models and bivariate normal models. Regressions involve relationships between fixed variables (that is, variables determined by the experimenter, for example, a number of trials or a number of objects in one trial, and so on.), whereas correlations in general are applied to investigate relationships between random variables (that is, variables that are beyond the experimenter's control). Despite this little nuance, the practical implementation of both methods is very similar.

In our example, based on the NILT dataset, we will try to predict the level of happiness of the respondents (`ruhappy`) from four variables: the age of the respondent (`rage`), personal income (`persinc2`), the number of people in the household (`househld`), and the number of ethnic groups that the respondent has regular contact with (`contegrp`). We believe that these variables may have some impact on the self-reported happiness of the respondents, but it is just a layperson's hypothesis, unsupported by any prior research. We will, however, make the assumption that we are reasonably justified in including all four predictors.

We may begin by creating a subset from the original NILT data to include all five variables of interest:

```
> reg.data <- subset(grel.data, select = c(ruhappy, rage, persinc2,
househld, contegrp))
> str(reg.data)
'data.frame':   1204 obs. of  5 variables:
 $ ruhappy : num  2 2 1 NA 1 2 3 3 1 3 ...
 $ rage    : num  44 86 26 72 40 44 76 23 89 56 ...
 $ persinc2: num  22100 6760 9880 NA 27300 ...
 $ househld: num  4 2 3 1 2 2 1 1 2 1 ...
 $ contegrp: num  3 0 3 2 2 2 0 2 0 3 ...
```

As the variables are numeric, we can firstly explore pair-wise Pearson's r correlations between all of them. To achieve this you may use a number of methods. First, you can use the psych package and its corr.test() function applied on all variables in our subset:

```
> library(psych)
> corr.test(reg.data[1:5], reg.data[1:5], use="complete", method="pearson",
alpha=.05)
Call:corr.test(x = reg.data[1:5], y = reg.data[1:5], use = "complete",
    method = "pearson", alpha = 0.05)
Correlation matrix
         ruhappy  rage persinc2 househld contegrp
ruhappy     1.00 -0.06    -0.08    -0.07     0.01
rage       -0.06  1.00    -0.04    -0.40    -0.28
persinc2   -0.08 -0.04     1.00     0.14     0.20
househld   -0.07 -0.40     0.14     1.00     0.19
contegrp    0.01 -0.28     0.20     0.19     1.00
Sample Size
[1] 876
Probability values  adjusted for multiple tests.
         ruhappy rage persinc2 househld contegrp
ruhappy     0.00  0.6     0.21     0.25        1
rage        0.60  0.0     1.00     0.00        0
persinc2    0.21  1.0     0.00     0.00        0
househld    0.25  0.0     0.00     0.00        0
contegrp    1.00  0.0     0.00     0.00        0
```

If you prefer, you may visualize the correlations with the `scatterplotMatrix()` function from the `car` package:

```
> library(car)
> scatterplotMatrix(reg.data[1:5], spread=FALSE, lty.smooth=2,
main="Scatterplot Matrix")
```

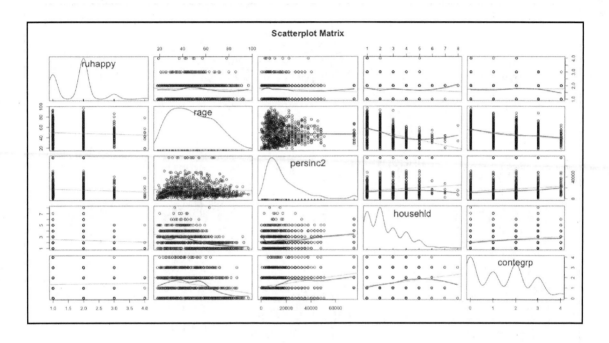

Or through the correlogram created with `corrgram()` available in the `corrgram` library:

```
> library(corrgram)
> corrgram(reg.data[1:5], order=TRUE, lower.panel=panel.shade,
  upper.panel=panel.pie, text.panel=panel.txt)
```

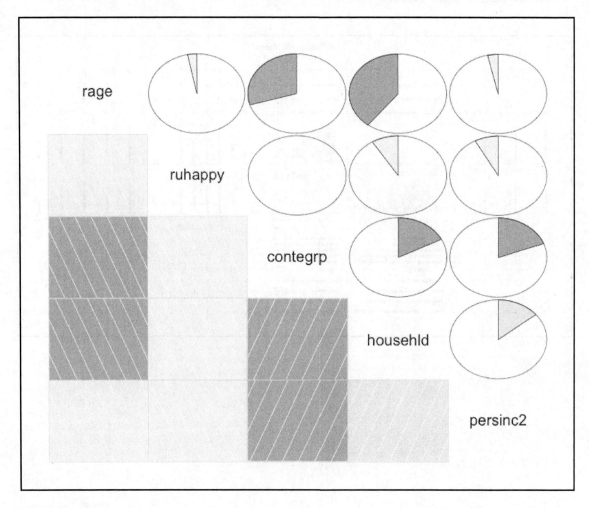

From what we've seen so far, the `ruhappy` variable is not correlated very strongly with any of the predictors, but it doesn't mean that we can't predict its value from the other four variables. We may now attempt to put all the variables into our regression model using the `lm()` function:

```
> attach(reg.data)
```

```
> regress1 <- lm(ruhappy~rage+persinc2+househld+contegrp)
> detach(reg.data)
> regress1
Call:
lm(formula = ruhappy ~ rage + persinc2 + househld + contegrp)
Coefficients:
  (Intercept)          rage      persinc2      househld      contegrp
 2.169521313  -0.004366977  -0.000004858  -0.036273587   0.005658574
```

A basic output of the `lm()` function provides us with the values of regression coefficients for each predictor, and the base intercept value of the criterion (the level of happiness), when all the predictors are equal to 0. You can observe, for instance, that the increase of the number of members in the household generally has a positive effect on the level of happiness; note here that the minus sign is confusing, because, as we pointed out before, the `ruhappy` variable is reverse-coded. The `summary()` function applied on the name of the R object that stores the linear model gives us a more detailed insight into our regression:

```
> summary(regress1)
Call:
lm(formula = ruhappy ~ rage + persinc2 + househld + contegrp)
Residuals:
    Min      1Q  Median      3Q     Max
-1.0256 -0.7280  0.1423  0.2553  3.3459
Coefficients:
                 Estimate    Std. Error  t value             Pr(>|t|)
(Intercept)   2.169521313   0.114745914   18.907 < 0.0000000000000002
rage         -0.004366977   0.001559900   -2.800              0.00523
persinc2     -0.000004858   0.000001907   -2.547              0.01104
househld     -0.036273587   0.020164265   -1.799              0.07237
contegrp      0.005658574   0.022231326    0.255              0.79914

(Intercept) ***
rage        **
persinc2    *
househld    .
contegrp
---
Signif. codes:  0 '***' 0.001 '**' 0.01 '*' 0.05 '.' 0.1 ' ' 1
Residual standard error: 0.7454 on 885 degrees of freedom
  (314 observations deleted due to missingness)
Multiple R-squared:  0.01855,   Adjusted R-squared:  0.01412
F-statistic: 4.182 on 4 and 885 DF,  p-value: 0.00232
```

The output clearly suggests that our model is indeed statistically significant with the overall p-value far below `0.05` and the two statistically significant independent variables of the age of respondents (`rage`) and personal income (`persinc2`). One variable-the number of people in the household (`househld`)-is just on the verge of significance. But, we have to admit, the model is not a very strong one, as the predictors explain only about 1.4% of the variation in the level of happiness (`ruhappy`). Feel free to select different variables from the data and use them as predictors of happiness.

Of course, there are many other methods in R that are applicable to regression analyses. For example, you may approach data without having any clear hypothesis in mind, and in this situation you could try the **stepwise regression** with the `stepAIC()` function in the `MASS` package to apply to several variables, and to discover potentially significant predictors. There are also special cases of regression such as a **logistic regression** or a **Poisson regression**, which attempt to predict either a binary dependent variable, or a count variable, from continuous and/or categorical predictors respectively. Both models can be obtained through the standard `glm()` function; however, in order to perform it, in the case of a logistic regression, the `family` argument must be set to `binomial`, whereas for a Poisson regression we require it to be set to `poisson`.

Data visualization packages

The topic of graphical data visualizations has recently evolved into a separate field that now covers a large array of methods and tools from such disciplines as statistics, graphic design, user experience, cognitive psychology, computer science, and many more. R gives you almost unlimited possibilities in how you can visualize the insights of your data. In the previous sections of this chapter, we've provided a few examples of data visualizations applicable to very specific tests and analyses, but they don't do justice to the vast graphical capabilities of the R language. In this section, we will only list a number of fantastic data visualization packages that R offers to its users, and we will let you explore them on your own depending on your interests and specific applications:

- **ggplot2**: An industry-standard static visualization package with its own grammar of graphics authored by Hadley Wickham
- **ggvis**: An interactive implementation of `ggplot2` grammar authored by Winston Chang and Hadley Wickham
- **shiny**: Allows users to create reactive web applications with R, developed and maintained by RStudio, `http://shiny.rstudio.com/`

- **lattice**: An R implementation of static Trellis graphics, great for multivariate visualizations, created by Deepayan Sarkar
- **rCharts**: Enables you to create and customize powerful interactive visualizations based on popular JavaScript libraries, for example `highcharts`, `morris`, and many more, developed by Ramnath Vaidyanathan, http://rcharts.io/
- **googleVis**: An R interface to the Google Charts API allowing users to create interactive visualization, such as well-known motion charts, authored by Markus Gesmann and Diego de Castillo with contributions from Joe Cheng
- **plotly:** Provides a way to translate static `ggplot2` charts into interactive web-based visualization, the product of a team effort with Carson Sievert as the maintainer, https://plot.ly/r/
- **htmlwidgets**: A framework for building interactive HTML visualizations based on JavaScript libraries, for example `leaflet` or `dygraphs`, authored by a number of contributors and maintained by JJ Allaire, http://www.htmlwidgets.org/

Apart from the amazing packages mentioned in the preceding list, let's not forget about the still extremely powerful static visualization capabilities of the core R installation. Paul Murrell's book on *R Graphics* (2nd edition) is a must-have text for all R expert wannabes who need a crash course in base R graphic facilities.

Summary

In this chapter we've revisited many concepts related to data management, data processing, transformations, and data analysis, using R programming language and a statistical environment. Our target was to enable you to familiarize yourself with major functions and R packages, which facilitate manipulation of data and hypothesis testing. Finally, we have mentioned a few words on the topic of static and interactive data visualizations and their nearly limitless applications.

This chapter was by no means inclusive of all available methods and techniques. We have merely scratched the surface of what's possible in R, but we also believe that the information provided in this chapter enabled you to either revise your existing R skills, or to identify potential gaps.

In the following chapters we will be building on these skills with almost exclusive focus on Big Data. In `Chapter 3`, *Unleashing the Power of R from Within* you will be exposed to a number of packages which allow R users to transform and process large amounts of data on a single machine.

3
Unleashing the Power of R from Within

In the first chapter we introduced you to a number of general terms and concepts related to Big Data. In Chapter 2, *Introduction to R Programming Language and Statistical Environment*, we presented you with several frequently used methods for data management, processing, and analysis using the R language and its statistical environment. In this chapter we will merge both topics and attempt to explain how you can use powerful mathematical and data modeling R packages in large datasets, without the need for distributed computing. After reading this chapter you should be able to:

- Understand R's traditional limitations for Big Data analytics and how they can be resolved
- Use R packages such as ff, ffbase, ffbase2, and bigmemory to enhance out-of-memory performance
- Apply statistical methods to large R objects through the biglm and ffbase packages
- Enhance the speed of data processing with R libraries supporting parallel computing
- Benefit from faster data manipulation methods available in the data.table package

Whenever possible, the methods presented in this chapter will be accompanied by practical examples containing real-world data. The samples of the datasets used in the examples are available for you to download from the Packt Publishing website. Of course, the samples are not large enough to be considered as Big Data; however we provide you with URL links to websites and online services where you can download full, large versions of the datasets, or you can mine the data and hence control its size yourself.

Traditional limitations of R

The usual scenario is simple. You've mined or collected unusually large amounts of data as part of your professional work, or university research, and you appreciate the flexibility of the R language and its ever-growing, rich landscape of useful and open-source libraries. So what next? Before too long you will be faced with two traditional limitations of R:

- Data must fit within the available RAM
- R is generally very slow compared to other languages

Out-of-memory data

The first of the claims against using R for Big Data is that the entire dataset you want to process has to be smaller than the amount of available RAM. Currently, most of the commercially sold, off-the-shelf personal computers are equipped with anything from 4GB to 16GB of RAM, meaning that these values will be the upper bounds of the size of your data which you will want to analyze with R. Of course, from these upper limits, you still need to deduct some additional memory resources for other processes to run simultaneously on your machine and provide extra RAM allowance for algorithms and computations you will like to perform during your analysis in the R environment. Realistically speaking, the size of your data shouldn't exceed a maximum of 50 to 60% of available memory, unless you want to see your machine become sluggish and unresponsive, which may also result in R and system crashes, and even potential data loss. At the moment, it all may seem pretty grim, and I won't blame you if you have suddenly become much less enthusiastic about integrating R with your currently used Big Data analytics workflows. However, there are already a number of solutions and workarounds available to R users who want to do some serious data crunching on their machines, even without turning to cloud computing platforms such as Microsoft Azure, Amazon EC2, or Google Cloud Platform. These ready-made solutions usually come as R packages, which you can simply download and install with your R distribution to enjoy some extra processing boost. We describe these libraries in the next part of this chapter, where we will also present their most useful functions and address some of their specific limitations. If, however, you want to learn how to set up R and RStudio in the Cloud, feel free to fast-forward to *Online Chapter, Pushing R Further* (https://www.packtpub.com/sites/default/files/downloads/5396_6457OS_PushingRFurther.pdf).

Processing speed

The second argument, which keeps R's antagonists going, is its processing speed. Although R's speed is still acceptable in some small-scale computations, it generally lags behind Python and, even more so, the C family of languages. There are several reasons why R cannot keep up with others.

First, R is considered to be an *interpreted language*, and, as such, its slower code execution comes by definition. It's also interesting that, despite a large bulk of the core R being written in C language (almost 39% according to `https://www.openhub.net/p/rproject`), a number of R functions created in C (or Fortran), are still much slower than in native code, and this may be partly due to poor memory management in R (for example, time spent on garbage collection, duplications, and vector allocation).

Second, the core R is *single-threaded*, meaning that the code of a function or a computation is processed line-by-line, one at a time, engaging a single CPU. There are, however, several methods, including some third-party packages, which allow multi-threading. We will explore these packages in more detail later in this chapter.

Third, the poor performance of the R code may come from the fact that the R language is *not formally defined*, and its processing speed largely depends on the designs of R implementations, rather than the R language itself. There is currently quite a lot of work being done to create new, much faster, alternative implementations and the recent release of Microsoft R Open distribution is an example of another, hopefully more performance-optimized, implementation of R.

Also we need to bear in mind that R is an open-source, community-run project with only a few R core development team members who are authorized to make any changes to the R internals. This puts serious constraints on how quickly poorly written parts of R code are altered. We also mustn't forget about one very important thing that underlies the whole development of the R language: it wasn't created to break computational speed records, but to provide statisticians and researchers (often with no programming or relevant IT skills) with a rich variety of robust and customizable data analysis and visualization techniques.

If you are interested in the particulars of the R language definition, and the details of its source code design, a book written by Hadley Wickham, *Advanced R* should serve as a top recommendation amongst several good references on that subject. You can purchase it from Amazon or visit its companion website at `http://adv-r.had.co.nz/`.

In the following section we will present several techniques for squeezing this extra power out of and from within R to allow data analytics of large datasets on a single computer.

To the memory limits and beyond

We will start off by introducing you to three very useful and versatile packages which facilitate out-of-memory data processing: `ff`, `ffbase`, and `ffbase2`.

Data transformations and aggregations with the ff and ffbase packages

Although the `ff` package authored by Adler, Glaser, Nenadic, Ochlschlagel, and Zucchini, is several years old it still proves to be a popular solution to large data processing with R. The title of the package *Memory-efficient storage of large data on disk and fast access functions* roughly explains what it does. It chunks the dataset, and stores it on a hard drive, while the `ff` data structure (or `ffdf` data frame), which is held in RAM, like the other R data structures, provides mapping to the partitioned dataset. The chunks of raw data are simply binary flat files in native encoding, whereas the `ff` objects keep the metadata, which describe and link to the created binary files. Creating `ff` structures and binary files from the raw data does not alter the original dataset in any way, so there is no risk that your data may get corrupted or lost. The `ff` package includes a number of general data-processing functions, which support the import of large datasets to R, their basic transformations such as recoding levels of factors, sampling, applying other functions to rows and columns, and setting various attributes to `ff` objects. The resulting data structures can be easily exported to TXT or CSV files.

The `ffbase` package authored by Edwin de Jonge, Jan Wijffels, and Jan van der Laan, on the other hand, extends the functionality of the original `ff` library by allowing users to apply a number of statistical and mathematical operations, including basic descriptive statistics, and other useful data transformations, manipulations, and aggregations such as creating subsets, performing cross-tabulations, merging `ff` objects, and transforming `ffdf` data frames, converting numeric `ff` vectors to factors, finding duplicated rows and missing values, and many more. Moreover, a very versatile `ffdfapply` function enables users to apply any function to the created binary flat files, for example, to easily calculate any statistic of interest for each level of a factor, and so on. The `ffbase` package also makes it possible to perform selected statistical models directly on `ff` objects such as classifications and regressions, least-angle regressions, random-forest classifications, and clustering. These techniques are available due to the `ffbase` package's connectivity with other third-party packages, supporting Big Data analytics such as `biglm`, `biglars`, `bigrf`, and `stream`.

In the following section, we will present several of the most widely used ff and ffbase functions, which you may wish to implement into your current workflows for Big Data processing and analytics.

We will be using a flights_sep_oct15.txt dataset (downloadable from the Packt Publishing website for this book), which contains all flights to and from all American airports in September and October 2015. The data have been obtained from the Bureau of Transportation Statistics (http://www.transtats.bts.gov/DL_SelectFields.asp?Table_ID =236&DB_Short_Name=On-Time) and we've selected 28 variables of interest that describe each flight, such as year, month, day of month, day of week, flight date, airline id, the names of the flight's origin and destination airports, departure and arrival times and delays, distance and air time of the flight, and several others. Feel free to mine as many months, years, or specific variables as you wish, but note that a complete year of data containing exactly the same 28 variables which we chose for our example will result in a file of slightly less than 1GB in size. The dataset used in this section (and the one you can download from Packt Publishing) is limited to two months only (951,111 rows in total), and hence its size is roughly 156MB (almost 19MB when compressed). This is, however, enough for us to guide you through some most interesting and relevant applications of the ff, ffbase, and ffbase2 packages. In addition to the main dataset, we also provide a small CSV file, which includes the full names of airlines to match with their IDs, contained within the AIRLINE_ID variable in the flight data. Also, during our tutorial, we will present statistics related to the elapsed time and used memory for each call for our example data, as well as for a 2GB version of the dataset which covers all flights to and from all American airports between January 2013 and December 2014. These benchmarks will be compared with similar calls performed using functions coming from the core R and other relevant third-party packages.

Before processing the data using packages related to ff, we first need to specify a path to a folder which will store our binary flat files-partitioned chunks of our original dataset. In your current working directory (which contains the data), you may explicitly create an additional folder directly from R console:

```
> system("mkdir ffdf")
```

If you are a Windows user, instead of the `system()` function you may need to use its equivalent `shell()` function to execute the commands through a shell.

Then, set the path to this newly created folder, which will store `ff` data chunks, for example:

```
> options(fftempdir = "/your_working_directory/ffdf")
```

Once this is done we may now upload the data as `ff` objects. Depending on the format of the data file, you may use either the `read.table.ffdf()` function or a convenience wrapper `read.csv.ffdf()` for CSV files. In addition to these functions, another package `ETLUtils`, created and maintained by Jan Wijffels (one of the `ffbase` package's original co-authors), extends the `ff` importing capabilities to include SQL databases such as Oracle, MySQL, PostgreSQL, and Hive, through functions which use DBI, RODBC, and RJDBC connections (more on these connections in Chapter 5, *R with Relational Database Management Systems (RDBMSs)*). Let's then import the data to R using a standard `read.table.ffdf()` function:

```
> flights.ff <- read.table.ffdf(file="flights_sep_oct15.txt",
                    sep=",", VERBOSE=TRUE,
                    header=TRUE, next.rows=100000,
                    colClasses=NA)
read.table.ffdf 1..100000 (100000)   csv-read=3.365sec ffdf-write=1.54sec
read.table.ffdf 100001..200000 (100000)   csv-read=3.596sec ffdf-write=0.595sec
read.table.ffdf 200001..300000 (100000)   csv-read=3.636sec ffdf-write=0.526sec
...
... #output truncated
...
read.table.ffdf 900001..951111 (51111)   csv-read=1.845sec ffdf-write=0.466sec
csv-read=34.24sec   ffdf-write=6.303sec   TOTAL=40.543sec
```

The `next.rows` argument sets how many rows of data will be assigned to each chunk. From the preceding output you can see that the data have been read in nine chunks, with the last part bringing in the remaining 51,111 cases of the original data. The output also gives us a basic estimate of the time spent on reading the data file, and writing its `ffdf` copies to a disk. In total, it took over 40 seconds to upload this relatively small dataset, and create the `ff` files in the previously specified folder.

The whole process of importing the data resulted in only one very small `ffdf` object (426.4 KB) being created in the R workspace, and 28 `ff` files of equal sizes (3.8 MB each) on a disk. It is also important here to mention that importing the data entailed only minimal costs in terms of RAM. We may now compare the `read.table.ffdf()` method with a standard `read.table()` procedure:

```
> flights.table <- read.table("flights_sep_oct15.txt",
                              sep=",", header=TRUE)
```

This more conventional method took just over 32 seconds to run; however it resulted in a much larger `data.frame` object created (101.9 MB) in the R workspace, and a little bit more memory usage. As we are working on a relatively small dataset of around 156 MB, the differences in RAM consumption will naturally be quite negligible. Let's then compare both approaches on much greater 2 GB-heavy data, which cover two full years (2013-2014) of flights (12,189,293 rows in total).

The `read.table.ffdf()` method took almost 456 seconds to import the dataset, in 23 chunks, creating ,as a result, just one `ffdf` R object of only 516.5 KB in size, and 28 `ff` data files (48.8 MB each, so nearly 1.37 GB in total). What's truly impressive is that the process involved a maximum of about 380 MB of RAM at most, which is generally just slightly above the base level of an R session in RStudio. The `read.table()` approach achieved a slightly faster import (441 seconds), but remember that this method does not include any writing to a disk, so obviously we will expect it to outperform the `ff` package on this measurement. The real difference, though, is in the resources used to complete the operation. The base `read.table()` function created one large `data.frame` object (1.3 GB) at huge memory costs; during the execution of this approach the RAM consumption oscillated between 2 GB and 3.6 GB, and at times it spiked up to as much as 4.85 GB. After the completion of the method, 4.13 GB of RAM was still in use and only an explicit call for the garbage collection (`gc()` function) lowered it to 1.47 GB-still more than four times higher than following the `read.table.ffdf()` application.

By this time you should see the obvious benefits of using the `ff` package for uploading large datasets to your R workspace. The question remains, however: What can you do with the `ff` or `ffdf` objects loaded to R?

You can begin by inspecting the `ffdf` data structure just as you will do with standard data frames in R:

```
> class(flights.ff)
[1] "ffdf"
> dim(flights.ff)
[1] 951111       28
> dimnames.ffdf(flights.ff)
[[1]]
NULL

[[2]]
 [1] "YEAR"                "MONTH"
 [3] "DAY_OF_MONTH"        "DAY_OF_WEEK"
 [5] "FL_DATE"             "UNIQUE_CARRIER"
 [7] "AIRLINE_ID"          "TAIL_NUM"
 [9] "FL_NUM"              "ORIGIN_AIRPORT_ID"
[11] "ORIGIN"              "ORIGIN_CITY_NAME"
[13] "ORIGIN_STATE_NM"     "ORIGIN_WAC"
[15] "DEST_AIRPORT_ID"     "DEST"
[17] "DEST_CITY_NAME"      "DEST_STATE_NM"
[19] "DEST_WAC"            "DEP_TIME"
[21] "DEP_DELAY"           "ARR_TIME"
[23] "ARR_DELAY"           "CANCELLED"
[25] "CANCELLATION_CODE"   "DIVERTED"
[27] "AIR_TIME"            "DISTANCE"
> str(flights.ff)
... #output truncated
```

The output of the last call will give you an understanding of how `ff` files on disk are mapped. The `ffdf` object is in fact a list with three components, which store virtual and physical attributes and the row names (in this case the `row.names` component is empty). The attributes hold metadata, which describe each variable and point to specific binary flat files.

We may now use the `read.csv.ffdf()` function to upload supplementary information with full names of airlines:

```
> airlines.ff <- read.csv.ffdf(file="airline_id.csv",
                               VERBOSE=TRUE, header=TRUE,
                               next.rows=100000,
                               colClasses=NA)
read.table.ffdf 1..1607 (1607)   csv-read=0.02sec ffdf-write=0.016sec
 csv-read=0.02sec    ffdf-write=0.016sec    TOTAL=0.036sec
```

We now have both datasets in R, so we can merge them by the `AIRLINE_ID` variable. As the names of variables differ, we first need to rename the `Code` variable in the `airlines.ff` object to `AIRLINE_ID` and the `Description` variable to `AIRLINE_NM`.

```
> names(airlines.ff) <- c("AIRLINE_ID", "AIRLINE_NM")
```

Let's merge both objects using the `merge.ffdf()` method:

```
> flights.data.ff <- merge.ffdf(flights.ff, airlines.ff, by="AIRLINE_ID")
```

The resulting `flights.data.ff` data frame is only 551.2 KB in size, and the merging process did not increase memory consumption. A similar operation executed on the 2 GB dataset with the use of the `merge.ffdf()` function from the `ffbase` package took just over 26 seconds to complete and, as a result of that, it created a `ffdf` data structure of 641.3 KB with minimal RAM costs. The large dataset previously uploaded to the R session with the standard `read.table()` function, and now merged with a small file containing names of airlines using the base `merge()` method, took more than 73 seconds to run, and increased the object size stored in RAM from 1.37 GB to 1.41 GB. What is even more striking is that, during this data merging, the memory usage peaked on a few occasions to 6.4 GB. It is clear to see how the `ff` approach can benefit Big Data processing and manipulation. The traditional core R methods, for example `read.table()` or `merge()`, applied on a dataset of just 2 GB, would produce out-of-memory errors on machines equipped with only 4 GB of RAM, and would most likely cause considerable problems even on PCs with 8 GB of RAM installed. The `ff` and `ffbase` packages provide, then, a very handy mechanism for avoiding memory-related issues at initial stages of large data processing.

With `ff` and `ffdf` objects you can use a number of base R functions without the need to transform them into native R data structures such as a `data.frame` or a `vector`. We saw that earlier when we applied the `names()` function to an `ffdf` data frame, to rename its variables. In a similar way, you can use `unique()` to extract all the names of the states for departing flights in our dataset:

```
> origin_st <- unique(flights.data.ff$ORIGIN_STATE_NM)
```

Alternatively, the `ffbase` package provides the `unique.ff()` function to apply to `ff` vectors. In the same way we are able to perform a cross-tabulation with the `table.ff()` method, for example the count of flights for each unique state of origin:

```
> orig_state_tab <- table.ff(flights.data.ff$ORIGIN_STATE_NM, exclude = NA)
```

When running table.ff() and table() functions on a ffdf structure, and a standard R data.frame object, respectively, you will see certain differences in how both functions perform. These differences are, again, best seen when processing a large dataset, for example 2 GB in size. The ff approach uses a maximum of 350 to 360 MB of RAM, but it takes almost 12 seconds to complete. The standard table() function is much faster on this data.frame object finishing the job in just over 1 second, but it increases the memory consumption by 700 MB, which, combined with the size of the data.frame already stored in RAM by R, and other earlier processes run on this object, uses up to 2.8 GB of available memory. It constitutes up to 10x greater RAM cost than through the ff and ffbase packages.

Following cross-tabulations, you may easily use other generic functions on ff and ffdf objects to get some basic descriptive statistics on your data, for example, mean(), quantile(), range(), and others. You should already know how to perform these simple operations if you read Chapter 2, *Introduction to R Programming Language and Statistical Environment*. Both ff and ffdf structures also work very well with functions contained in third-party packages such as describe() from the Hmisc package; however, in these situations, we need to explicitly tell R to treat our ffdf object as a standard data.frame using the as.data.frame.ffdf() function from the ff package:

```
> library(Hmisc)
> describe(as.data.frame.ffdf(flights.data.ff$DISTANCE))
      n  missing  unique    Info     Mean     .05     .10     .25
 951111        0    1241       1    816.2     168     224     370
    .50      .75     .90     .95
    641     1050    1721    2239
lowest :   31   36   67   68   69
highest: 4243 4502 4817 4962 4983
```

Again, if you run describe(), or a core R summary() function, on the *big* data with over 12,000,000 rows using an as.data.frame.ffdf() wrapper on a ffdf object, the memory consumption is relatively small (a spike of up to 920 MB for describe()), compared with describe() applied on a standard, large data.frame (a maximum memory usage of 5.2 GB). This time there is also barely any difference in the processing speed.

The ff-approach allows other data manipulation methods. For example, it is possible to convert a numeric ff vector to a factor ff using the cut.ff() function. In our example we will transform the DAY_OF_WEEK numeric variable to a new factor variable called WEEKDAY:

```
> flights.data.ff$WEEKDAY <- cut.ff(flights.data.ff$DAY_OF_WEEK,
+                             breaks = 7,
+                             labels = c("Monday", "Tuesday",
+                                        "Wednesday", "Thursday",
+                                        "Friday", "Saturday",
+                                        "Sunday"))
```

The preceding code only marginally engages the machine's resources. Even when run on a large dataset there is as little as 357 MB of RAM usage. A similar code applied on the standard data.frame object may use up to 4.4 GB of memory.

What makes the ff and ffbase packages even better tools, is their ability to perform quite complex dataaggregations. The ffdfdply() function deserves at least a few minutes of your attention. It allows us to carry out a *split-apply-combine* type of operation on an ffdf object. During the *apply* part of the process, you may specify any function (FUN parameter) to use in order to aggregate the data and store it as a separate ffdf object. In our flights example we will calculate a mean departure delay for each city of origin by calling summaryBy() from the doBy package in the FUN argument:

```
> DepDelayByOrigCity <- ffdfdply(flights.data.ff,
+                          split = flights.data.ff$ORIGIN_CITY_NAME,
+                          FUN=function(x) {
+                              summaryBy(DEP_DELAY~ORIGIN_CITY_NAME,
+                              data=x, FUN=mean, na.rm=TRUE) }
+                          )
... #output truncated
```

The output provides a verbose explanation of how the splits are created, and presents us with informative details on the progress of the function execution (these have been removed from the listing above to save space). Depending on the size of your raw data, the output will contain more, or fewer, splits and will take a longer, or shorter, time to run. Performed on the large dataset of 2 GB, ffdfdply() took 181 seconds to finalize the aggregation. During this time the memory usage fluctuated between 250 MB and 300 MB, and it reached 459 MB only for a very short time. You can compare these results with the original summaryBy() performance on a large data.frame object loaded with the generic core R approach (read.table()).

summaryBy() was much faster than its implementation on an ffdf with ffdfdply(), as it took only 5.6 seconds to complete, but the RAM consumption momentarily sky-rocketed to reach 4.85 GB. As our raw data are split between several binary flat files through the ff package, many operations performed on an ffdf object will obviously be much slower than those run directly in RAM on a standard data.frame. The ff-approach through the ffdfdply() function requires that each very small partition of the data is initially extracted to RAM, where it is processed with a specific function (FUN) applied on the selected data. The result of this computation is then finally appended to the output ffdf object, in our case it was the DepDelayByOrigCity object. It doesn't surprise us that all these tasks may take a relatively long time to complete, but the question to answer here is: Which strategy of data aggregation are you more likely to choose? If you are dealing with out-of-memory data, are you able (or allowed) to compromise on the speed of processing?

The output object of the preceding aggregation is another ffdf structure. You can convert it to a standard data.frame through the previously introduced as.data.frame.ffdf():

```
> plot1.df <- as.data.frame.ffdf(DepDelayByOrigCity)
> str(plot1.df)
'data.frame':   305 obs. of  2 variables:
 $ ORIGIN_CITY_NAME: Factor w/ 305 levels "Abilene, TX",..: 13 41 53 56 121 93 147 162 181 26 ...
 $ DEP_DELAY.mean  : num  4.52 6.35 5.51 5.15 5.62 ...
```

Now the data.frame is small enough to be easily used with all functions available in core R or third-party packages. For example you may want to sort the cities from which the flights departed based on the average departure delay in the descending order:

```
> plot1.df <- orderBy(~-DEP_DELAY.mean, data=plot1.df)
> plot1.df
            ORIGIN_CITY_NAME DEP_DELAY.mean
198              Pago Pago, TT     49.11764706
286             Adak Island, AK     21.23529412
289            Christiansted, VI     20.46875000
268       North Bend/Coos Bay, OR     15.68055556
271              Plattsburgh, NY     15.63333333
302                Nantucket, MA     13.00952381
209    Scranton/Wilkes-Barre, PA     12.38565022
... #output truncated
```

Such prepared data can now be very conveniently used for visualizations or further data crunching with standard R techniques.

The `ff` and `ffbase` packages also support subsetting of `ffdf` objects through the `subset.ffdf()` method. It takes similar arguments to the generic `subset()` function:

```
> subs1.ff <- subset.ffdf(flights.data.ff, CANCELLED == 1,
+                   select = c(FL_DATE, AIRLINE_ID,
+                              ORIGIN_CITY_NAME,
+                              ORIGIN_STATE_NM,
+                              DEST_CITY_NAME,
+                              DEST_STATE_NM,
+                              CANCELLATION_CODE))
```

In the preceding code, we have specified that we want to subset all records with cancelled flights only, and that in our new `subs1.ff` `ffdf` object we wish to include only selected (`select` argument) variables from the original `flights.data.ff` object. If applied on the `ffdf` object, which maps to the large 2 GB-heavy data, the `subset.ffdf()` function consumes only minimal amounts of memory. As with preceding examples, the generic `subset()` executed on the `data.frame` object was more RAM-hungry; it used over 0.7 GB of available memory.

The newly created `ffdf` object may be exported to a flat data file; in fact it will be saved as seven separate `ff` files, one for each variable (that is column) in the `subs1.ff` object:

```
> save.ffdf(subs1.ff)
```

By default the `save.ffdf()` function saves the flat files to a new folder called `ffdb`, which will be created automatically in your working directory. You can, of course, change the name of the destination folder by altering the `dir` argument (by default set to `"./ffdb"`). The resulting flat files are stored with filenames in the format:
`<ffdf_name>$<variable_name>.ff`.

The exported `ff` files can be loaded back to the R session using the `load.ffdf()` command. If you want to see how it works, remove the `subs1.ff` object from your workspace and type the correct path to the destination folder with `ff` files in the `load.ffdf()` function:

```
> rm(subs1.ff)
> load.ffdf("~/Desktop/data/ffdb")
```

A new *old* `subs1.ff` object should appear in the environment with all its default metadata and `ffdf` features. But you may also want to export it into CSV or TXT files. This functionality is also possible through the `ff` package. The `write.csv.ffdf()` or `write.table.ffdf()` expressions will accomplish this task for you:

```
> write.csv.ffdf(subs1.ff, "subset1.csv", VERBOSE = TRUE)
```

So far in this section, we have presented numerous applications of several functions from the `ff` and `ffbase` packages, which you might find very useful when processing datasets that are larger than the available RAM resources. By a rule of thumb, and following the recommendations given by the authors of `ff` and `ffbase`, the packages will generally benefit workflows executed on data up to 10 times bigger than the memory capacity. This threshold seems very reasonable. Based on our extensive testing, all data manipulations, transformations, and aggregations performed on `ff` or `ffdf` objects and discussed in this section used a maximum of only 425.3 MB of RAM the final value of the memory resources assigned to the R session processes at the end of the R script run on flat files (the `ff`-approach). On the other hand, the same data processing activities, but utilizing a more generic approach, and executed on a `data.frame` object which was loaded to R with the `read.table()` function, and as result stored in RAM, consumed as much as 3.7 GB of memory with the moment of execution of the last line of the R code. This almost ten-fold difference makes sense. The value of RAM usage in the `data.frame` approach will have probably been even greater if we hadn't used the garbage collection call (`gc()`) very frequently.

Unfortunately the `ff` and `ffbase` packages are not without their own issues and limitations. In the preceding examples, we've shown that in the `ff` approach, data input is not as fast as when uploaded to the R session, even using generic `read.table()` or `read.csv()`. Remember that, in order to create an `ff` or `ffdf` object, containing mapping, to the raw data, the original dataset needs to be loaded in chunks, and their content copied to binary flat files. The processes of chunking, mapping, and writing to `ff` files may take considerably more time than simply loading to a `data.frame` using core R functions. Later in this chapter you will learn much faster methods of getting the data into R, using, for example, the `fread()` function from the `data.table` package. Second, we've already explained that some functions may be slower in execution on `ffdf` objects than their counterparts on standard data frames. Again, data first have to be retrieved from chunks and only then may a function or an operation be applied to this particular, small part of the data.

The output of the function is consequently appended, one-by-one, to a new `ff` or `ffdf` object. These several processes extend the time spent on the execution of a function. Third, the filenames of chunks are quite confusing, and definitely not user-friendly. Assuming we will like to find the name of the `ff` file where the `AIRLINE_ID` variable is stored we can obtain it as follows:

```
> basename(filename(flights.data.ff$AIRLINE_ID))
[1] "ffdfe9c4f870103.ff"
```

This file naming convention makes working with flat files very difficult, especially when moving data around.

Despite these flaws, the `ff` and `ffbase` packages offer an interesting alternative to transformations and aggregation of out-of-memory data in R. The `ffbase` library is currently being re-developed by Edwin de Jonge to include grammar and internal functions from a very popular data manipulation package `dplyr`. The re-branded version of `ffbase`, now called `ffbase2`, contains a set of transformation functions such as `summarize()`, `group_by()`, `filter()`, or `arrange()` that are known from the `dplyr` package, but will also be applicable to `ffdf` objects. As the `ffbase2` library is still under development, its current version can only be installed from Edwin de Jonge's GitHub repository at `https://github.com/edwindj/ffbase2` and its functionality is still very limited, but you may find some of its methods operational. In order to install and load this package you must have the recent version of the `devtools` library installed and ready-to-use in your RStudio:

```
> install.packages("devtools")
> devtools::install_github("edwindj/ffbase2")
> library(ffbase2)
```

The use of `ff` and `ffbase` packages is not just limited to data transformations and aggregations. In the following section, you will learn how to perform more complex Big Data modeling and analytics operations on `ffdf` objects.

Generalized linear models with the ff and ffbase packages

If you followed the preceding section, you should already know how to transform and aggregate data contained within the `ffdf` objects. But the `ffbase` package also allows us to run a **Generalized Linear** Model (**GLM**) through its `bigglm.ffdf()` function.

> In case you have forgotten, GLM is simply the generalization of general linear models. In other words, GLMs try to determine a linear relationship between a **dependent variable** (*Y*) and **predictor variables** (*Xi*), but they differ from general linear models in two important aspects. Firstly, GLMs don't assume that the dependent variable comes from a normal distribution; it may well be generated from a **non-continuous distribution,** for example if it represents one of only two (**binomial**) or more (**multinomial**) possible discrete outcomes, or when the model predicts the count (**Poisson**), for example the number of children in the household (we know this value has to be an integer and we also assume the distribution of this variable to be highly skewed). The second reason for the implementation of GLM is that the predicted relationship may not in fact be linear. For example, the relationship between personal income and the age of a person is not linear in nature. In generalized linear models, the values of the response variables are predicted from a linear combination of predictors with the help of a **link function**. The use of an appropriate link function depends on the assumed distribution of the dependent variable (more precisely distribution family of its residuals); for example, for the Gaussian distribution you will use the **identity link** function, but for the binomial distribution (for example in the logistic regression) you will have a choice between the **logit**, **probit**, and **cloglog** functions. There is a large selection of good resources which provide detailed explanations of how to implement GLMs in R. Recommended online introductory GLM tutorials are available at: `http://data.princeton.edu/R/glms.html` and `http://www.statmethods.net/advstats/glm.html`. The `glm()` function from the `stats` package (within the base R installation) is the most frequently used method for fitting GLM and its documentation can offer further insights into how GLM are operationalized in R.

In order to use the `bigglm.ffdf()` function you need to install the `biglm` package, which is called by the `ffbase` library. Please use standard `install.packages("biglm")` and `library(biglm)` statements to do this.

Thomas Lumley's `biglm` package provides two main methods that create either linear (`biglm()`) or generalized linear (`bigglm()`) objects from data larger than available memory resources. Apart from standard support for data frames, `biglm` provides connectivity to databases through `SQLite` and `RODBC` connections. The `ffbase` package supplies just a wrapper function in the form of the `bigglm.ffdf()` method, that can be applied to `ffdf` objects. The following presents a logistic regression example as a GLM application of this function.

Logistic regression example with ffbase and biglm

In **logistic regression** we predict a binomial or multinomial outcome of a categorical variable, for example, you may want to predict either a yes or no answer, win or lose, healthy or sick, or one of the levels of the ordered factor, for example, very happy, somewhat happy, neither happy nor unhappy, somewhat unhappy, or very unhappy. In the following example, we will use a dataset called the *Chronic Kidney Disease Dataset*, which is available from the excellent **Machine Learning Depository** maintained by the University of California Irvine at http://archive.ics.uci.edu/ml/index.html. Although the dataset is very small (it contains merely 400 rows), you will learn how to expand it to out-of-memory size through `ff` and `ffbase` functions. First, however, let's load the data in. Because of the small size of the file, we can upload it straight to R's workspace, but remember that for large files you will need to use `read.table.ffdf()` or `read.csv.ffdf()` (or their variants) to map the `ffdf` objects to the raw data stored on a disk. Another issue with our Chronic Kidney Disease Dataset is that it comes in the **Attribute-Relation File Format (ARFF)** format, so we have to use a third-party package called `RWeka` to load this type of file into R:

```
> library(RWeka)
> ckd <- read.arff("ckd_full.arff")
> str(ckd)
'data.frame':   400 obs. of  25 variables:
 $ age   : num  48 7 62 48 51 60 68 24 52 53 ...
 $ bp    : num  80 50 80 70 80 90 70 NA 100 90 ...
 ... #output truncated
 $ class: Factor w/ 2 levels "ckd","notckd": 1 1 1 1 1 1 1 1 1 1 ...
> levels(ckd$class)
[1] "ckd"      "notckd"
```

At this stage, we may already define what we will like to predict in our logistic regression model. We simply want to obtain the coefficients predicting the level of `class` variable- whether an individual suffers from a chronic kidney disease, or not, based on several chosen predictor variables such as age (`age` variable), blood pressure (`bp`), blood glucose random (`bgr`), and others. The `class` variable is therefore binomially coded as 1 = `ckd` (denoting chronic kidney disease) and 2 = `notckd` (no chronic kidney disease). To improve the clarity of our model, we will first transform the factor into a numeric variable and recode the values, so that 0 will now stand for `notckd`, and 1 will denote `ckd`. It is easier to think of a lack of disease as the base value (hence 0 assigned to the `class` variable for those individuals who do not have chronic kidney disease). We will also keep values of this variable as numeric, and without labels, for simplicity. We can achieve this transformation by recoding the existing levels of the `class` variable into the recommended ones, using `recoder()` function from the `ETLUtils` package. Following the recoding transformation we may want to obtain the class distribution for both levels of the variable:

```
> ckd$class <- as.numeric(ckd$class)
> library(ETLUtils)
> ckd$class <- recoder(ckd$class, from = c(1,2), to=c(1,0))
> table(ckd$class)
  0   1
150 250
```

We are now dealing with a standard, small `data.frame` object and we could now easily run a typical generalized linear model algorithm in the form of logistic regression with the `glm()` function:

```
> model0 <- glm(class ~ age + bp + bgr + bu + rbcc + wbcc + hemo, data =
ckd, family=binomial(link = "logit"), na.action = na.omit)
> model0
Call:  glm(formula = class ~ age + bp + bgr + bu + rbcc + wbcc + hemo,
    family = binomial(link = "logit"), data = ckd, na.action = na.omit)
Coefficients:
(Intercept)          age           bp          bgr
 12.6368996    0.0238911    0.1227932    0.0792933
         bu         rbcc         wbcc         hemo
  0.0026834   -0.9518575    0.0002601   -2.3246914
Degrees of Freedom: 236 Total (i.e. Null);  229 Residual
   (163 observations deleted due to missingness)
Null Deviance:      325.9
Residual Deviance: 35.12    AIC: 51.12
```

From the function structure, and the output, you can see that we have used seven different predictor variables for our logistic regression model0 model. Because we were interested in predicting a binomial dependent variable (class), we set the family argument to binomial with a link function defined as logit for two outcomes (or 1), and we also assumed that the errors were logistically distributed. The simple output of the model presents the values of coefficients without any suggestions as to their probability, but with a number of other relevant statistics such as degrees of freedom, null and residual deviances, and the **Akaike Information Criterion (AIC)**, which informs us of the quality of our model, and may help with the correct model selection (generally the lower the better). The summary() function, applied to the model, will provide us with more detailed information about our regression:

```
> summary(model0)
Call:
glm(formula = class ~ age + bp + bgr + bu + rbcc + wbcc + hemo,
    family = binomial(link = "logit"), data = ckd, na.action = na.omit)
Deviance Residuals:
    Min      1Q   Median      3Q      Max
-1.3531 -0.0441 -0.0025  0.0016  3.2946
Coefficients:
              Estimate Std. Error z value Pr(>|z|)
(Intercept) 12.6368996  6.2190363   2.032 0.042157 *
age          0.0238911  0.0327356   0.730 0.465500
bp           0.1227932  0.0591324   2.077 0.037840 *
bgr          0.0792933  0.0212243   3.736 0.000187 ***
bu           0.0026834  0.0296290   0.091 0.927838
rbcc        -0.9518575  0.8656291  -1.100 0.271501
wbcc         0.0002601  0.0002432   1.070 0.284725
hemo        -2.3246914  0.6404712  -3.630 0.000284 ***
---
Signif. codes:  0 '***' 0.001 '**' 0.01 '*' 0.05 '.' 0.1 ' ' 1
(Dispersion parameter for binomial family taken to be 1)
    Null deviance: 325.910  on 236  degrees of freedom
Residual deviance:  35.121  on 229  degrees of freedom
  (163 observations deleted due to missingness)
AIC: 51.121
Number of Fisher Scoring iterations: 9
```

From the output, we can find three statistically significant predictor variables: `bp` (blood pressure), `bgr` (blood glucose random), and `hemo` (haemoglobin level). Of course the generic `glm()` function was performed on a `data.frame` object with only 400 observations. But the `ffbase` package allows the implementation of generalized linear models on `ffdf` objects mapped to much larger, out-of-memory raw data. Let's firstly re-run logistic regression on a small `ffdf` object created from our `ckd` data frame. This is quite straightforward:

```
> library(ffbase)
> options(fftempdir = "~/ffdf")
> ckd.ff <- as.ffdf(ckd)
> dimnames(ckd.ff)
[[1]]
NULL
[[2]]
 [1] "age"   "bp"    "sg"    "al"    "su"    "rbc"   "pc"
 [8] "pcc"   "ba"    "bgr"   "bu"    "sc"    "sod"   "pot"
[15] "hemo"  "pcv"   "wbcc"  "rbcc"  "htn"   "dm"    "cad"
[22] "appet" "pe"    "ane"   "class"
```

Before transforming `data.frame` into `ffdf`, make sure that you set your `fftempdir` to a directory where you want to keep your temporary `ff` files as shown in the preceding listing. Because our newly created `ckd.ff` object maps to very small data, you can run either the previously used `glm()` function or `bigglm.ffdf()` from the `ffbase` package. Both outputs should be almost identical. We will use `bigglm.ffdf()` as you already know how to apply a standard `glm()` function:

```
> model1 <- bigglm.ffdf(class ~ age + bp + bgr + bu + rbcc + wbcc + hemo,
data = ckd.ff, family=binomial(link = "logit"), na.action = na.exclude)
> summary(model1)
Large data regression model: bigglm(class ~ age + bp + bgr + bu + rbcc +
wbcc + hemo, data = ckd.ff,
    family = binomial(link = "logit"), na.action = na.exclude)
Sample size =  400
failed to converge after 8  iterations
              Coef    (95%      CI)     SE        p
(Intercept) 12.6330  0.3275  24.9384  6.1527  0.0401
age          0.0239 -0.0410   0.0888  0.0324  0.4616
bp           0.1228  0.0061   0.2394  0.0583  0.0353
bgr          0.0793  0.0375   0.1210  0.0209  0.0001
bu           0.0027 -0.0558   0.0611  0.0292  0.9268
rbcc        -0.9515 -2.6587   0.7557  0.8536  0.2650
wbcc         0.0003 -0.0002   0.0007  0.0002  0.2780
hemo        -2.3239 -3.5804  -1.0675  0.6282  0.0002
```

The only difference between both methods is that `bigglm.ffdf()` does not identify which of the predictors were found to be significant, so users need to screen *p values* visually, and decide which variables fall below the significance level. Similarly, the `bigglm.ffdf()` function can be applied on large out-of-memory datasets. In the following example we will expand the size of our sample to 8,000,000 and also to 80,000,000 observations, but you may try to make it even bigger if you like.

Also, remember that if you have concerns about the performance of your machine. or you do not have at least 20GB of free hard drive space available on your computer, we advise you not to run the following scripts. Should you still want to run them, you may wait until *Online Chapter, Pushing R Further* (https://www.packtpub.com/sites/default/files/downloads/5396_6457OS_PushingRFurther.pdf) where you will be shown how to build and deploy a much more powerful virtual machine that will allow heavier data processing.

The increase of the data size will simply mean that you will have to wait for the algorithm to do its job for much longer. In the first example, the processing time will be relatively short, but for the second, 10 times larger data, we decided to keep the waiting time to a reasonable two to four minutes, depending on your machine's architecture and processing power. But before we apply logistic regression to our data, let's see how you can expand the data with the `ffbase` package:

```
> ckd.ff$id <- ffseq_len(nrow(ckd.ff))
```

The first operation will simply append a new `id` variable (an `ff` vector with a row number for each record) to our `ckd.ff` object. Let's now create a separate `ffdf` object, which will hold 20,000 times (or 200,000 for 80,000,000 cases) more data than the original file. For that purpose, we will use the `expand.ffgrid()` function from the `ffbase` package:

```
> system.time(big.ckd.ff <- expand.ffgrid(ckd.ff$id, ff(1:20000)))
```

It takes roughly just over 2 seconds to create a new `ffdf` data frame with 8,000,000 rows and around 27 seconds for data with 80,000,000 observations. In order to expand the original data, we just need to merge this newly created data frame with the `ckd.ff` object based on the common variable (`id`). Therefore, we first need to rename the columns in the `big.ckd.ff` with 8,000,000 observations, and assign an `id` name to the `Var1` column:

```
> colnames(big.ckd.ff) <- c("id","expand")
> big.ckd.ff <- merge.ffdf(big.ckd.ff, ckd.ff, by.x="id", by.y="id",
all.x=TRUE, all.y=FALSE)
```

The merging of `ffdf` objects may take some time. In our case it took 29 seconds for 8,000,000 rows and 227 seconds (3 mins 47 secs) when we merged the data with 80,000,000 cases. At this stage, you may also want to see what happened with the original `ff` files created in your `fftempdir` directory. All files became much larger; in fact, our `big.ckd.ff` data frame now maps to raw data of 1.16 GB and 11.62 GB in size for 8,000,000 and 80,000,000 rows respectively. During the process of merging, our R session has also additionally consumed over 700 MB of RAM (on top of around 500 to 600 MB RAM already used by the R session in earlier computations and data transformations in this tutorial). This example can, therefore, serve as another piece of evidence supporting the claim that the `ff` and `ffbase` approaches are suitable for data up to 10 times the amount of RAM installed on a machine.

Finally, we are ready to perform a logistic regression on our large data, in the same way we did with an `ffdf` object containing only 400 observations:

```
> model2 <- bigglm.ffdf(class ~ age + bp + bgr + bu + rbcc + wbcc + hemo,
  data = big.ckd.ff, family=binomial(), na.action = na.exclude)
> summary(model2)
Large data regression model: bigglm(class ~ age + bp + bgr + bu + rbcc +
wbcc + hemo, data = big.ckd.ff, family = binomial(), na.action =
na.exclude)
Sample size =  8000000
failed to converge after 8   iterations
            Coef    (95%      CI)     SE p
(Intercept) 12.6330 12.5460 12.7200 0.0435 0
age          0.0239  0.0234  0.0243 0.0002 0
bp           0.1228  0.1219  0.1236 0.0004 0
bgr          0.0793  0.0790  0.0796 0.0001 0
bu           0.0027  0.0023  0.0031 0.0002 0
rbcc        -0.9515 -0.9635 -0.9394 0.0060 0
wbcc         0.0003  0.0003  0.0003 0.0000 0
hemo        -2.3239 -2.3328 -2.3150 0.0044 0
```

This job took around 181 seconds to complete for the smaller `ffdf` object, but almost 1,920 seconds (32 minutes) when processing larger data with 80,000,000 cases. Although obtaining logistic regression model parameters using the `bigglm.ffdf()` approach is not the fastest process in the world, it only consumed minimal (100 to 200MB) memory resources for either size of the data. The ability to estimate logistic regression coefficients through the `bigglm.ffdf()` function comes at the expence of slow processing speed. Another shortcoming of the `ffbase` package, which is still present in its current version, is that for large `ffdf` objects it doesn't seem to display *p values* for each predictor; all are set to zero.

An explicit call to retrieve particular values of a predictor (for example the `age` variable) does not help either:

```
> summary(model2)$mat[2,]
        Coef              (95%             CI)              SE
0.02387827036   0.02373321074   0.02402332998   0.00007252981
              p
0.00000000000
```

There is also a known issue with model convergence after a specific number of **Fisher scoring iterations**. As you look at the preceding outputs, you will notice that they include a warning message indicating that they *failed to converge after 8 iterations*. Unfortunately, the documentation of the `ffbase` or `biglm` packages does not explain how to set the `maxit` argument properly for different sizes of data, and types of predicted variables. It is a somewhat hit-and-miss affair. Based on my experiences, it is also recommended to adjust the `chunksize` parameter of the `bigglm.ffdf()` function-generally try to increase the value (by default set to 5000) for larger data. The following implementation of the `bigglm.ffdf()` command works on our 8,000,000-strong dataset, but because of the increased number of iterations, the processing time expanded to 706 seconds (almost four times longer than previously):

```
> model2 <- bigglm.ffdf(class ~ age + bp + bgr + bu + rbcc + wbcc + hemo,
data = big.ckd.ff, family=binomial(), na.action = na.exclude, sandwich =
TRUE, chunksize = 20000, maxit = 40)
> summary(model2)
Large data regression model: bigglm(class ~ age + bp + bgr + bu + rbcc +
wbcc + hemo, data = big.ckd.ff,
    family = binomial(), na.action = na.exclude, sandwich = TRUE,
    chunksize = 20000, maxit = 40)
Sample size =  8000000
              Coef     (95%       CI)     SE  p
(Intercept)  12.6369  12.5332  12.7406  0.0518  0
age           0.0239   0.0235   0.0243  0.0002  0
bp            0.1228   0.1223   0.1233  0.0002  0
bgr           0.0793   0.0789   0.0796  0.0002  0
bu            0.0027   0.0023   0.0030  0.0002  0
rbcc         -0.9519  -0.9589  -0.9448  0.0035  0
wbcc          0.0003   0.0003   0.0003  0.0000  0
hemo         -2.3247  -2.3375  -2.3118  0.0064  0
Sandwich (model-robust) standard errors
```

A good compromise, however, is to increase `chunksize` and simultaneously lower the value of maximum **Fisher iterations** and not use the `sandwich` parameter at all. It is also important here to warn you that inflating `chunksize` may increase the usage of RAM-so make sure to find the right balance when adjusting the arguments of the `bigglm.ffdf()` function. The following call runs for around three minutes on data with 8,000,000 rows (and almost 30 minutes for 80,000,000 observations) and does not throw the warning message:

```
> model2 <- bigglm.ffdf(class ~ age + bp + bgr + bu + rbcc + wbcc + hemo,
data = big.ckd.ff, family=binomial(), na.action = na.exclude, chunksize =
100000, maxit = 20)
> summary(model2)
Large data regression model: bigglm(class ~ age + bp + bgr + bu + rbcc +
wbcc + hemo, data = big.ckd.ff,
    family = binomial(), na.action = na.exclude, chunksize = 100000, maxit
= 20)
Sample size =  8000000
              Coef    (95%      CI)      SE p
(Intercept) 12.6369 12.5489 12.7249 0.0440 0
age          0.0239  0.0234  0.0244 0.0002 0
bp           0.1228  0.1220  0.1236 0.0004 0
bgr          0.0793  0.0790  0.0796 0.0002 0
bu           0.0027  0.0023  0.0031 0.0002 0
rbcc        -0.9519 -0.9641 -0.9396 0.0061 0
wbcc         0.0003  0.0003  0.0003 0.0000 0
hemo        -2.3247 -2.3337 -2.3156 0.0045 0
```

Overall the `ff/ffdf` approach allows easy data manipulations and flexible aggregations for datasets up to 10 times bigger than the available RAM resources. The *crunched* data can then be conveniently transformed into a standard `data.frame` object for further processing, analysis, or visualizations. The `ffbase` package extends basic `ff` functionalities, and offers several very useful descriptive statistics, and even more complex analytics, in the form of the `bigglm.ffdf()` function. The major and obvious drawback of the `ff/ffdf` method is that, owing to the data being processed on a disk, the jobs are generally run much slowly, but this might be a fair price to pay for the ability to manipulate out-of-memory data on a single machine using R.

In the next section we will present another approach, which allows processing of larger-than-memory data through the `bigmemory` package and its extensions.

Expanding memory with the bigmemory package

As previously stated, the usual importing of data into R using the `read.table()` family of functions is limited by the available RAM. In fact, however, we always need to allow additional memory resources for further data manipulations, and other R objects that we create in the process, during the same R session. The use of `read.table()` functions also results in a memory overhead of anything from 30% to 100% of the original data size. It simply means that an import of a 1 GB file to an R workspace requires roughly 1.3 GB to 2 GB of available RAM to succeed. Fortunately, you already know how to deal with much large data using the `ff/ffdf` approach. The `bigmemory` package, authored by Michael J. Kane and John W. Emerson (http://www.bigmemory.org), offers an alternative solution, but again, it's not without its own limitations and disadvantages which we will address in this section. In general, `bigmemory` facilitates data import, processing, and analysis by allocating the S4 class objects (matrices) to shared memory and using memory-mapped files. It also provides a C++ framework for the development of tools and functions, allowing greater processing speed and memory management. All this sounds good, but one aspect should probably alarm you straightaway. The issue is that the `bigmemory` package supports only matrices, and from Chapter 2, *Introduction to R Programming Language and Statistical Environment* we know that matrices can hold only one type of data. Therefore, if your dataset includes a mix of character, numeric, or logical variables, they cannot be used together in `bigmemory`. There are, however, several ways of dealing with this problem. You can:

- Mine and collect your data using methods that only code responses as numeric values (and keep the labels or factor levels in a separate file if necessary), or
- If your original dataset is within the range of the available RAM, import it using `read.table()` or, even better, through the `data.table` package (which we will present later in this chapter) and transform the classes of variables to numeric only, or
- For out-of-memory data, use the `ff` package, change the classes of variables to numeric, and save the resulting data to another file on a disk.

The list of options is not inclusive of all relevant methods, but, from my personal experience, I found myself using the preceding strategies more often than others. Let's now prepare some data for the import and further processing with the `bigmemory` package using the second method.

In this part we will be using an interesting governmental dataset *National Energy Efficiency Data – Framework (NEED)* provided by the Department of Energy & Climate Change in the United Kingdom. The public use dataset and documentation files are available at `https://www.gov.uk/government/statistics/national-energy-efficiency-data-framework-need-anonymised-data-2014`, but you can also download the data and the look-up tables, with value labels, from Packt Publishing's website for this book. The public use file, which we are going to use here, is very small (7.8 MB) as it only contains a representative sample of 49,815 records, but you are encouraged to test the R code on a much larger sample of 4,086,448 cases (which you can obtain from UK Data Service at `https://discover.ukdataservice.ac.uk/catalogue/?sn=7518`. In short, the data contain information on annual electricity and gas consumption, energy efficiency characteristics, and socio-demographic details of UK-based households over several years, from 2005 until 2012.

Once you save the data to your working directory, you may import the public use file through a standard `read.csv()` command:

```
> need0 <- read.csv("need_puf14.csv", header = TRUE, sep = ",")
> str(need0)
'data.frame':   49815 obs. of  50 variables:
 $ HH_ID           : int  1 2 3 4 5 6 7 8 9 10 ...
 $ REGION          : Factor w/ 10 levels "E12000001","E12000002",..: 7 2 2 5 3 7 6 5 7 3 ...
 $ IMD_ENG         : int  1 4 4 1 1 2 3 5 4 2 ...
...#output truncated
```

Notice that our data contains a number of categorical variables (`factors`). You can extract `class` information for all the variables in the data using the following snippet:

```
> classes <- unlist(lapply(colnames(need0), function(x) {
+     class(need0[,x])
+ }))
> classes
[1] "integer" "factor"  "integer" "integer" "integer" "factor"
[7] "integer" "factor"  "integer" "factor"  "integer" "factor"
```

Based on the contents of the `classes` object, we can now identify indices of factors and use this information to convert them to integers with a `for()` loop statement:

```
> ind <- which(classes=="factor")
> for(i in ind) {need0[,i] <- as.integer(need0[, i])}
> str(need0)
'data.frame':   49815 obs. of  50 variables:
 $ HH_ID           : int  1 2 3 4 5 6 7 8 9 10 ...
 $ REGION          : int  7 2 2 5 3 7 6 5 7 3 ...
 $ IMD_ENG         : int  1 4 4 1 1 2 3 5 4 2 ...
...#output truncated
```

You can now export the `need0 data.frame` to an external file on a disk and import it back using the `read.big.matrix()` function from the `bigmemory` package:

```
> write.table(need0, "need_data.csv", sep = ",",
+             row.names = FALSE, col.names = TRUE)
> need.mat <- read.big.matrix("need_data.csv", header = TRUE,
+                             sep = ",", type = "double",
+                             backingfile = "need_data.bin",
+                             descriptorfile = "need_data.desc")
> need.mat
An object of class "big.matrix"
Slot "address":
<pointer: 0x108b5cb00>
```

The `read.big.matrix()` call comes with two useful arguments: `backingfile` and `descriptorfile`. The first is responsible for holding the raw data on a disk, the latter keeps all metadata describing the data (hence the name). By mapping the resulting object to data stored on a disk, they allow users to import large, out-of-memory data into R or attach a cached `big.matrix` without explicitly reading the whole data again when needed. If you didn't set these two arguments, the `bigmemory` package will still import the file; however, the data will not be stored in a backing file, and hence it will take much longer to load the data in the future. This difference can be easily noticed in the following experiment, in which we compared the time taken and RAM used for importing a large sample of over 4 million records of NEED data using `read.big.matrix()` implementations with and without `backingfile` and `descriptorfile` arguments, and the standard `read.csv()` approach. On each occasion the R session was started from fresh and every time the initial memory usage was found to be the same:

```
> gc()
          used (Mb) gc trigger (Mb) max used (Mb)
Ncells 431186 23.1     750400 40.1    592000 31.7
Vcells 650931  5.0    1308461 10.0    868739  6.7
```

The first attempt included the following call:

```
> need.big1 <- read.big.matrix("need_big.csv", header = TRUE,
+                              sep = ",", type = "double")
```

It took 102 seconds to create the need.big1 object-a big.matrix of only 664 bytes in size. The RAM usage did not increase significantly above the base max used values. No files were created in the working directory as a result of this call.

The second trial contained the backingfile and descriptorfile arguments as follows:

```
> need.big2 <- read.big.matrix("need_big.csv", header = TRUE,
+                              sep = ",", type = "double",
+                              backingfile = "need_big.bin",
+                              descriptorfile = "need_big.desc")
```

Just as in the first trial, the newly created big.matrix was only 664 bytes light and the memory usage remained at the base level. It also took exactly 102 seconds to import the data, however two new files named as specified in the backingfile and descriptorfile emerged in the working directory. The first was 1.73 GB-heavy and the latter contained only 1 KB of metadata. The second step of this trial involved the attachment of big.matrix using the reference to the descriptor file with either the attach.resource() or attach.big.matrix() functions:

```
> rm(list=ls())
> need.big2b <- attach.big.matrix("need_big.desc")
```

This operation took only 0.001 seconds to complete, which clearly shows how useful caching through backingfile and descriptorfile can be, especially if you need to import the same data a few times, or in several separate, but parallel R sessions.

The third trial compared the previous two with a standard read.csv() function performance. The following call was used:

```
> need.big3 <- read.csv("need_big.csv", header = TRUE,
+                       sep = ",")
```

As anticipated, this statement engaged the memory resources the most of all three trials, creating a `data.frame` object of 841.8 MB in size, and using quite substantial amount of RAM in the process:

```
> gc()
            used    (Mb)  gc trigger      (Mb)   max used      (Mb)
Ncells     447877   24.0     3638156     194.3    4541754     242.6
Vcells  113035196  862.4   356369969    2718.9  354642630    2705.8
```

R also spent almost 191 seconds completing the job, much longer than in the previous attempts. The winner is obvious, and, depending on your needs, you can either choose the `read.big.matrix()` implementation with or without data caching. Personally, I prefer to store a copy of the data through `backingfile` and `descriptorfile` in case I have to import `big.matrix` again. Although it takes a rather generous slice of hard drive space, it saves a great amount of time, especially when you deal with a multi-GB dataset. But reading and writing big matrices are not the only *selling* points of the `bigmemory` package. The library can also perform quite an impressive number of data management and analytical tasks. First of all, you can apply generic functions such as `dim()`, `dimnames()`, or `head()` to the `big.matrix` object:

```
> dim(need.mat)
[1] 49815    50
> dimnames(need.mat)
[[1]]
NULL
[[2]]
 [1] "HH_ID"           "REGION"             "IMD_ENG"
 [4] "IMD_WALES"       "Gcons2005"          "Gcons2005Valid"
 [7] "Gcons2006"       "Gcons2006Valid"     "Gcons2007"
[10] "Gcons2007Valid"  "Gcons2008"          "Gcons2008Valid"
... #output truncated
> head(need.mat)
     HH_ID REGION IMD_ENG IMD_WALES Gcons2005 Gcons2005Valid
[1,]     1      7       1        NA     35000              5
[2,]     2      2       4        NA     19000              5
[3,]     3      2       4        NA     22500              5
[4,]     4      5       1        NA     21000              5
[5,]     5      3       1        NA        NA              3
[6,]     6      7       2        NA        NA              4
...#output truncated
```

Unleashing the Power of R from Within

The `describe()` function prints a description of the backing file, just like the content of a file created with the `descriptorfile` argument. Some basic functions such as `ncol()` and `nrow()` have their `bigmemory` implementation as well. But more descriptive statistics, and some serious modeling, can be achieved through two supplementary packages that use big matrices created by the `bigmemory` package: `bigtabulate` and `biganalytics`. You can easily obtain contingency tables through `bigtable()` and `bigtabulate()` commands (but the base `table()` will work too), for example:

```
> library(bigtabulate)
> library(biganalytics)
> bigtable(need.mat, c("PROP_AGE"))
   101   102   103   104   105   106
 13335  7512  8975  9856  5243  4894
> bigtabulate(need.mat, c("PROP_AGE", "PROP_TYPE"))
     101  102  103  104  105  106
101 1506 2787 1514 5192  320 2016
102  815 3854  605  995  623  620
103  856 3084  697 1134 1641 1563
104 1737 1790  861 1344 1721 2403
105 1439  760  388  550  589 1517
106 1557  704  465  642  261 1265
```

In the preceding listing we first obtained a frequency table of properties belonging to specific age bands (`PROP_AGE` variable), and then a contingency table of property age band (`PROP_AGE`) by property type (`PROP_TYPE`).

Functions such as `summary()` or `bigtsummary()` allow users to calculate basic descriptive statistics about variables or table summaries when conditioned on other variables for example:

```
> summary(need.mat[, "Econs2012"])
   Min. 1st Qu.  Median    Mean 3rd Qu.    Max.    NA's
    100    2100    3250    3972    4950   25000     102
> sum1 <- bigtsummary(need.mat, c(39, 40), cols = 35, na.rm = TRUE)
> sum1[1:length(sum1)]
[[1]]
     min   max    mean        sd NAs
[1,] 100 25000  6586.7  4533.779  21

[[2]]
     min   max      mean        sd NAs
[1,] 100 25000  5513.041  3567.643   6
[[3]]
     min   max      mean        sd NAs
[1,] 100 25000  5223.626  3424.481   1
...#output truncated
```

In the first call we have only printed statistics simply describing the total annual electricity consumption in 2012, whereas in the second call we have obtained descriptive statistics for each level of crossed factors: property, age, and property type, using indices of variables rather than their names. More stat functions such as `colmean()`, `colsum()`, `colmin()`, `colmax()`, `colsd()`, and others are also available from the `biganalytics` package.

The `bigmemory` approach can also perform a split-apply-combine type of operation, similar to **MapReduce** known from **Hadoop**, which we will address very thoroughly in Chapter 4, *Hadoop and MapReduce Framework for R*. For example, we may want to split the electricity consumption in 2012 (`Econs2012`) for each level of the electricity efficiency band (`EE_BAND`) and then calculate the mean electricity consumption for each band using `sapply()` function:

```
> need.bands <- bigsplit(need.mat, ccols = "EE_BAND", splitcol =
"Econs2012")
> sapply(need.bands, mean, na.rm=TRUE)
       1        2        3        4        5        6
2739.937 3441.517 3921.660 4379.745 5368.460 5596.172
```

A similar job, run on a large `big.matrix`, with over 4 millions records, took only one second to run and used as little as 68MB of RAM.

Moreover, `bigmemory` is also capable of running more complex modeling jobs such as generalized linear models through its `bigglm.big.matrix()` function (which makes use of the previously introduced `biglm` package of Thomas Lumley) and memory efficient **k-means clustering** with the `bigkmeans()` function. In the following snippet, we will attempt to perform a multiple linear regression predicting the electricity consumption in 2012 from a number of predictors:

```
> library(biglm)
> regress1 <- bigglm.big.matrix(Econs2012~PROP_AGE + FLOOR_AREA_BAND +
CWI_YEAR + BOILER_YEAR, data = need.mat, fc = c("PROP_AGE",
"FLOOR_AREA_BAND"))
> summary(regress1)
Large data regression model: bigglm(formula = formula, data =
getNextDataFrame, chunksize = chunksize,
    ...)
Sample size =   49815
                   Coef        (95%        CI)        SE
(Intercept)    10547.0626  -7373.4056  28467.5307  8960.2341
PROP_AGE102     -119.0011   -196.1487    -41.8535    38.5738
PROP_AGE103     -139.3637   -212.7222    -66.0052    36.6792
PROP_AGE104     -127.8420   -201.2732    -54.4109    36.7156
PROP_AGE105     -285.7916   -375.4228   -196.1605    44.8156
PROP_AGE106     -269.9347   -441.3419    -98.5275    85.7036
```

```
FLOOR_AREA_BAND2   1041.7280    967.8755   1115.5806    36.9263
FLOOR_AREA_BAND3   1966.9689   1888.4783   2045.4594    39.2453
FLOOR_AREA_BAND4   3676.7904   3571.4992   3782.0816    52.6456
CWI_YEAR              10.7261      4.6191     16.8331     3.0535
BOILER_YEAR          -14.8711    -22.2074     -7.5348     3.6681
                         p
(Intercept)         0.2392
PROP_AGE102         0.0020
PROP_AGE103         0.0001
PROP_AGE104         0.0005
PROP_AGE105         0.0000
PROP_AGE106         0.0016
FLOOR_AREA_BAND2    0.0000
FLOOR_AREA_BAND3    0.0000
FLOOR_AREA_BAND4    0.0000
CWI_YEAR            0.0004
BOILER_YEAR         0.0001
```

Depending on the original value labelling, the function allows us to indicate which predictors are factors (Thomas Lumley's `fc` argument). The implementation is quite efficient; it ran for 22 seconds and used only 70 MB of RAM for the dataset with over 4 million cases.

There are two other packages that are part of the `bigmemory` project. The `bigalgebra` library, as the name suggests, allows matrix algebra on `big.matrix` objects, whereas the `synchronicity` package provides a set of functions supporting synchronisation through **mutexes,** and thus can be used for multiple thread processes. A package named `bigpca`, authored and maintained by Nicholas Cooper, offers fast scalable **Principle Components Analysis (PCA)** and **Singular Value Decompositions (SVD)** on `big.matrix` objects. It also supports multi-core implementation of the `apply` functionality (through `bmcapply()` function) and convenient transposing. The data can be imported quickly using a reference to the descriptor file, for example:

```
> need.mat2 <- get.big.matrix("need_data.desc")
> prv.big.matrix(need.mat2)
Big matrix with: 49815 rows, 50 columns
 - data type: numeric
Loading required package: BiocInstaller

        colnames
   row#       HH_ID      REGION   .....   BOILER_YEAR
      1           1           7     ...            NA
      2           2           2     ...            NA
      3           3           2     ...          2004
   ....         ...         ...     ...           ...
  49815       49815           2     ...          2010
```

The `bigpca` package also enables easy subsetting of big matrices for example:

```
> need.subset <- big.select(need.mat2, select.cols = c(35, 37:50),
+                           pref = "sub")
> prv.big.matrix(need.subset)
Big matrix; 'sub.RData', with: 49815 rows, 15 columns
 - data type: numeric
Loading required package: BiocInstaller
        colnames
   row#    Econs2012    E7Flag2012    .....    BOILER_YEAR
      1         6300            NA      ...             NA
      2         3000            NA      ...             NA
      3         4700            NA      ...           2004
    ....          ...           ...     ...            ...
  49815         2200            NA      ...           2010
```

The `big.select()` function conveniently creates RData, backing, and descriptor files with names specified in the `pref` argument. The principal component analysis, and singular value decomposition, on `big.matrix` objects can be achieved through the `big.PCA()` and `svd()` functions respectively. As these methods go beyond the scope of this book, feel free to follow the examples provided in the help files and the manual of the `bigpca` package.

In general, the `bigmemory` approach allows fast and memory-efficient management and processing of large (or even massive and out-of-memory) matrices. The only serious limitation derives from the definition of a matrix as an R data structure, only holding one type of data. If the original dataset contains various classes of variables and is larger than the available RAM, the only way of converting differing types into a single class will be either through the `ff` package or by first moving data to a large enough server, or a database, and eventually importing the processed data in the required format back to R. However, once this initial step is complete, the `bigmemory` family of packages offers quite an impressive array of data manipulation and processing functions. These can be further extended and speeded up by the multicore support offered by some of the R packages, which allow parallel computing.

Parallel R

In this part of of the chapter, we will introduce you to the concept of parallelism in R. More precisely, we will focus here almost entirely on *explicit* methods for parallel computation, in which users are capable of controlling the parallelization on a single machine. In *Online Chapter, Pushing R Further* (https://www.packtpub.com/sites/default/files/downloads/5396_6457OS_PushingRFurther.pdf) you will practice some of these methods on much larger clusters of commodity hardware through popular cloud computing platforms such as Amazon EC2 or Microsoft Azure. In Chapter 4, *Hadoop and MapReduce Framework for R* you will learn much more about the **MapReduce** approach in R (through the HadoopStreaming, Rhipe, and RHadoop packages)-an abstraction of parallelism for distributed files systems such as **Hadoop**.

Our motivation for parallel computing in R comes from the simple fact that many data-processing operations tend to be very similar, and some of them are extremely time-consuming, especially when using for loops on large datasets or when computing models with multiple different parameters. It is generally accepted that a function, which runs for a few minutes while computing an **embarrassingly parallel** problem, should be spread across several cores, if possible, to reduce the processing time. Another reason to implement parallelism in your data manipulation tasks in R is that most currently available, commercially-sold PCs are equipped with more than one CPU. R, however, by default, uses only one, so it is worth making the most of the available architecture, and forcing R to explicitly delegate some work to other cores. Also, in many cloud-computing solutions (which we present in *Online Chapter, Pushing R Further* https://www.packtpub.com/sites/default/files/downloads/5396_6457OS_PushingRFurther.pdf), users can benefit from multiple cores allowing their algorithms to complete faster.

There are a number of good online resources, which elaborate on parallel computing in R. The obvious destination is *CRAN Task View – High Performance Computing* available at https://cran.r-project.org/web/views/HighPerformanceComputing.html. The Task View lists all major packages, known for supporting parallelism in R, and provides brief descriptions about their most essential functionalities. Manuals, vignettes, and tutorials for specific R packages, which are accessible through CRAN, offer deep insights into particulars of their functions and used syntax. A concise book *Parallel R* written by Q. Ethan McCallum and Stephen Weston (published by O'Reilly Media) delivers a good overview of the most popular parallel packages in R.

Chapter 3

From bigmemory to faster computations

First, let's pick up where we left off. In the previous section we ran through a number of `bigmemory` functions which helped us make the most of available RAM. The library also contains several useful functions optimized for objects of the S4 class and provides *fast* computations such as `colmean()` or `colrange()` and many others. In this section, we will compare the performance of these functions with their base R and parallel implementations by using several methods and R packages that support parallelism.

For testing purposes, we will try to obtain mean values for each column in the larger version (over 4 mln rows) of the *National Energy Efficiency Data-Framework* (NEED), which we used previously (at this stage it does not really matter that some variables should in fact be factors, rather than numeric, and so arithmetic means are irrelevant-we just do it to compare the performance of specific methods). If you wish, you may run the scripts on a smaller sample provided in the file bundle on this book's website-but don't expect significant performance gains owing to the limited size of the dataset.

If you execute the following code on a smaller sample of NEED data, make sure to use the descriptor file `need_data.desc` instead of the `need_big.desc`. Alternatively, you may obtain the full dataset as explained earlier and create a large `big.matrix` object with the name of the descriptor file set to `need_big.desc`.

We will begin from the `bigmemory` functions to set the performance base levels for comparison. As explained previously, you can use the descriptor file to import the data into R quickly and then use the `colmean()` function from the `biganalytics` package to calculate mean values for each column:

```
> library(bigmemory)
> library(biganalytics)
> need.big.bm <- attach.resource("need_big.desc")
> meanbig.bm1 <- colmean(need.big.bm, na.rm = TRUE)
> meanbig.bm1
           HH_ID          REGION         IMD_ENG       IMD_WALES
     2043224.500000        5.664102        2.907485        2.793685
          FP_ENG    EPC_INS_DATE        Gcons2005  Gcons2005Valid
        7.839505        1.738521    17704.752048        4.620305
...#output truncated
```

It took 6.22 seconds to run this function on my computer. In the following examples, we will review other more optimized methods, including selected functions that support parallelism.

An apply() example with the big.matrix object

You can also obtain the same output by invoking the `apply()` function from the same `biganalytics` package. The `apply()` function from `biganalytics` is a generalization of the base R `apply()` function that additionally supports the S4 class object of `big.matrix` type. Apart from this subtle difference, both functions are identical. As the `big.matrix` object is a custom data structure, the `apply()` function deals differently with the memory overhead associated with extracting data from this S4 class, and, for that reason, we may expect that this implementation will actually run slower than `colmean()`:

```
> meanbig.bm2 <- apply(need.big.bm, 2, mean, na.rm=TRUE)
> meanbig.bm2
           HH_ID           REGION          IMD_ENG         IMD_WALES
     2043224.500000        5.664102        2.907485        2.793685
          FP_ENG      EPC_INS_DATE         Gcons2005   Gcons2005Valid
...#output truncated
```

The preceding process completed in 8.21 seconds, so almost 2 seconds slower than with the `colmean()` method.

> **The apply() family of functions**
> For those of you who are not familiar with the `apply()` family of functions, it is probably a good moment to pause for a minute, and read this concise, but to-the-point, review of `apply()` functions available at http://faculty.nps.edu/sebuttre/home/R/apply.html. In general, `apply()` methods can save quite a lot of your precious data-processing time, and you don't need to write loops to calculate the same statistics over all columns, rows, or other dimensions of your data structures.

A for() loop example with the ffdf object

Going back to our mean estimations for each column and performance comparison between available methods, let's see whether there is any improvement in processing speed if we wanted to apply a `for()` loop on the 4-million-row NEED file imported through the `ff` package which we described earlier. Assuming that the relevant `ffdf` object has been already created in the R environment, we may now run the following code:

```
> meanbig.ff <- list()
> for(i in 1:ncol(need.big.ff)) {
+     meanbig.ff[[i]] <- mean.ff(need.big.ff[[i]], na.rm=TRUE)
+ }
> meanbig.ff
```

```
[[1]]
[1] 2043224

[[2]]
[1] 5.664102

[[3]]
[1] 2.907485
...#output truncated
```

In the preceding snippet, we've used the mean.ff() function, which is simply a the S3 method for the class ff that derives from the generic mean() function in the base R. The for() loop completed in 7.72 seconds, providing only a slight improvement when compared to the performance of apply() used on the big.matrix object, but it was still slower than colmean() through the biganalytics package. How can we then boost the performance in a more significant way?

Using apply() and for() loop examples on a data.frame

One main reason why we have been using the bigmemory and ff/ffdf packages extensively in this chapter is their ability to process out-of-memory data directly from the R console. But by mapping their custom data structures to raw data stored on disk, we consciously compromise the performance of our operations. For datasets that can fit within the RAM boundaries, if you have the comfort of working on a large server, or if you are using some of the cloud computing services (and you will when you get to *Online Chapter, Pushing R Further* https://www.packtpub.com/sites/default/files/downloads/5396_6457OS_PushingRFurther.pdf), you may also import the data directly to the physical memory, and use either one of the generic functions such as colMeans() or for() loops on a created data frame object:

```
> for(i in 1:ncol(need.big.df)) {
+    x1[i] <- mean(need.big.df[,i], na.rm = TRUE) }
> x1
[[1]]
[1] 2043224
[[2]]
[1] 5.664102
[[3]]
[1] 2.907485
...#output truncated
```

The `for()` loop with the base `mean()` function with its completion time of 5.24 seconds, was faster than any other method presented previously. The catch is that there was quite a substantial peak of memory usage and we are now holding a large object in the workspace. But `colMeans()` approach from base R beats all others with only 2.9 seconds:

```
> x2 <- colMeans(need.big.df, na.rm = TRUE)
> x2
           HH_ID          REGION         IMD_ENG        IMD_WALES
   2043224.500000        5.664102        2.907485         2.793685
          FP_ENG    EPC_INS_DATE        Gcons2005   Gcons2005Valid
         7.839505        1.738521    17704.752048         4.620305
...#output truncated
```

However, as the `colMeans()` function is in fact equivalent to the `apply()` method, let's test to see if the simplified version of `apply()` in the form of `sapply()` can keep up the pace:

```
> x3 <- sapply(need.big.df, mean, na.rm = TRUE)
> x3
           HH_ID          REGION         IMD_ENG        IMD_WALES
   2043224.500000        5.664102        2.907485         2.793685
...#output truncated
```

It took almost 5.2 seconds for `sapply()` to calculate the means for all columns in our data, which is much slower than through `colMeans()`. We may try to optimize the speed of the `apply()` function by parallelizing its execution explicitly through the `parallel` package.

A parallel package example

The `parallel` package has come as an integral part of the core R installation since the R 2.14.0 version, and it has been built on two other popular R packages that support parallel data processing: `multicore` (authored by Simon Urbanek) and `snow` (*Simple Network of Workstations*, created by Tierney, Rossini, Li, and Sevcikova). In fact, `multicore` has already been discontinued and removed from the CRAN repository, as the `parallel` package took over all its essential components. The `snow` package is still available on CRAN, and it may be useful in certain, but limited, circumstances. The `parallel` library, however, extends their functionalities by allowing greater support for random-number generation. The package can be suitable for parallelizing repetitive jobs on unrelated chunks of data, and computations that do not need to communicate with, and between, one another.

The computational model adopted by `parallel` is similar to approaches known from earlier packages for `snow`, and it's based on the relationship between *master* and *worker* processes. The details of this model can be found in a short R manual dedicated to the `parallel` package and is available at http://stat.ethz.ch/R-manual/Rdevel/library/parallel/doc/parallel.pdf.

In order to perform any parallel processing jobs, it might first be advisable to know how many physical CPUs (or cores) are available on the machine that runs R. This can be achieved in the parallel package with the `detectCores()` function:

```
> library(parallel)
> detectCores()
[1] 4
```

It is important here to be aware that the returned number of cores may not be equal to the actual number of available logical cores (for example in Windows), or CPU accessible to a specific user on restricted multi-user systems.

The `parallel` package, by default, facilitates clusters communicating over two types of sockets: **SOCK** and **FORK**. The SOCK cluster, operationalized through the `makePSOCKcluster()` function, is simply an enhanced implementation of the `makeSOCKcluster()` command known from the `snow` package. The FORK cluster, on the other hand, originates from the `multicore` package and allows the creation of multiple R processes by copying the master process completely including R GUI elements such as an R console and devices. The forking is generally available on most non-Windows R distributions. Other types of clusters can be created using the `snow` package (for example through **MPI** or **NWS** connections) or with `makeCluster()` in the `parallel` package, which will call `snow`, provided it's included in the search path. In `parallel`, you may use either `makePSOCKcluster()` or `makeCluster()` functions to create a SOCK cluster:

```
> cl <- makeCluster(3, type = "SOCK")
> cl
socket cluster with 3 nodes on host 'localhost'
```

Generally, it is advisable to create clusters with *n* number of nodes, where n = `detectCores()` - 1, hence three nodes in our example. This approach allows us to benefit from multi-threading, without putting an excessive pressure on other processes or applications that may be run in parallel.

The `parallel` package allows the execution of `apply()` operations on each node in the cluster through the `clusterApply()` function and parallelized implementations of the `apply()` family of functions known from the `multicore` and `snow` packages: `parLapply()`, `parSapply()`, and `parApply()`.

```
> meanbig <- clusterApply(cl, need.big.df, fun=mean, na.rm=TRUE)
> meanbig
[[1]]
[1] 2043224

[[2]]
[1] 5.664102
[[3]]
[1] 2.907485
```

Unfortunately, the `clusterApply()` approach is quite slow and completes the job in 13.74 seconds. The `parSapply()` implementation returning the output is much faster and returns the output in 6.71 seconds:

```
> meanbig2 <- parSapply(cl, need.big.df, FUN = mean, na.rm=TRUE)
> meanbig2
          HH_ID          REGION         IMD_ENG       IMD_WALES
  2043224.500000        5.664102        2.907485        2.793685
         FP_ENG    EPC_INS_DATE        Gcons2005  Gcons2005Valid
        7.839505        1.738521    17704.752048        4.620305
```

In addition to the `apply()` operations presented earlier, the parallel package contains the `mclapply()` function, which is a parallelized version of `lapply()` relying on forking (not available on Windows unless `mc.cores = 1`). In the following example we will compare the performance of `mclapply()` with differing number of cores (the `mc.cores` argument from 1 to 4):

```
> meanbig3 <- mclapply(need.big.df, FUN = mean, na.rm = TRUE, mc.cores = 1)
> meanbig3
$HH_ID
[1] 2043224
$REGION
[1] 5.664102
$IMD_ENG
[1] 2.907485
```

The table below presents average timings of evaluation of the same `mclapply()` expression with a differing number of `mc.cores` (from 1 to 4):

mc.cores	time (in seconds)
1	4.71
2	4.14
3	3.51
4	3.10

It is clear that the increase in cores correlates with the better performance. Note, however, that because the parallel implementation of `mclapply()` initializes several processes which share the same GUI, it is advisable not to run it in the R GUI or embedded environments, otherwise your machine (and R sessions) may become unresponsive, cause chaos, or even crash. For larger datasets, several parallel R sessions may rapidly increase the memory usage and its pressure, so please be extremely careful when implementing the parallelized `apply()` family of functions into your data processing workflows.

Once the parallel jobs are complete it is a good habit to close all connections with the following statement:

```
> stopCluster(cl)
```

The previously mentioned R manual on `parallel` is available from http://stat.ethz.ch/R-manual/R-devel/library/parallel/doc/parallel.pdf and presents two very good frequent applications of the package: in **bootstrapping** and **maximum-likelihood estimations**. Please feel free to visit the manual and run through the given examples.

A foreach package example

The `foreach()` package, authored by Revolution Analytics, Rich Calaway, and Steve Weston, offers an alternative method of implementing `for()` loops, but without the need to use the loop counter explicitly. It also supports the parallel execution of loops through the `doParallel` backend and the `parallel` package. Sticking to our example, with mean estimates for each column of the data, we may apply `foreach()` in the following manner:

```
> library(foreach)
> library(parallel)
> library(doParallel)
```

```
> cl <- makeCluster(3, type = "SOCK")
> registerDoParallel(cl)
> x4 <- foreach(i = 1:ncol(need.big.df)) %dopar% mean(need.big.df[,i],
na.rm=TRUE)
> x4
[[1]]
[1] 2043224

[[2]]
[1] 5.664102
[[3]]
[1] 2.907485
...#output truncated
> stopCluster(cl)
```

The job took 5.2 seconds to complete. In the first part of the listing, we have created a three-node cluster `cl`, which we registered with the `foreach` package using `registerDoParallel()` from the `doParallel` library-a parallel backend. You've also probably noticed that the above `foreach()` statement contains an unfamiliar piece of syntax: `%dopar%`. It is a binary operator that evaluates an R expression (`mean(...)`) in parallel in an environment created by the `foreach` object. If you wished to run the same call sequentially, you could use the `%do%` operator instead. In fact, both implementations return the output of our mean calculations within a very similar time, but the actual timings will obviously depend on the specific computation and available architecture. The `foreach()` function contains a number of other useful settings. For example, you can present the output as a vector, matrix, or in any other way, defined by a function set in the `.combine` argument. In the following code snippet, we use `foreach()` with an `%do%` operator and a `.combine` argument set to concatenate the values (that is, to present them as a vector):

```
> x5 <- foreach(i = 1:ncol(need.big.df), .combine = "c") %do%
mean(need.big.df[,i], na.rm=TRUE)
> x5
[1]  2043224.500000        5.664102        2.907485        2.793685
[5]        7.839505        1.738521    17704.752048        4.620305
[9]    17012.460418        4.641852    16457.487437        4.662537
...#output truncated
```

The `foreach` package is still pretty new to the R community, and it is expected that more functionalities will be added within the next several months. Steve Weston's guide *Using The foreach Package* (available from https://cran.r-project.org/web/packages/foreach/vignettes/foreach.pdf) contains several simple, and slightly more complex, applications of specific parameters (and their values) which can be set in the `foreach()` function.

The future of parallel processing in R

In the preceding section, we have introduced you to some basics of parallel computing currently available from within R, on a single machine. R is probably not the ideal solution for parallelized operations, but a number of more recent approaches may potentially revolutionize the way R implements parallelism.

Utilizing Graphics Processing Units with R

Graphics Processing Units (GPUs) are specialized, high-performance electronic circuits that are designed for efficient and fast memory management in computationally demanding tasks, such as image and video rendering, dynamic gameplay, simulations (both 3D or virtual and also statistical), and many others. Although they are still rarely used in general calculations, a growing number of researchers benefit from GPU acceleration when carrying out repetitive, embarrassingly parallel, computations over multiple parameters. The major disadvantage of GPUs, however, is that they don't generally support Big Data analytics on a single machine owing to their limited access to RAM. Again, it depends what one means by Big Data, and also, their application in the processing of large datasets relies on the architecture in place. On average, however, parallel computing through GPU can be up to 12 times as fast, compared to parallel jobs using standard CPUs. The largest companies manufacturing GPU are Intel, NVIDIA, and AMD, and you are probably familiar with some of their products if you ever built a PC yourself, or at least played some computer games.

R also supports parallel computing through GPUs, but obviously you can only make the most of it if your machine is equipped with one of the leading GPUs for example NVIDIA CUDA. If you don't own one, you can quite cheaply create a cloud-computing cluster, for example, on Amazon Elastic Cloud Computing (EC2), which will include graphics processing units. We will show you how to deploy such an EC2 instance with R and RStudio Server installed in *Online Chapter, Pushing R Further* (https://www.packtpub.com/sites/default/files/downloads/5396_6457OS_PushingRFurther.pdf).

As the GPUs need to be programmed, and hence many R users may struggle with their configuration, there are several R packages that facilitate working with CUDA-compatible GPU. One of them is the `gputools` package authored by Buckner, Seligman, and Wilson. The package requires a recent version of the NVIDIA CUDA toolkit and it contains a set of GPU-optimized statistical methods such as (but not limited to) fitting generalized linear models (the `gpuGlm()` function), performing hierarchical clustering for vectors (`gpuHclust()`), computing distances between vectors (`gpuDist()`), calculating Pearson or Kendall correlation coefficients (`gpuCor()`), and estimating t-tests (`gpuTtest()`).

The `gputools` package can also implement **fastICA algorithm** (*Fast Independent Component Analysis algorithm* created by Prof. Aapo Hyvarinen from University of Helsinki, `http://www.cs.helsinki.fi/u/ahyvarin/`) through `CULA Tools` (`http://www.culatools.com/`)-a collection of GPU-supported linear algebra libraries for parallel computing. More details on the `gputools` package are available from the following sources:

- `https://cran.r-project.org/web/packages/gputools/index.html`- the `gputools` CRAN website with links to manuals and source code
- `https://github.com/nullsatz/gputools/wiki`-the `gputools` GitHub project repo
- `http://brainarray.mbni.med.umich.edu/brainarray/rgpgpu/`- the `gputools` website run by Microarray Lab at the Molecular and Behavioral Neuroscience Institute, University of Michigan

Apart from `gputools`, R packages, also provide an interface with **OpenCL**-a programming language framework used for operating heterogeneous computational platforms based on a variety of CPU, GPU, and other accelerator devices. The `OpenCL` package, developed and maintained by Simon Urbanek, allows R users to identify and retrieve a list of **OpenCL** devices and to execute kernel code that has been compiled for **OpenCL** directly from the R console.

On the other hand, the `gpuR` package, created and maintained by Charles Determan Jr., simplifies this task by providing users with ready-made custom `gpu` and `vcl` classes which function as wrappers for common R data structures such as `vector` or `matrix`. Without any prior knowledge of **OpenCL**, R users can easily perform a number of statistical methods through the `gpuR` package such as estimating row and column sums and arithmetic means, comparing elements of `gpuvector` and `vector` objects, calculating covariance and cross-product on `gpuMatrix` and `vclMatrix`, distance matrix estimations, eigenvalues computations, and many others.

CRAN High-Performance Computing Task View
(`https://cran.r-project.org/web/views/HighPerformanceComputing.html`) lists a few more specialized R packages, which support GPU acceleration. It doesn't, however, mention the `rpud` package, which was removed from CRAN in late October 2015 owing to maintenance issues. The package, however, has been quite successful in performing several GPU-optimized statistical methods such as hierarchical cluster analysis and classification tasks.

Although it is not available on CRAN, its most recent version can be downloaded from the developers' website at `http://www.r-tutor.com/content/download`. The `http://www.r-tutor.com/gpu-computing` site contains a number of practical applications of GPU-accelerated functions included in the `rpud` package. Also, the **NVIDIA CUDA Zone**-a blog run by CUDA developers, presents very good tutorials on the implementation of selected `rpud` methods using R and Cloud computing (for example `http://devblogs.nvidia.com/parallelforall/gpu-accelerated-r-cloud-teraproc-cluster-service/`).

Multi-threading with Microsoft R Open distribution

The acquisition of Revolution Analytics by Microsoft in summer 2015 sent a clear signal to the R community that the famous Redmond-based tech giant was soon going to re-package the already good and Big-Data-friendly **Revolution R Open** (**RRO**) distribution and enhance it by equipping it with more powerful capabilities. When writing this chapter, the **Microsoft R Open** (**MRO**) distribution, based on the previous RRO version, is only a few days old, but it has already energized R users. Unfortunately, it's too new to be incorporated into this book, as it requires quite extensive testing to assess the validity of Microsoft's claims in terms of MRO's performance.

According to MRO's developers, Microsoft R Open provides access to the multi-threaded **Math Kernel Library** (**MKL**) giving R computations an impressive boost across a spectrum of mathematical and statistical calculations. Diagrams and comparisons of performance benchmarks available at `https://mran.revolutionanalytics.com/documents/rro/multithread/` clearly indicate that MRO with as little as 1 core can significantly increase computation speed for a variety of operations. MRO equipped with four cores may make them run up to 48 times faster (for a matrix multiplication) compared with the R distribution obtained from CRAN. Depending on the type of algorithm used during performance testing, the Microsoft RO distribution excelled in two areas of matrix calculation and matrix functions; these are where MRO recorded the greatest performance gains. The programming functions designed for loops, recursions, or control flow generally performed at the same speed as in CRAN R. Microsoft R Open is supported on 64-bit platforms only including Windows 7.X, Linux (Ubuntu, CentOS, Red Hat, SUS, and others.), and Mac OS X (10.9+), and can be installed from `https://mran.revolutionanalytics.com/download/`.

Parallel machine learning with H2O and R

In this section we will only very briefly mention the mere existence of **H2O** (http://www.h2o.ai/)-a fast and scalable platform for parallel and Big Data machine learning algorithms. The platform is also supported by the R language through the h2o package (authored by Aiello, Kraljevic, and Maj with contributions from the actual developers of the **H2O.ai**team), which provides an interface for the **H2O** open-source ML engine. This exciting collaboration is only mentioned here as we dedicate a large part of Chapter 8, *Machine Learning methods for Big Data in R* to a detailed discussion and a number of practical tutorials of the H2O platform and R applied to a real-world, Big Data issue.

Boosting R performance with the data.table package and other tools

The following two sections present several methods of enhancing the speed of data processing in R. The larger part is devoted to the excellent data.table package, which allows convenient and fast data transformations. At the very end of this section we also direct you to other sources, that elaborate, in more detail, on the particulars of faster and better optimized R code.

Fast data import and manipulation with the data.table package

In a chapter devoted to optimized and faster data processing in the R environment, we simply must spare a few pages for one, extremely efficient and flexible package called data.table. The package, developed by Dowle, Srinivasan, Short, and Lianoglou with further contributions from Saporta and Antonyan, took the primitive R data.frame concept one (huge) step forward and has made the lives of many R users so much easier since its release to the community.

The `data.table` library offers (very) fast subsetting, joins, data transformations, and aggregations as well as enhanced support for fast date extraction and data file import. But this is not all, it also has other great *selling* points, for example:

- A very convenient and easy chaining of operations
- Key setting functionality allowing (even!) faster aggregations
- Smooth transition between `data.table` and `data.frame` (if it can't find a `data.table` function it uses the base R expression and applies it on a `data.frame`, so users don't have to convert between data structures explicitly)
- Really easy-to-learn, natural syntax.

Oh, did I say it's fast? How about any drawbacks? It still stores data and all created `data.table` objects in RAM, but owing to its better memory management, it engages RAM only for processes, and only on specific subsets (rows, columns) of the data that have to be manipulated. The truth is that `data.table` can in fact save you a bit of cash if you work with large datasets. As its computing time is much shorter than when using standard base R functions on data frames, it will significantly reduce your time and bill for cloud-computing solutions. But instead of reading about its features, why don't you try to experience them first hand by following the introductory tutorial?

Data import with data.table

In order to show you real performance gains with `data.table` we will be using the flight dataset, which we have already explored when discussing the `ff/ffdf` approach at the beginning of this chapter. As you probably remember, we were comparing the speed of processing and memory consumption between the `ff/ffbase` packages and base R functions such as `read.table()` or `read.csv()` performed on a smaller dataset with two months of flights (you can download this from Packt Publishing's website for this book), and a bigger, almost 2 GB two-year dataset for which we were also giving performance benchmarks just as a reference. In this section, we will be quoting only the estimates for the larger, 2 GB file with 12,189,293 observations, but feel free to follow the examples by running the same code on smaller data (just remember to specify the name of the file correctly).

The `data.table` package imports the data through its fast file reader `fread()` function:

```
> library(data.table)
> flightsDT <- fread("flights_1314.txt", stringsAsFactors = TRUE)
Read 12189293 rows and 28 (of 28) columns from 1.862 GB file in 00:00:29
```

Unleashing the Power of R from Within

It's quite spectacular that operation which took 456 seconds using `read.table.ffdf()` and 441 seconds with `read.table()` was slashed to a mere 29 seconds in `fread()`. The `fread()` function also comes with a large number of additional arguments users can set; for example, they may `select` or `drop` specific columns, define standard separators between read columns (`sep`) or even within columns (`sep2`), specify the number of rows to read (`nrows`), the number of rows to skip (`skip`), define column classes (`colClasses`), and many others. The resulting object is both a `data.table` and a `data.frame` allowing the flexibility of syntax and applications depending on users' specific needs:

```
> str(flightsDT)
Classes 'data.table' and 'data.frame':  12189293 obs. of   28 variables:
 $ YEAR              : int  2013 2013 2013 2013 2013 2013 2013 2013 2013 2013 ...
 $ MONTH             : int  1 1 1 1 1 1 1 1 1 ...
 $ DAY_OF_MONTH      : int  17 18 19 20 21 22 23 24 25 26 ...
 $ DAY_OF_WEEK       : int  4 5 6 7 1 2 3 4 5 6 ...
 ...#output truncated
```

This flexibility of `data.table` semantics is best noticed in data transformations such as subsetting and aggregations.

Lightning-fast subsets and aggregations on data.table

Datatables can be subsetted and aggregated using their indexing operators surrounded by square `[]` brackets and with the following default format:

```
> DT[i, j, by]
```

The structure of this call can be compared to a standard SQL query where the `i` operator stands for WHERE, `j` denotes SELECT, and `by` can be simply translated to the SQL GROUPBY statement. In the most basic form we may subset specific rows, which match set conditions as in the example below:

```
> subset1.DT <- flightsDT[
+   YEAR == 2013L & DEP_TIME >= 1200L & DEP_TIME < 1700L,
+ ]
> str(subset1.DT)
Classes 'data.table' and 'data.frame':  1933463 obs. of   28 variables:
 $ YEAR              : int  2013 2013 2013 2013 2013 2013 2013 2013 2013 2013 ...
 $ MONTH             : int  1 1 1 1 1 1 1 1 1 ...
 $ DAY_OF_MONTH      : int  30 3 4 5 6 7 8 9 10 11 ...
 $ DAY_OF_WEEK       : int  3 4 5 6 7 1 2 3 4 5 ...
 ...#output truncated
```

[120]

The task took only 1.19 seconds to complete, and the new subset (a `data.table` and a `data.frame`) contains all 2,013 flights which departed in the afternoon between 12:00 and 16:59. In the same way we can perform simple aggregations in which we may even calculate other arbitrary statistics. In the following example we will estimate the total delay (`TotDelay`) for each December flight and the average departure delay (`AvgDepDelay`) for all December flights. Additionally, we will group the results by the state of the flight origin (`ORIGIN_STATE_NM`):

```
> subset2.DT <- flightsDT[
+ MONTH == 12L,
+ .(TotDelay = ARR_DELAY - DEP_DELAY,
+   AvgDepDelay = mean(DEP_DELAY, na.rm = TRUE)),
+ by = .(ORIGIN_STATE_NM)
+ ]
> subset2.DT
        ORIGIN_STATE_NM TotDelay AvgDepDelay
     1:        New York        1    11.85232
     2:        New York      -42    11.85232
     3:        New York      -16    11.85232
     4:        New York      -32    11.85232
     5:        New York       -2    11.85232
    ---
993918:        Delaware       11    11.73494
993919:        Delaware       -3    11.73494
993920:        Delaware        0    11.73494
993921:        Delaware       -5    11.73494
993922:        Delaware        4    11.73494
```

By indicating the names of columns in the `j` parameter you can easily extract variables of interest, for example:

```
> subset3.DT <- flightsDT[, .(MONTH, DEST)]
> str(subset3.DT)
Classes 'data.table' and 'data.frame':  12189293 obs. of  2 variables:
 $ MONTH: int  1 1 1 1 1 1 1 1 1 1 ...
 $ DEST : Factor w/ 332 levels "ABE","ABI","ABQ",..: 174 174 174 174 174 174 174 174 174 174 ...
```

As in the previous listing, we may now quickly aggregate any statistic in `j` by group, specified in the `by` operator:

```
> agg1.DT <- flightsDT[, .(SumCancel = sum(CANCELLED),
+                          MeanArrDelay = mean(ARR_DELAY, na.rm = TRUE)),
+                      by = .(ORIGIN_CITY_NAME)
+]
> agg1.DT
         ORIGIN_CITY_NAME SumCancel MeanArrDelay
```

```
     1:   Dallas/Fort Worth, TX    13980    10.0953618
     2:          New York, NY      12264     6.0086474
     3:       Minneapolis, MN       2976     3.6519174
     4:     Raleigh/Durham, NC      2082     5.8777458
     5:          Billings, MT         75    -1.0170240
   ---
   325:        Devils Lake, ND        10    13.6372240
   326:           Hyannis, MA          1    -0.6933333
   327:              Hays, KS         16    -3.0204082
   328:          Meridian, MS          3     9.8698630
   329: Hattiesburg/Laurel, MS         2    10.0434783
```

In the preceding snippet we simply calculated the number of cancelled flights, which were supposed to depart from each city, and the mean arrival delay for all remaining connections, which flew from specific locations.

The resulting `data.table` may be sorted using `order()` just as in base R. However, the `data.table` package offers an internally-optimized implementation of the `order()` function, which performs much faster on large datasets than its generic counterpart. We will now compare both implementations by sorting arrival delay values (ARR_DELAY) in the decreasing order for all flights in our data (12,189,293 observations):

```
> system.time(flightsDT[base::order(-ARR_DELAY)])
   user  system elapsed
 15.493   0.732  16.310
> system.time(flightsDT[order(-ARR_DELAY)])
   user  system elapsed
  4.669   0.401   5.075
```

The `data.table` implementation is clearly at least 3x faster than when R was forced to use the base `order()` function. The package contains a number of other shortcuts that speed up data processing; for example `.N` can be used to produce fast frequency calculations:

```
> agg2.DT <- flightsDT[, .N, by = ORIGIN_STATE_NM]
> agg2.DT
          ORIGIN_STATE_NM       N
     1:             Texas 1463283
     2:          New York  553855
     3:         Minnesota  268206
     4:    North Carolina  390979
     5:           Montana   34372
     6:              Utah  227066
     7:          Virginia  356462
...#output truncated
```

R spent only 0.098 seconds estimating the counts of flights from each state. Compared to the `table()` approach on a `data.frame` (1.14 seconds), the `data.table` package offers roughly a ten-fold speedup:

```
> agg2.df <- as.data.frame(table(flightsDT$ORIGIN_STATE_NM))
> agg2.df
             Var1    Freq
1         Alabama   66695
2          Alaska   74589
3         Arizona  383253
4        Arkansas   59792
5      California 1495110
6        Colorado  498276
7     Connecticut   44041
...#output truncated
```

Chaining, more complex aggregations, and pivot tables with data.table

One of the smartest things about `data.table` is that users can easily chain several fast operations into one expression reducing both programming and computing time, for example:

```
> agg3.DT <- flightsDT[, .N, by = ORIGIN_STATE_NM]
+                     [order(-N)]
> agg3.DT
      ORIGIN_STATE_NM       N
1:         California 1495110
2:              Texas 1463283
3:            Florida  871200
4:           Illinois  806230
5:            Georgia  802243
...#output truncated
```

Chaining is especially useful in more complex aggregations. In the following example we want to calculate the arithmetic mean (set in the `j` index) for all December flights (`i` index) on departure and arrival delay variables as indicated by the `.SDcols` parameter, group the results by the `ORIGIN_STATE_NM`, `DEST_STATE_NM`, and `DAY_OF_WEEK` variables, and finally sort the output firstly by `DAY_OF_WEEK` (in ascending order) and then by `DEP_DELAY` and `ARR_DELAY` (both in descending order):

```
> agg4.DT <- flightsDT[MONTH == 12L,
+                      lapply(.SD, mean, na.rm = TRUE),
+                      by = .(ORIGIN_STATE_NM,
```

```
+                              DEST_STATE_NM,
+                              DAY_OF_WEEK),
+               .SDcols = c("DEP_DELAY", "ARR_DELAY")]
+            [order(DAY_OF_WEEK, -DEP_DELAY, -ARR_DELAY)]
> head(agg4.DT, n=5)
   ORIGIN_STATE_NM        DEST_STATE_NM DAY_OF_WEEK DEP_DELAY
1:       Louisiana             Kentucky           1  111.6667
2:            Ohio       South Carolina           1  106.0000
3:          Alaska                Texas           1  103.0000
4:    Pennsylvania U.S. Virgin Islands           1   92.0000
5:         Indiana            Tennessee           1   90.0000
   ARR_DELAY
1:  108.0000
2:  104.3333
3:   93.0000
4:   88.5000
5:   82.7500
```

Note that the expression looks very tidy and contains multiple data-manipulation techniques. It is also extremely fast even on a large dataset; the preceding statement was executed in only 0.13 seconds. In order to replicate this aggregation in the base R, we would probably need to create a separate function, to process the call in stages, but it is very likely that the performance of this approach will be much worse. The data.table package through elegant chaining allows users to shift their focus from programming to true data science.

The chaining of operations may also be achieved through custom-built functions. In the following example, we want to create a delay function that will calculate TOT_DELAY for each flight in the dataset and we also want to attach this variable to our main dataset using the := operator. Second, based on the newly-created TOT_DELAY variable, the function will compute MEAN_DELAY for each DAY_OF_MONTH:

```
> delay <- function(DT) {
+   DT[, TOT_DELAY := ARR_DELAY - DEP_DELAY]
+   DT[, .(MEAN_DELAY = mean(TOT_DELAY, na.rm = TRUE)),
+        by = DAY_OF_MONTH]
+ }
> delay.DT <- delay(flightsDT)
> names(flightsDT)
 [1] "YEAR"              "MONTH"              "DAY_OF_MONTH"
 [4] "DAY_OF_WEEK"       "FL_DATE"            "UNIQUE_CARRIER"
 [7] "AIRLINE_ID"        "TAIL_NUM"           "FL_NUM"
[10] "ORIGIN_AIRPORT_ID" "ORIGIN"             "ORIGIN_CITY_NAME"
[13] "ORIGIN_STATE_NM"   "ORIGIN_WAC"         "DEST_AIRPORT_ID"
[16] "DEST"              "DEST_CITY_NAME"     "DEST_STATE_NM"
[19] "DEST_WAC"          "DEP_TIME"           "DEP_DELAY"
```

```
[22] "ARR_TIME"           "ARR_DELAY"       "CANCELLED"
[25] "CANCELLATION_CODE"  "DIVERTED"        "AIR_TIME"
[28] "DISTANCE"           "TOT_DELAY"
> head(delay.DT)
   DAY_OF_MONTH MEAN_DELAY
1:           17  -3.235925
2:           18  -3.369053
3:           19  -3.439632
4:           20  -3.976177
5:           21  -3.229061
6:           22  -3.166394
```

The := operator added the TOT_DELAY variable to the original data stored in the flightsDT object and the delay function computed the requested MEAN_DELAY by DAY_OF_MONTH and stored the results in delay.DT of the data.table package.

We should also mention here a very useful **casting** implementation through the dcast.data.table() function which allows rapid pivot tables, for example:

```
> agg5.DT <- dcast.data.table(flightsDT,
                              UNIQUE_CARRIER~MONTH,
                              fun.aggregate = mean,
                              value.var = "TOT_DELAY",
                              na.rm=TRUE)
> agg5.DT
   UNIQUE_CARRIER          1          2          3          4
1:             9E -5.0604885 -4.3035723 -3.9639291 -3.97549369
2:             AA -5.4185064 -4.7585774 -4.8034317 -3.52661907
3:             AS -3.6258772 -3.9423103 -2.5658019 -1.42202236
4:             B6 -3.0075933 -1.6200692 -2.9693669 -2.58131086
5:             DL -4.9593914 -5.1271330 -4.9993318 -4.78600081
...#output truncated
```

In the preceding output we have obtained a pivot table with mean values of TOT_DELAY for each carrier and month in our data. The fun.aggregate argument may take multiple functions, and similarly, the value.var parameter may now also refer to multiple columns.

In this section we have presented several common applications of fast data transformations available in the data.table package. This is, however, not inclusive of all functionalities this great package can offer. For more examples (especially on chaining, joins, key setting, and many others) and tutorials, please visit the data.table GitHub repository at https://github.com/Rdatatable/data.table/wiki. The CRAN page for the package (https://cran.r-project.org/web/packages/data.table/index.html) contains several references to comprehensive manuals elaborating on different aspects of data manipulation with data.table.

Writing better R code

Finally, in the last section of this chapter we will direct you to several good sources that can assist you in writing better optimised and faster R code. The best primary resource of knowledge on this subject is the previously mentioned book by Hadley Wickham *Advanced R*, and more specifically its chapter on code optimization. It includes very informative, but still pretty concise and approachable, sections on profiling and benchmarking tools, which can be used to test the performance of R scripts. Wickham also shares a number of tips and tricks on how to organize the code, minimise the workload, compile the functions, and use other techniques such as R interfaces for compiled code. The online version of the chapter is available from Wickham's personal website at
http://adv-r.had.co.nz/Profiling.html.

Another great source of information in this field is Norman Matloff's book titled *The Art of R Programming*. It consists of several comprehensive chapters dedicated to performance enhancements in processing speed and memory consumption, interfacing R to other languages-most predominantly C/C++ and Python, and also introducing readers to parallel techniques available in R through Hadley Wickham's `snow` package, compiled code and GPU programming amongst others. Besides, Matloff includes essential details on code debugging and rectifying issues with specific programming methods in R.

Unfortunately, the code optimization goes beyond the scope of this book, but the contents of both Wickham's and Matloff's publications cover this gap in a very comprehensive way. There are also a number of good web-based resources on specific high-performance computing approaches available in R and the CRAN Task View
https://cran.r-project.org/web/views/HighPerformanceComputing.html
should serve well as the index of the most essential packages and tools on that subject. Also, most of the packages, that were either referenced in the preceding sections of this chapter, or are listed in the High-Performance Computing Task View on CRAN, contain well-written manuals, and at least vignettes, addressing most important concepts and common applications.

Summary

In this chapter we have began our journey through the meanders of Big Data analytics with R. First, we introduced you to the structure, definition, and major limitations of the R programming language hoping that this may clarify why traditionally R was an unlikely choice for a Big Data analyst. But then we showed you how some of these concerns can be quite easily dispelled by using several powerful R packages which facilitate processing and analysis of large datasets.

We have spent a large proportion of this chapter on approaches, that allow out-of-memory data management, first through the `ff` and `ffbase` packages, and later by presenting methods contained within the `bigmemory` package and other libraries that support operations and analytics on `big.matrix` objects.

In the second part of the chapter we moved on to methods that can potentially boost the performance of your R code. We explored several applications of parallel computing through the `parallel` and `foreach` packages and you learnt how to calculate statistics using the `apply()` family of functions and `for()` loops. We also provided you with a gentle introduction to GPU computing, a new highly-optimized Microsoft R Open distribution, which supports multi-threading, and we have also mentioned the H2O platform for fast and scalable machine learning for Big Data (which we discuss in detail in Chapter 8, *Machine Learning methods for Big Data in R*).

We ended this chapter with an introductory tutorial on the `data.table` package-a highly efficient and popular package for fast data manipulation in R.

Further if you want to know how we can take Big Data outside of the limitation of a single machine and to deploy and configure instances and clusters on leading Cloud computing platforms, such as Amazon Elastic Cloud Computing (EC2), Microsoft Azure, and Google Cloud, you can go through the *Online Chapter, Pushing R Further* available at https://www.packtpub.com/sites/default/files/downloads/5396_6457OS_PushingRFurther.pdf

4
Hadoop and MapReduce Framework for R

In this chapter we are entering the diverse world of Big Data tools and applications that can be relatively easily integrated with the R language. In this chapter, we will present you with a set of guidelines and tips on the following topics:

- Deploying cloud-based virtual machines with Hadoop, the ready-to-use **Hadoop Distributed File System (HDFS)**, and MapReduce frameworks
- Configuring your instance/virtual machine to include essential libraries and useful supplementary tools for data management in HDFS
- Managing HDFS using shell/Terminal commands and running a simple MapReduce word count in Java for comparison
- Integrating R statistical environment with Hadoop on a single-node cluster
- Managing files in HDFS and run simple MapReduce jobs using the rhadoop bundle of R packages
- Carrying out more complex MapReduce tasks on large-scale electricity meter readings datasets on a multi-node HDInsight cluster on Microsoft Azure

However, just before we dive into practical tutorials, let's build on our Hadoop and MapReduce introduction from Chapter 1, *The Era of Big Data* and familiarize ourselves with how they are suited for Big Data processing and analysis.

Hadoop architecture

Apache Hadoop is an open source, integrated framework for Big Data processing and management, which can be relatively easy to deploy on commodity hardware. Hadoop can also be defined as an ecosystem of tools and methods that allow distributed storage and analytics of massive amounts of structured and unstructured data. In this section, we will present an array of tools, frameworks, and applications that come as integral parts of the Hadoop ecosystem and are responsible for a variety of data management and processing purposes.

Hadoop Distributed File System

As explained in `Chapter 1`, *The Era of Big Data*, **Hadoop Distributed File System** (**HDFS**) derives from the original Google File System presented in 2003 in a paper titled *The Google file system* authored by Ghemawat, Gobioff, and Leung. The architecture and design of current HDFS (based on the Apache Hadoop 2.7.2 release) are explained thoroughly in the HDFS Architecture Guide available at the Apache website at `http://hadoop.apache.org/docs/r2.7.2/hadoop-project-dist/hadoop-hdfs/HdfsDesign.html`. In a nutshell, we can characterize HDFS in the following few statements:

- It's a distributed, highly-accessible (can be accessed from multiple applications using a spectrum of programming languages) file system designed to run on a commodity hardware.
- HDFS is tuned and configured to support large or massive files by providing high data bandwidth and enormous scalability (up to thousands of nodes in a cluster).
- It's a fault-tolerant and robust distributed file system—this is achieved through its innovative ability of storing each file as a sequence of blocks (typically 64 MB in size each) and replicating them across numerous nodes.
- HDFS contains a number of safety measures to minimize the risk of data loss; for example, upon initialization, the **NameNode** restarts in the Safemode state, which ensures that each block is replicated accordingly (that is it has a minimum number of *replicas*) across **DataNodes**. If some of the copies of specific blocks are missing, the NameNode replicates them to other DataNodes.
- The file system's metadata stored in the NameNode is consistently checked and updated by the EditLog.

- In case of hardware failure or any other damage to central data structures of HDFS, it's configured to provide cluster rebalancing by moving data from one DataNode to another if required or to create additional replicas of certain blocks in high-demand scenarios. It also ensures data integrity by replicating *healthy* blocks of their corrupted copies and creating multiple copies of Edit Log in case of metadata disk failure.
- Data stored in HDFS can be easily managed by a variety of applications and it also supports automatic file back up and possible restoration.

During practical tutorials in this chapter you will learn how to manage files and directories on HDFS using Linux shell/Terminal commands and the R language.

MapReduce framework

As much as HDFS is important for data storage and file management, it cannot offer its users any real insights into the stored data. The MapReduce framework, on the other hand, is responsible for writing applications that process and analyze huge amounts of data in a distributed (or parallel) manner.

Simply speaking, the MapReduce job consists of a number of stages in which the data stored in HDFS is processed. Typically, a user initially designs a MapReduce application by defining a format of the input data, which will then be split into separate, independent chunks across HDFS and processed using a Mapper function in parallel.

Once the Mapper maps the data according to the user's requirements, it returns a key-value pair as an output, which becomes a new input for a Reducer function. The role of the Reducer is then to perform an aggregation or calculation of a specific statistic requested by the user. It again returns a key-value pair as the final output of the MapReduce job.

It is important to know that in most cases, the storage of data in HDFS and its processing through the MapReduce framework takes place on the same nodes. This allows the Hadoop framework to schedule the tasks more effectively and provide extremely high data bandwidth across all nodes in the cluster.

Secondly, the usual feature of the MapReduce job is that it very often changes the types of input and output formats depending on user's requirements. Users can freely create their own custom format of the requested data, which allows enormous flexibility in working with a large array of different data objects and formats.

Finally, each MapReduce job can contain numerous Mappers, and users can also specify a number of Reducer tasks for each job. Besides, MapReduce applications may include a Combiner function, which summarizes the output records from the Mapper before passing them into the Reducer function. However, in our MapReduce tutorials in this book we will only be using Mappers and Reducers.

A simple MapReduce word count example

In order to better understand operations of the MapReduce framework, it might be advisable to go through a very basic example of the word count algorithm applied to three simple sentences.

In this exercise, our task is to calculate how many times each word is repeated across all three following sentences:

Simon is a friend of Becky.
Becky is a friend of Ann.
Ann is not a friend of Simon.

Ignoring special characters, each sentence is a string of words, which we will extract and split during the MapReduce job. Let's also assume that all sentences constitute one single data file, which is chunked into data blocks replicated three times on DataNodes in our HDFS:

1. The Mapper of the MapReduce application is written in such a way that it recognizes the format of the input file (that is, our sentences) and passes each line of the text through the mapping function. In our case, Mapper will split the sentences into individual words (**keys**) and will assign the **value** 1 to each occurrence of a word. This step is visually presented in the following diagram:

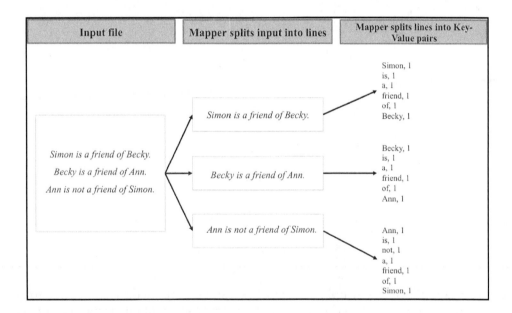

2. The extracted words are now sorted and shuffled into blocks that store the same keys. The Reducer can now sum the values for each key and return the final output in the form of an aggregated set of key-value pairs as displayed in the following diagram:

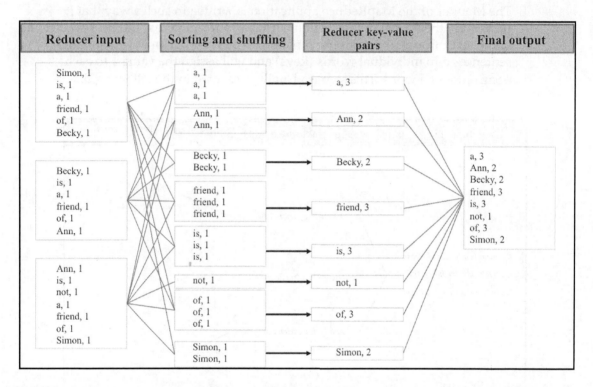

As stated earlier, this overly simplified workflow can be expanded further to include other Mappers, Combiners, and/or Reducers, and will most likely contain far more complex transformations, data aggregations, and statistical methods.

Other Hadoop native tools

HDFS and MapReduce are not the only components included in the Apache Hadoop framework. It also contains other open source tools and applications that support distributed computing and help users in managing the Hadoop cluster, scheduling its data processing tasks, obtaining performance metrics, manipulating data, and many others.

The Hortonworks website, http://hortonworks.com/hadoop/, includes a very useful graphical visualization of the diversity of Hadoop's native ecosystem:

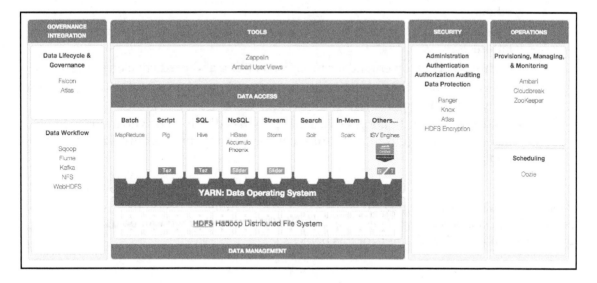

The preceding diagram also presents how all these tools are related to one another and what specific aspects of the data product cycle they participate in.

You can see that the basis of this framework resides in the previously described HDFS. On top of that we've got **Yet Another Resource Negotiator** (**YARN**), which fulfills a role for the Hadoop operating system. YARN controls job scheduling and manages all cluster resources. It also includes a **Resource Manager User Interface** through which users can access **MapReduce Job Tracker** providing all fine-grained application metrics in real time. We will be using Resource Manager UI during all tutorials in this chapter.

YARN supervises the activities and measures the performance of a collection of frameworks and tools that store, process, and analyze different types of structured and unstructured data. You have already got to know one of them, the **MapReduce** model, but there is also a number of other tools which you might find useful in your Big Data analytics workflows; they are listed as follows:

- **Hive**: An SQL-based, highly-scalable, distributed **Relational Database Management System** (**RDBMS**) allowing users to explore and analyze huge amounts of data using interactive SQL queries
- **HBase**: A low-latency, fault-tolerant, NoSQL database optimized for real-time access to data of differing structures and schemas

- **Storm**: A fast, real-time analytics framework
- **Spark**: A high-velocity in-memory data processing and analytics engine with a selection of add-on libraries for machine learning, graph analysis, and many others
- **Cassandra**: A highly reliable and scalable database with no single points of failure
- **Mah**out: A Hadoop engine for machine learning and predictive analytics
- Ambari: A web-based cluster management and monitoring tool with support for numerous Hadoop frameworks and applications
- Zookeper: A coordination service that provides distributed synchronization by maintaining configuration information and naming for other distributed Hadoop applications

The preceding list does not include many other extremely useful, open-source Apache projects that have been designed to work with Hadoop framework and its tools at different stages of Big Data processing. These are, however, the tools that we will be using throughout the next few chapters of this book.

Learning Hadoop

Data processing in Hadoop has recently become one of the major and most influential topics in the Big Data community. There are a large number of online and printed resources with information on how to implement MapReduce applications in Java, C, or other programming languages. As of March 2016, the StackOverflow website contained over 25,000 questions related to any type of Hadoop distribution and their related tools. Amongst several very good books on Hadoop, three titles are highly recommended:

- Gunarathne, T. (2015). *Hadoop MapReduce v2 Cookbook*, 2nd Edition. Packt Publishing.
- Grover, M., Malaska, T., Seidman, J., and Shapira, G. (2015). *Hadoop Application Architecture*, 1st Edition. USA: O'Reilly
- White, T. (2015). *Hadoop: The Definitive Guide*, 4th Edition. USA: O'Reilly

In addition to these, the Apache Hadoop website at `http://hadoop.apache.org/` is a very good starting point, and it provides a vast collection of documentation and reference guides for each specific Hadoop-related Apache project.

Last but not least, the websites of leading companies that offer commercial distributions of Hadoop, for example, Cloudera or Hortonworks, contain comprehensive explanations of native Apache Hadoop tools and other proprietary Hadoop-related applications. They also present a number of showcases and tutorials based on their own Hadoop distributions.

However, if you are interested in learning how to integrate the R language with Hadoop, proceed to the next section of this book, where we will present how to create a single-node Hadoop that can be used for Big Data processing and analytics using R.

A single-node Hadoop in Cloud

Hopefully by now you should have obtained an understanding of what outcomes you can achieve by running MapReduce jobs in Hadoop or by using other Hadoop components. In this chapter, we will put theory into practice.

We will begin by creating a Linux-based virtual machine with a pre-installed Hortonworks distribution of Hadoop through Microsoft Azure. The reason why we opt for a pre-installed, ready-to-use Hadoop is because this book is not about Hadoop *per se,* and we also want you to start implementing MapReduce jobs in the R language as soon as possible.

Once you have your Hadoop virtual machine configured and prepared for Big Data crunching we will present you with a simple word count example initially carried out in Java. This example will serve as a comparison for a similar job run in R.

Finally, we will perform a word count task in the R language. Before that, however, we will guide you through some additional configuration operations and we will explain how to resolve certain common installation issues when preparing our Hadoop virtual machine for data processing in the R language.

> Note that apart from the expected R scripts, and some statistical concepts and standard installation guidelines, this part of the chapter will also contain some Java snippets and a large amount of shell/Terminal commands. This clearly shows you how varied a typical toolbox of modern data scientists is in order to allow them to process and analyze unusually large or complex datasets. In later chapters, you will also get to know a variety of other tools and approaches including (and not limited to) SQL queries or certain flavors of NoSQL statements.

Let's now get on with our plan and deploy a new virtual machine with Hadoop.

Deploying Hortonworks Sandbox on Azure

In the preceding part of this chapter, we introduced you to some basics of Hadoop architecture and we also mentioned a few words about commercial Hadoop distributions such as Cloudera, Hortonworks, and MapR. In this section, we will create a virtual machine with a Hortonworks Hadoop release called **Sandbox**. We won't be installing Hortonworks Sandbox from scratch; in fact we will use a ready-made Hadoop image available for Microsoft Azure customers. Also, our Hadoop will operate just on a single machine (so we will create a single-node cluster) and you can also deploy a similar setup locally on your own PC using one of the virtualization tools and Sandbox installation guidelines available at http://hortonworks.com/products/hortonworks-sandbox/#install.

Hortonworks Sandbox is a free trial version of the fully licensed enterprise Hadoop Data Platform and it comes with a number of tools and functionalities typical for a standard Apache Hadoop distribution. We will explore some of them in the following sections of this chapter and we will also install other useful tools that will help us monitor all activities and processes on our virtual machine:

1. Login to your free/trial or full version of Microsoft Azure using your personal credentials. Upon successful log, in you should be re-directed automatically to the Azure Portal. If not, go to `https://portal.azure.com/`:

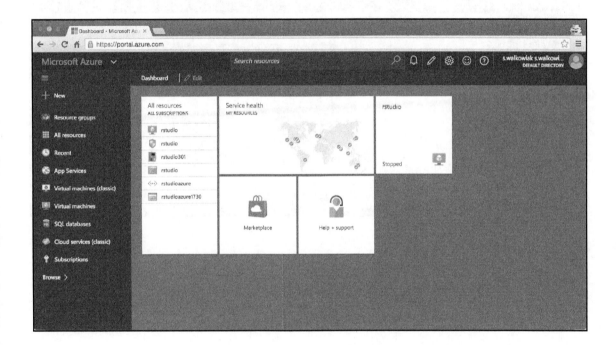

2. Click on the **Marketplace** icon located in the main dashboard pane. A new menu panel will appear in the main window, as shown in the following screenshot:

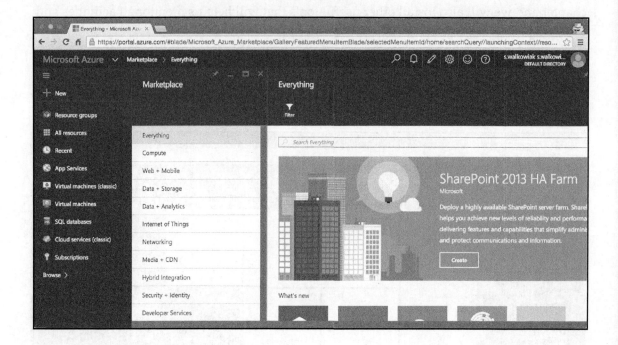

3. You can now either scroll down and find an icon depicting a green elephant with a caption that reads **Hortonworks Sandbox** or type `hortonworks` in the **Search Everything** search bar located in the top section of the pane. Choose **Hortonworks Sandbox with HDP 2.3.2**:

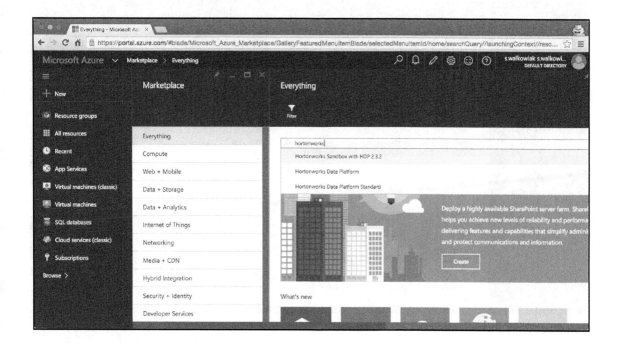

Hadoop and MapReduce Framework for R

4. The new panel with additional information on Hortonworks Sandbox should appear. Take a minute to read through and once you are ready select **Resource Manager** as the deployment model and click the **Create** button:

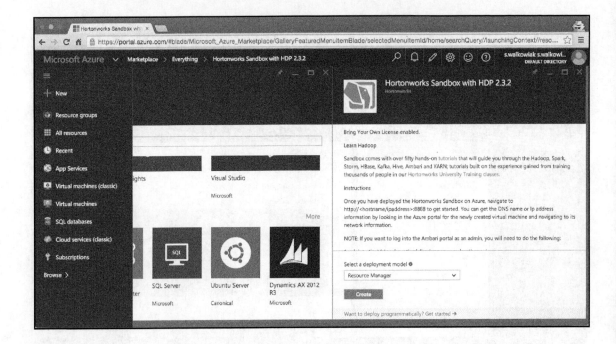

Chapter 4

5. Two new panes should open now. They will guide you through the process of setting up and configuring your new virtual machine with Sandbox:

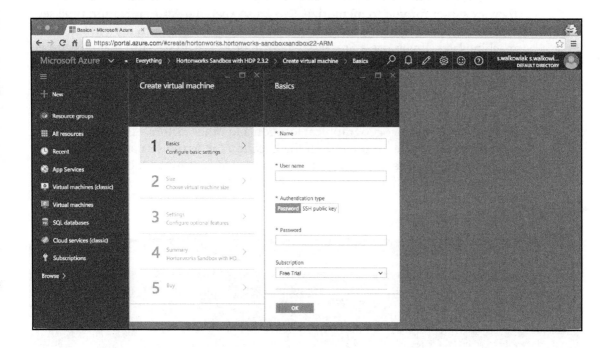

Hadoop and MapReduce Framework for R

6. Complete the form associated with the **Basics** tab. This allows us to configure some basic settings of our machine. I will call it `swalkohadoop` and the username will be set to `swalko`, but you should probably choose other names to ensure that they are easier for you to memorize. I will also keep the **Authentication type** set to **Password** and will provide a **Password** to be able to log in to my virtual machine later. If you use a Free Trial version of Azure, make sure that it is chosen as your preferred **Subscription**. As explained earlier, the fees are dependent on several different factors and additional charges may apply if you go beyond the limits indicated in the free/trial offering, so make sure you always read terms and conditions, and if in doubt, contact Microsoft Azure support to obtain current pricing based on your geographical location and other factors. Also, you need to name your **Resource group** and set the **Location** for Azure servers to be used I'm going to stick to **West Europe** as it is default for my account. Once you are happy with your basic configuration settings, click the **OK** button at the bottom of the form to proceed:

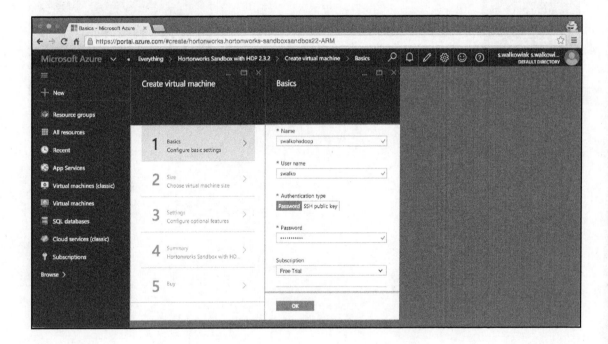

Chapter 4

7. You may now select the size and specification of your virtual machine. The Azure portal provides you with a recommended Azure infrastructure to choose and its provisional or estimated pricing. Remember that the pricing may not reflect applicable discounts for your specific subscriptions and it is also based on the constant (24/7) use of the machine. We will go with the recommended size **A5 Standard** with two cores, 14 GB of RAM and four data disks, but you may review and choose other machines depending on your subscription and budget limitations. Select your preferred instance by clicking on its information box and confirm your choice by pressing the **Select** button:

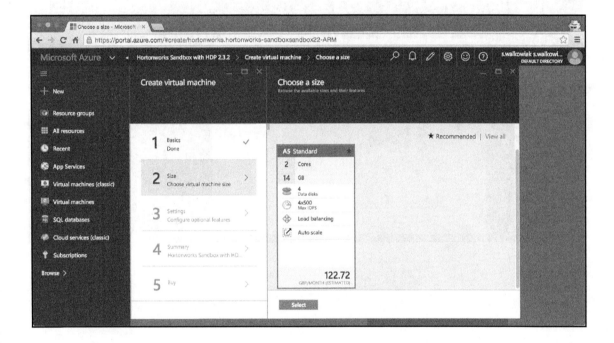

8. At this stage you will be provided with an additional panel of optional features and settings to configure. Most of them, such as **Storage account** name, **Network** details, for example, **Virtual network** name or **Public IP address,** will be populated automatically for you, as shown in the following screenshot:

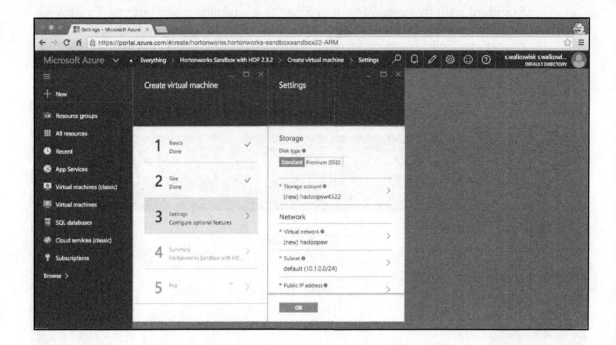

Scroll down to the bottom of the **Settings** pane and click on the **Availability set** option. This allows you to add another virtual machine to the availability set as a precaution and to provide redundancy for your data processing task or application in case of a maintenance issue, and other problems. It's recommended to do it, but because we are only testing the setup we will select **None** to skip this step:

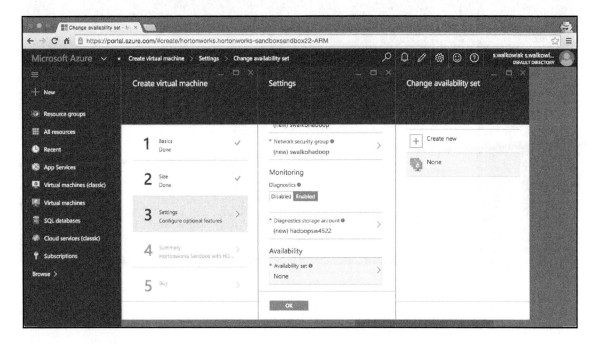

Accept all optional settings by clicking on the OK button.

Hadoop and MapReduce Framework for R

9. The penultimate phase allows you to review all details about your virtual machine. If you are happy with all settings, click **OK** to proceed:

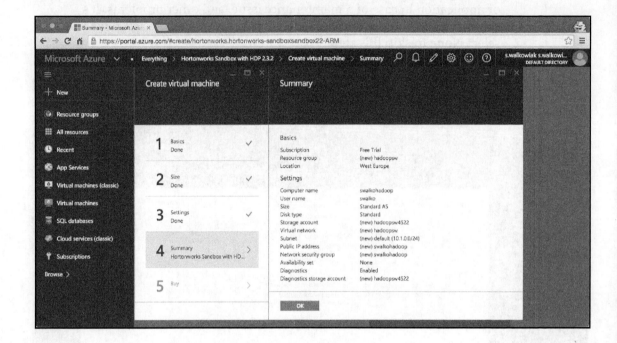

10. The this step confirms the final pricing for our setup. Depending on your subscription, your newly created machine may be included in your available credit, but may also incur additional charges. The virtual machine that I'm going to create will cost me 16.5 pence per hour. Of course, these estimates may vary depending on your specific location and many other factors, as explained before. If you want to proceed, read the terms of use and click the **Purchase** button to confirm:

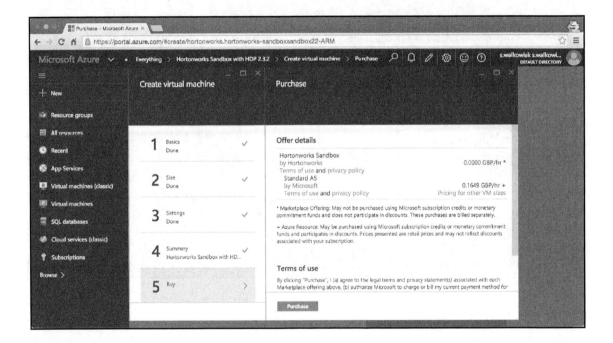

11. Azure will now start deploying the virtual machine. You should have also received a notification to your dashboard confirming that the deployment had indeed started. It may take several minutes for the machine and all its resources to be fully operational and ready to use:

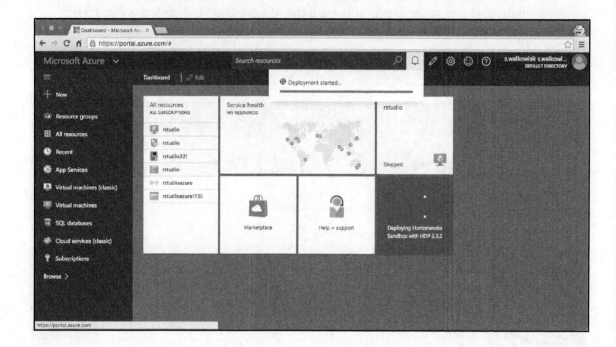

12. After a few minutes, the Azure Portal will notify you about the status of the job. If it is successful you will be re-directed to the **Settings** panel, where you can access all vital details of your virtual machine and monitor its performance:

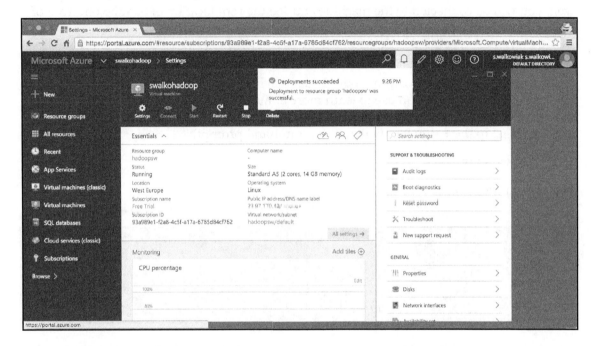

We have now successfully deployed a Linux-based virtual machine with Hortonworks Sandbox installed. We can try connecting to the machine using a simple `ssh` command from the shell/Terminal editor and by providing our credentials, which we set up in step 6. We will also need our **Public IP address** for the created machine you can obtain it from the main panel with settings called **Essentials** or by choosing the **Properties** option in the **Settings** screen. Open a new SSH/Terminal window and type the following:

```
ssh <username>@<IPaddress>
```

In my case, the preceding command should look as follows (adjust `<username>` and `<IPaddress>` for your machine):

```
ssh swalko@23.97.179.42
```

You will be informed that **the authenticity of host '<IPaddress>' can't be established** and then asked whether **you are sure you want to continue connecting (yes/no)?**. Proceed by typing `yes`. At this stage, the IP address of your machine will be added to the list of known hosts and you will be asked to provide a password for the specified username. If all goes well, you will be connected to the machine and see the root location of Sandbox in front of the command prompt, for example:

```
[swalko@sandbox ~]$
```

It is now a good moment to explore our virtual machine and revise some of the most useful Linux commands that we can use to learn more about the infrastructure at our disposal. In order to check your current version of Hadoop installed on the machine, use the following commands (skip the $ sign it only denotes a command prompt):

```
$ hadoop version
```

This should produce an output similar to the following screenshot:

```
[swalko@sandbox ~]$ hadoop version
Hadoop 2.7.1.2.3.2.0-2950
Subversion git@github.com:hortonworks/hadoop.git -r 5cc60e0003e33aa98205f18bccaeaf36cb193c1c
Compiled by jenkins on 2015-09-30T18:08Z
Compiled with protoc 2.5.0
From source with checksum 69a3bf8c667267c2c252a54fbbf23d
This command was run using /usr/hdp/2.3.2.0-2950/hadoop/lib/hadoop-common-2.7.1.2.3.2.0-2950.jar
```

It is also recommended that we check the location of our Hadoop folders. You can use either the `which` or `whereis` command, for example:

```
$ which hadoop
/usr/bin/hadoop
$ whereis hadoop
hadoop: /usr/bin/hadoop /etc/hadoop
```

Also, it is essential that you know the version and location of your **Java Runtime Environment (JRE)**, which will be used during data processing directly in Hadoop and also when deploying MapReduce jobs from an R environment:

```
$ java -version
java version "1.7.0_91"
OpenJDK Runtime Environment (rhel-2.6.2.2.el6_7-x86_64 u91-b00)
OpenJDK 64-Bit Server VM (build 24.91.b01, mixed mode)
$ which java
/usr/lib/jvm/java-1.7.0-openjdk-1.7.0.91.x86_64/bin/java
```

Of course, we also can't forget about checking the actual version of Linux and its kernel used in our machine:

```
$ cat /etc/system-release
CentOS release 6.7 (Final)
$ uname -r
2.6.32-573.7.1.el6.x86_64
```

Once you obtain the basic information about your machine you will very often find yourself wanting to monitor available resources when implementing MapReduce jobs in Hadoop. There is a number of ways to do it. Firstly, you can obtain some insights from a chart available through the Azure Portal, which allows visual inspection and monitoring of how your virtual machine behaves. You can access these functions by selecting the name of the virtual machine of interest from the **All resources** tab on the left-hand menu bar and clicking on the main chart below the **Essentials** of the selected machine. A new panel called **Metric** will appear in which you can specify visualized measurements by selecting the **Edit chart** button and choosing one or several available metrics from the list. Tick the ones that you find relevant, scroll down to the bottom of the list, and click the **Save** button to apply the changes:

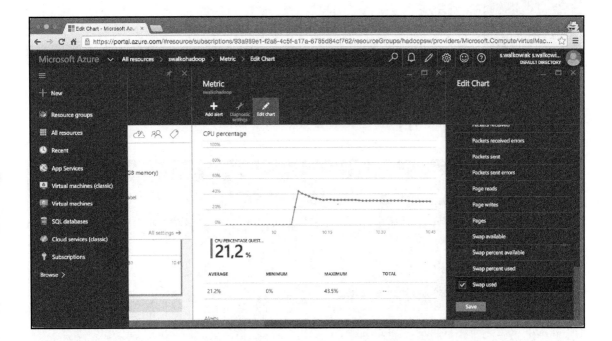

Hadoop and MapReduce Framework for R

The applied changes will appear after a few seconds on the newly created **Metric** chart, as shown in the following screenshot:

Of course, you may also use a number of Linux commands to obtain more quantitative details about available resources and processes. To check the size and usage of hard drives on your machine, you can type the following two commands:

```
[swalko@sandbox ~]$ df
Filesystem            1K-blocks      Used Available Use% Mounted on
/dev/mapper/vg_sandbox-lv_root
                       44717136  13962276  28476668  33% /
tmpfs                   7194736         0   7194736   0% /dev/shm
/dev/sda1                487652     30214    431838   7% /boot
/dev/sdb1             139203560   4255236 127870540   4% /mnt/resource
[swalko@sandbox ~]$ df -H
Filesystem            Size  Used Avail Use% Mounted on
/dev/mapper/vg_sandbox-lv_root
                       46G   15G   30G  33% /
tmpfs                 7.4G     0  7.4G   0% /dev/shm
/dev/sda1             500M   31M  443M   7% /boot
/dev/sdb1             143G  4.4G  131G   4% /mnt/resource
```

They are almost identical, but the second call gives us a more human-friendly conversion to megabytes, gigabytes, or terabytes.

Obtaining RAM information is indispensable in measuring the performance of applications and the efficiency of your code while processing large amounts of data. You can find out basic hardware details about installed RAM on your machine through the `dmidecode` command:

```
$ sudo dmidecode -t memory
```

Depending on the number of modules installed and their configuration, the command may return quite a long output. It is more useful, however, to determine the actual usage of memory resources during the processing jobs. In order to do this, you can use a generic `free` command:

```
[swalko@sandbox ~]$ free -m
               total       used       free     shared    buffers     cached
Mem:           14052       5244       8807          7         31        319
-/+ buffers/cache:          4894       9158
Swap:           4095          0       4095
```

The `free` command provides only the most essential and overall memory allowance and usage. For more detailed information, we have to look at the `meminfo` virtual file:

```
[swalko@sandbox ~]$ cat /proc/meminfo
MemTotal:       14389476 kB
MemFree:         8852180 kB
Buffers:           32964 kB
Cached:           328420 kB
SwapCached:            0 kB
Active:          5076224 kB
Inactive:         214640 kB
Active(anon):    4928756 kB
Inactive(anon):     8028 kB
Active(file):     147468 kB
```

Again, as the output is quite long, we've decided to include only a short snapshot of returned memory metrics. A similar output can be achieved using the `vmstat` command with the `-s` option, as follows:

```
$ vmstat -s
```

For more in-depth monitoring of memory usage for specific processes we recommend the `top` command or an additional tool called `htop`. Both return roughly the same metrics, `htop` being, however, a little bit more human-readable:

```
$ top
```

```
top - 11:19:22 up  1:20,  2 users,  load average: 2.37, 2.56, 2.16
Tasks: 193 total,   1 running, 192 sleeping,   0 stopped,   0 zombie
Cpu(s):  1.0%us,  0.2%sy,  0.0%ni, 98.8%id,  0.0%wa,  0.0%hi,  0.0%si,  0.0%st
Mem:  14389476k total,  5403360k used,  8986116k free,    33952k buffers
Swap:  4194300k total,        0k used,  4194300k free,   329864k cached

  PID USER      PR  NI  VIRT  RES  SHR S %CPU %MEM    TIME+  COMMAND
 1794 ranger    20   0 2477m 725m  14m S  1.3  5.2   1:55.44 java
 3012 oozie     20   0 2812m 513m  25m S  1.0  3.7   3:22.19 java
 3365 yarn      20   0 1038m 275m  24m S  0.7  2.0   1:32.65 java
49848 swalko    20   0 15028 1380  984 R  0.7  0.0   0:00.07 top
 1263 root      20   0  401m  17m 4396 S  0.3  0.1   0:07.14 python
 1796 atlas     20   0 2378m 399m  20m S  0.3  2.8   0:56.35 java
 2111 ranger    20   0 4882m 144m  12m S  0.3  1.0   0:14.27 java
 2296 hdfs      20   0 1627m 345m  24m S  0.3  2.5   1:37.63 java
 2299 hdfs      20   0  988m 323m  24m S  0.3  2.3   0:44.48 java
 2682 root      20   0 1167m  52m 4688 S  0.3  0.4   4:11.42 python2.6
 3381 yarn      20   0 1157m 257m  24m S  0.3  1.8   0:50.09 java
 5689 hue       20   0  205m  39m 5192 S  0.3  0.3   0:15.36 python2.6
    1 root     20   0 19232 1528 1232 S  0.0  0.0   0:02.11 init
    2 root     20   0     0    0    0 S  0.0  0.0   0:00.00 kthreadd
    3 root     RT   0     0    0    0 S  0.0  0.0   0:00.05 migration/0
```

The preceding output has been truncated. You can quit it by pressing the letter q on your keyboard.

The `htop` tool is not included in the standard Linux distribution on our machine so it has to be installed separately. Knowing that our Linux is a RHEL/CentOS 6.7 release, we firstly need to install and enable **RepoForge** (**RPMForge**) repositories for our 64-bit system. They can be found and downloaded from the following website, `http://pkgs.repoforge.org/rpmforge-release/`. The most recent RepoForge repositories for a 64-bit RHEL/CentOS 6.x were released on 20th March 2013, and this is the version we are going to install:

```
$ wget http://pkgs.repoforge.org/rpmforge-release/rpmforge-release-0.5.3-1.el6.rf.x86_64.rpm
...
$ rpm -ihv rpmforge-release-0.5.3-1.el6.rf.x86_64.rpm
...
```

With the RepoForge repositories installed, we can now use a standard `yum` command to install the `htop` tool:

```
$ sudo yum install htop
```

Provide your password and confirm that you are willing to install `htop` when prompted. After a few seconds, the tool should be installed and ready to use on your machine:

```
Total download size: 75 k
Installed size: 161 k
Is this ok [y/N]: y
Downloading Packages:
htop-1.0.1-2.el6.x86_64.rpm                               | 75 kB     00:00
Running rpm_check_debug
Running Transaction Test
Transaction Test Succeeded
Running Transaction
  Installing : htop-1.0.1-2.el6.x86_64                                    1/1
  Verifying  : htop-1.0.1-2.el6.x86_64                                    1/1

Installed:
  htop.x86_64 0:1.0.1-2.el6

Complete!
```

You can now run `htop`:

```
$ htop
```

The `htop` tool provides user-friendly access to resources for each process, including RAM information, CPU percentage use, and workload information for each available core amongst other metrics. You can quit the `htop` window by pressing the letter q on your keyboard.

In this section, you've learned how to create and configure a fully operational Linux virtual machine with a trial version of Hadoop distribution offered by Hortonworks Sandbox. We have also revised several frequently used Linux commands and tools which allow us to obtain basic information about the deployed infrastructure and to monitor processes and their usage of available resources on the machine.

In the following section, we will introduce you to essential Hadoop commands that will enable you to manage data files within HDFS and perform a simple MapReduce task in Java to obtain word count information, as described in the first part of this chapter.

A word count example in Hadoop using Java

Earlier in this chapter, we explained how the HDFS and MapReduce frameworks work by giving you an example of a very simplified word count task applied to a few random sentences. In this section, you will implement a similar word count MapReduce job yourself, but this time on a much larger data set that contains all Mark Twain's works, available at Project Gutenberg.

> **Project Gutenberg,** https://www.gutenberg.org/, is an open access book repository that contains over 51,000 free ebooks written and published by some of the world's most famous authors. Users can search books and browse the catalogue based on several criteria. If specific books are freely-available in their countries, users can also download these ebooks and read them (or analyze them) on their own devices. Project Gutenberg is a non-profit initiative and donations are always welcome. The ebooks may be provided for free, but there is a cost behind their preparation and the hosting of the site.

> To see all available books authored by Mark Twain, visit: https://www.gutenberg.org/ebooks/author/53. Depending on the copyright access terms and conditions in your country, you may be able to download all Twain's works available through Project Gutenberg by visiting: https://www.gutenberg.org/ebooks/3200. We will be using the Plain Text UTF-8 format as our input file for further analysis. If you are sure you can use Twain's books freely, download the file as it is to any known or easily memorable location on your personal computer and rename the file twain_data.txt. The file is not very large (only 15.3 MB), but will provide us with a good example of how the Hadoop ecosystem works in practice.

In this section, we will also take a very brief detour from the R language and carry out our first MapReduce job in **Java**Hadoop's native language. The understanding of Java scripts may pose some difficulties for readers and data analysts without any Java knowledge; however, we believe it will be interesting to compare how Java-operated MapReduce differs from its implementation using the R language. We will try to explain step-by-step processes and actions you should take in order to run Java MapReduce successfully. Packt Publishing's website for this book allows you to download all required files, including the `jar` MapReduce job package, which contains Java classes created for Mapper and Reducer. You can then follow the example using the downloaded data file from Project Gutenberg and the instructions provided in this section.

Before we attempt to perform any MapReduce tasks on the data, we should learn how to manage HDFS and move our data set from the local machine to our virtual machine deployed in this chapter. Let's firstly create a folder in our current default directory called `data`, which will store our data file:

```
$ mkdir data
```

You can now inspect the contents of the default directory for the main user:

```
$ ls
data rpmforge-release-0.5.3-1.el6.rf.x86_64.rpm
```

As you can see, it also contains the RepoForge repositories that we downloaded when installing the `htop` tool. If you want, you can delete this file:

```
$ rm rpmforge*
$ ls
data
```

It's also worth knowing the actual path to the current directory where the new `data` folder has been created. You can obtain it using the `pwd` command:

```
$ pwd
/home/swalko
```

Once we've created a new folder for our data and we know the actual path to the `data` folder, the easiest way to move the data from a local to virtual machine is by using the following command in a new shell/Terminal window:

```
$ scp -r ~/Desktop/twain_data.txt swalko@23.97.179.42:~/data
```

Of course, you may need to amend the preceding line depending on the location where you stored the data file, your actual username to access the virtual machine, its IP address, and the path to the `data` folder. You will also need to provide your password for the specified username and then wait a couple of seconds for the file to transfer to the virtual machine. At this stage, we have only copied the file from our PC to the created virtual machine the file is still not accessible in the HDFS. For the time being, you can only find it in the `data` folder for the main user (`swalko` in our case), for example:

```
[swalko@sandbox ~]$ cd data
[swalko@sandbox data]$ ls
twain_data.txt
```

In order to copy the file from the user's area of the virtual machine to the HDFS, we need to create a new folder for data files, but before this happens, we have to make sure that there is a new directory created on HDFS for our user `swalko`.

Firstly, let's inspect which users have permissions to write to HDFS:

```
$ hadoop fs -ls /
```

```
Found 9 items
drwxrwxrwx   - yarn    hadoop          0 2015-10-27 14:39 /app-logs
drwxr-xr-x   - hdfs    hdfs            0 2015-10-27 15:16 /apps
drwxr-xr-x   - hdfs    hdfs            0 2015-10-27 15:04 /demo
drwxr-xr-x   - hdfs    hdfs            0 2015-10-27 14:39 /hdp
drwxr-xr-x   - mapred  hdfs            0 2015-10-27 14:39 /mapred
drwxrwxrwx   - mapred  hadoop          0 2015-10-27 14:40 /mr-history
drwxr-xr-x   - hdfs    hdfs            0 2015-10-27 15:09 /ranger
drwxrwxrwx   - hdfs    hdfs            0 2015-10-27 14:54 /tmp
drwxr-xr-x   - hdfs    hdfs            0 2015-10-27 15:18 /user
```

We need to create a new directory for `swalko` through the `hdfs` user, as follows:

```
$ sudo -u hdfs hadoop fs -mkdir /user/swalko
```

You will be asked again to type your default password. If you try accessing the newly created directory as `swalko` it is highly likely that you still won't be able to do so, for example:

```
$ hadoop fs -ls /user
ls: Permission denied: user=swalko, access=READ_EXECUTE,
inode="/user":hdfs:hdfs:drwx------
```

However, you can check through the `hdfs` user whether the new directory has in fact been created:

```
$ sudo -u hdfs hadoop fs -ls /user
Found 2 items
drwx------   - hdfs hdfs          0 2016-02-29 18:42 /user/hdfs
drwxr-xr-x   - hdfs hdfs          0 2016-02-29 18:47 /user/swalko
... //(output truncated)
```

In order to make `swalko` able to write files to the `/user/swalko` directory on HDFS we need to explicitly assign permissions to this directory using the following command:

```
$ sudo -u hdfs hadoop fs -chown swalko:swalko /user/swalko
```

Provide your password when prompted and run the line below to copy the data from the `swalko` user's area in the `data` folder to the `/user/swalko` folder on HDFS:

```
$ hadoop fs -copyFromLocal data/twain_data.txt /user/swalko/twain_data.txt
```

Check whether the file has been copied correctly to the `/user/swalko` directory on HDFS:

```
$ hadoop fs -ls /user/swalko
```

```
[swalko@sandbox ~]$ hadoop fs -ls /user/swalko
Found 1 items
-rw-r--r--   3 swalko swalko   16013935 2016-02-29 23:43 /user/swalko/twain_data.txt
```

Once we have the data in the HDFS we can now download the `jar` package along with the Java classes for the word count MapReduce task from Packt Publishing's website for this chapter and transfer it from the personal computer to our virtual machine in the same way as we did with the data file several steps earlier. Open a new Terminal window and type the following line, making sure that you alter the paths, username, and IP address accordingly:

```
$ scp -r ~/Desktop/wordcount swalko@23.97.179.42:~/wordcount
```

The preceding code will copy the contents of the `wordcount` folder to your local area (but not HDFS) of the virtual machine. It should include the `dist` folder (with a `WordCount.jar` file) and the `src` folder (with three `*.java` files).

One of the Java files is a Mapper (`WordMapper.java`) used in our simple WordCount MapReduce job to split sentences into words and assign the value 1 to each word that occurs in the text file. At this stage, our MapReduce task ignores the fact that some words are repeated throughout the text of Mark Twain's works. It is the Reducer's (`SumReducer.java`) responsibility to sum all 1 for each specific word and give us the final output with the aggregated number of how many times a word occurred in the text. In the Mapper, we have also added a couple of functions that clean up our data a little bit. We have removed a number of special characters (for example, punctuation characters) and we have transformed all words to lowercase. If you want to inspect the code, feel free to open the included Java files in any Java programming IDE (for example, *Eclipse*) or a Java-friendly text editor (for example, *Notepad++* for Windows, *TextWrangler* for Mac OS, or *Gedit*, *Emacs*, and *Vim* for Linux).

Before we execute the MapReduce job on our data stored in HDFS, move to the wordcount folder in your local area on the virtual machine:

```
$ cd wordcount
```

Now run the following line to start the MapReduce task (you may need to change the paths, the username of your HDFS directory, and some other things):

```
$ hadoop jar dist/WordCount.jar WordCount /user/swalko /user/swalko/out
```

In the preceding code, we explicitly tell Hadoop to execute the `jar` file stored at `dist/WordCount.jar` on the user's local area, which will subsequently perform a `WordCount` MapReduce job on data stored in the `/user/swalko` directory on HDFS. We have also specified the location on HDFS for MapReduce output files in our case, it will be `/user/swalko/out`. Once your MapReduce job starts you should expect the following output appear in your shell/Terminal window:

```
16/03/02 13:13:20 INFO impl.TimelineClientImpl: Timeline service address: http://sandbox.hortonworks.com:8188/ws/v1/timeline/
16/03/02 13:13:21 INFO client.RMProxy: Connecting to ResourceManager at sandbox.hortonworks.com/10.1.0.4:8050
16/03/02 13:13:21 WARN mapreduce.JobResourceUploader: Hadoop command-line option parsing not performed. Implement the Tool interface and execute your application with ToolRunner to remedy this.
16/03/02 13:13:23 INFO input.FileInputFormat: Total input paths to process : 1
16/03/02 13:13:23 INFO mapreduce.JobSubmitter: number of splits:1
16/03/02 13:13:24 INFO mapreduce.JobSubmitter: Submitting tokens for job: job_1456920375766_0002
16/03/02 13:13:25 INFO impl.YarnClientImpl: Submitted application application_1456920375766_0002
16/03/02 13:13:25 INFO mapreduce.Job: The url to track the job: http://sandbox.hortonworks.com:8088/proxy/application_1456920375766_0002/
```

Following the advice given in the last line of the output, you can monitor the progress of the job at `http://<IPAddress>:8088`, where `<IPAddress>` is the IP address of your virtual machine, so in our case, the URL should read as follows: `http://23.97.179.42:8088`. The Hadoop application tracking page is part of the Resource Manager and it includes essential information about the progress of the job and its resources:

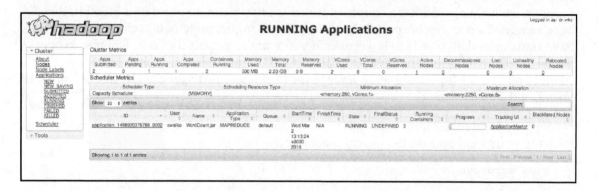

You can click on the application ID link (`ID` column) for more details on the particular MapReduce job that is currently running:

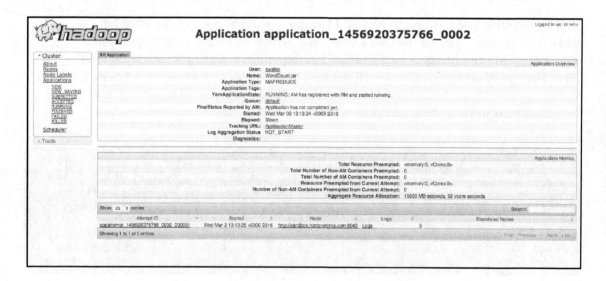

Chapter 4

In the meantime, the Terminal window should return the real-time progress of Mapper and Reducer:

```
16/03/02 13:13:25 INFO mapreduce.Job: Running job: job_1456920375766_0002
16/03/02 13:13:39 INFO mapreduce.Job: Job job_1456920375766_0002 running in uber mode : false
16/03/02 13:13:39 INFO mapreduce.Job:  map 0% reduce 0%
16/03/02 13:13:55 INFO mapreduce.Job:  map 47% reduce 0%
16/03/02 13:13:59 INFO mapreduce.Job:  map 56% reduce 0%
16/03/02 13:14:08 INFO mapreduce.Job:  map 59% reduce 0%
16/03/02 13:14:11 INFO mapreduce.Job:  map 67% reduce 0%
16/03/02 13:14:16 INFO mapreduce.Job:  map 100% reduce 0%
16/03/02 13:15:15 INFO mapreduce.Job:  map 100% reduce 7%
16/03/02 13:15:19 INFO mapreduce.Job:  map 100% reduce 17%
16/03/02 13:15:23 INFO mapreduce.Job:  map 100% reduce 27%
16/03/02 13:15:27 INFO mapreduce.Job:  map 100% reduce 30%
16/03/02 13:15:29 INFO mapreduce.Job:  map 100% reduce 43%
16/03/02 13:15:31 INFO mapreduce.Job:  map 100% reduce 50%
16/03/02 13:15:34 INFO mapreduce.Job:  map 100% reduce 60%
16/03/02 13:15:36 INFO mapreduce.Job:  map 100% reduce 61%
16/03/02 13:15:38 INFO mapreduce.Job:  map 100% reduce 67%
16/03/02 13:15:39 INFO mapreduce.Job:  map 100% reduce 76%
16/03/02 13:15:42 INFO mapreduce.Job:  map 100% reduce 78%
16/03/02 13:15:46 INFO mapreduce.Job:  map 100% reduce 80%
16/03/02 13:16:11 INFO mapreduce.Job:  map 100% reduce 90%
16/03/02 13:16:12 INFO mapreduce.Job:  map 100% reduce 100%
16/03/02 13:16:15 INFO mapreduce.Job: Job job_1456920375766_0002 completed successfully
```

You can also see that the last line of the output informs us that the job has been completed successfully. You can double-check the status on the application tracking page which confirms that the MapReduce job succeeded:

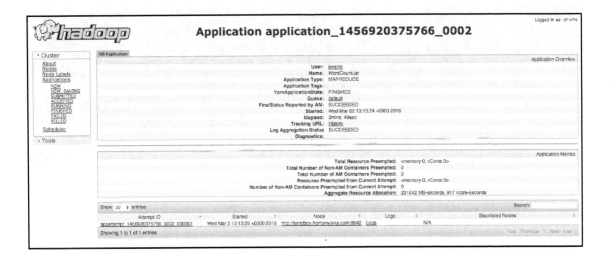

The Terminal window also returns numerous metrics related to file system counters, job counters, and the MapReduce framework. The last three sets of metrics indicate any potential shuffle errors and determine the size of input and output files:

```
        Shuffle Errors
                BAD_ID=0
                CONNECTION=0
                IO_ERROR=0
                WRONG_LENGTH=0
                WRONG_MAP=0
                WRONG_REDUCE=0
        File Input Format Counters
                Bytes Read=16013935
        File Output Format Counters
                Bytes Written=922308
[swalko@sandbox wordcount]$
```

From the bottom two lines of the Terminal output we can imply that our MapReduce job reduced the data from 16,013,935 bytes (~16 MB) to 922,308 bytes (~0.9 MB). Let's then find our output data files in HDFS and inspect their structure.

In order to locate the MapReduce output data, we can list the contents of the output folder, /user/swalko/out, which we created on HDFS when starting the MapReduce job with a jar file:

```
$ hadoop fs -ls /user/swalko/out
```

```
Found 11 items
-rw-r--r--   3 swalko swalko          0 2016-03-02 13:16 /user/swalko/out/_SUCCESS
-rw-r--r--   3 swalko swalko      91509 2016-03-02 13:15 /user/swalko/out/part-r-00000
-rw-r--r--   3 swalko swalko      92086 2016-03-02 13:15 /user/swalko/out/part-r-00001
-rw-r--r--   3 swalko swalko      91091 2016-03-02 13:15 /user/swalko/out/part-r-00002
-rw-r--r--   3 swalko swalko      92762 2016-03-02 13:15 /user/swalko/out/part-r-00003
-rw-r--r--   3 swalko swalko      91470 2016-03-02 13:15 /user/swalko/out/part-r-00004
-rw-r--r--   3 swalko swalko      91541 2016-03-02 13:15 /user/swalko/out/part-r-00005
-rw-r--r--   3 swalko swalko      93344 2016-03-02 13:15 /user/swalko/out/part-r-00006
-rw-r--r--   3 swalko swalko      93144 2016-03-02 13:15 /user/swalko/out/part-r-00007
-rw-r--r--   3 swalko swalko      92432 2016-03-02 13:16 /user/swalko/out/part-r-00008
-rw-r--r--   3 swalko swalko      92929 2016-03-02 13:16 /user/swalko/out/part-r-00009
```

This MapReduce job created 10 separate output files containing the result of operations executed by Mapper and Reducer (a confirmation of the number of output files can be found in the `Map-Reduce Framework` part of the Terminal job processing output where the `Merged Map outputs` metric is shown). The number of output files can be controlled by adjusting the `job.setNumReduceTasks(10);` statement in the `WordCount.java` file. Remember, however, that if you alter this value you will have to package all Java files into a new `jar` file before you run another MapReduce job.

All output files are named as `part-r-xxxxx`, where x is the identifier of an individual output file, for example, `part-r-00000` for the first output file. You may want to inspect only a small chunk (for example, the top 20 records) of each file using the following command:

```
$ hadoop fs -cat /user/swalko/out/part-r-00000 | head -n 20
```

```
[swalko@sandbox wordcount]$ hadoop fs -cat /user/swalko/out/part-r-00000 | head -n 20
00half  1
03      6
1       361
100000000       1
104851  1
113     1
1150    3
12      113
12000000        4
1204    1
1222    2
12th    15
131     1
140     1
1547    1
1592    1
1600    5
1600000 1
162000  2
1673    3
cat: Unable to write to output stream.
```

Most of the time, however, you will want to create a single output file that you can transfer to a separate server or to your personal computer for further processing or analysis. You can merge all output files into one document and move it from HDFS to the user's local area with the following line of code:

```
$ hadoop fs -getmerge /user/swalko/out wordcount.txt
```

The `-getmerge` option allows us to merge all data files from a specified directory on HDFS (for example, `/user/swalko/out`) and transfer them as a single file, for example, `wordcount.txt`, to your current directory on a local file system. Remember that before running the MapReduce job we advised you to move your current directory to the `wordcount` folder using the `cd wordcount` command. Therefore, you should now be able to find the `wordcount.txt` output file in this folder:

```
[swalko@sandbox wordcount]$ ls
dist  src  wordcount.txt
```

If you wish, you can now download the output file from the virtual machine to the desktop area of your personal computer. Open a new Terminal window and type the following (make sure to adjust the username, IP address, and paths accordingly to reflect your actual values):

```
$ scp -r swalko@23.97.179.42:~/wordcount/wordcount.txt
~/Desktop/wordcount_final.txt
```

Finally, we can remove the `/user/swalko/out` folder which contains all 10 chunks of the MapReduce output files on HDFS:

```
$ hadoop fs -rm -r /user/swalko/out
```

Hopefully, after reading this section, you will be able to perform a simple MapReduce task on a newly deployed virtual machine with a Hortonworks distribution of Hadoop using Hadoop's native Java. We have also introduced you to essential Linux shell/Terminal commands frequently used for data management in HDFS. More specific and advanced Hadoop, HDFS, Yarn, and MapReduce commands can be found in the Apache Hadoop reference documents:

- `http://hadoop.apache.org/docs/current/hadoop-project-dist/hadoop-common/CommandsManual.html` for **Hadoop commands**
- `http://hadoop.apache.org/docs/current/hadoop-mapreduce-client/hadoop-mapreduce-client-core/MapredCommands.html` for **MapReduce commands**
- `http://hadoop.apache.org/docs/current/hadoop-project-dist/hadoop-hdfs/HDFSCommands.html` for **HDFS commands**

If you feel like you need to brush up on your Linux shell/Terminal skills, you can find some good tutorials on the Internet:

- `http://linuxcommand.org/lc3_learning_the_shell.php`, or
- `http://ryanstutorials.net/linuxtutorial/`

In the next section of this chapter, we will implement the same MapReduce job on exactly the same data, but this time using the R language.

A word count example in Hadoop using the R language

Knowing that our virtual machine is Linux RedHat/CentOS 6, this part of the chapter will firstly focus on essential procedures to configure the virtual machine and install an appropriate version of an RStudio Server environment. The following steps will guide you through this process.

RStudio Server on a Linux RedHat/CentOS virtual machine

1. Just like in the preceding chapter, in order to download and install the correct version of RStudio Server, visit `https://www.rstudio.com/products/rstudio/download-server/` and select the distribution that is suitable for your machine. In our case, it will be a 64-bit RedHat/CentOS 6 and 7. Log in to your virtual machine using shell/Terminal or an SSH client such as **PuTTY** we will use a standard Terminal window:

```
$ ssh swalko@23.97.179.42
```

2. Go to your home directory and make sure to install the core R files:

```
$ cd ..
$ sudo yum install R
```

Currently installed Linux libraries and R dependencies will be updated and you will be asked to confirm whether you want to download a set of files required for the core R installation. Type *y* and press *Enter*:

```
Transaction Summary
=================================================================
Install       52 Package(s)

Total download size: 135 M
Installed size: 307 M
Is this ok [y/N]:
```

3. It may take several minutes to download, install, and verify all core R files and packages. Once complete, Terminal will return an output that should resemble the following screenshot:

```
Installed:
  R.x86_64 0:3.2.3-4.el6

Dependency Installed:
  R-core.x86_64 0:3.2.3-4.el6                            R-core-devel.x86_64 0:3.2.3-4.el6
  R-devel.x86_64 0:3.2.3-4.el6                           R-java.x86_64 0:3.2.3-4.el6
  R-java-devel.x86_64 0:3.2.3-4.el6                      blas.x86_64 0:3.2.1-4.el6
  blas-devel.x86_64 0:3.2.1-4.el6                        bzip2-devel.x86_64 0:1.0.5-7.el6_0
  desktop-file-utils.x86_64 0:0.15-9.el6                 fontconfig-devel.x86_64 0:2.8.0-5.el6
  freetype-devel.x86_64 0:2.3.11-15.el6_6.1              gcc-gfortran.x86_64 0:4.4.7-16.el6
  kpathsea.x86_64 0:2007-60.el6_7                        lapack.x86_64 0:3.2.1-4.el6
  lapack-devel.x86_64 0:3.2.1-4.el6                      libRmath.x86_64 0:3.2.3-4.el6
  libRmath-devel.x86_64 0:3.2.3-4.el6                    libX11-devel.x86_64 0:1.6.0-6.el6
  libXau-devel.x86_64 0:1.0.6-4.el6                      libXft-devel.x86_64 0:2.3.1-2.el6
  libXmu.x86_64 0:1.1.1-2.el6                            libXrender-devel.x86_64 0:0.9.8-2.1.el6
  libgfortran.x86_64 0:4.4.7-16.el6                      libicu-devel.x86_64 0:4.2.1-12.el6
  libxcb-devel.x86_64 0:1.9.1-3.el6                      netpbm.x86_64 0:10.47.05-11.el6
  netpbm-progs.x86_64 0:10.47.05-11.el6                  pcre-devel.x86_64 0:7.8-7.el6
  psutils.x86_64 0:1.17-34.el6                           tcl.x86_64 1:8.5.7-6.el6
  tcl-devel.x86_64 1:8.5.7-6.el6                         tex-preview.noarch 0:11.85-10.el6
  texinfo.x86_64 0:4.13a-8.el6                           texinfo-tex.x86_64 0:4.13a-8.el6
  texlive.x86_64 0:2007-60.el6_7                         texlive-dvips.x86_64 0:2007-60.el6_7
  texlive-latex.x86_64 0:2007-60.el6_7                   texlive-texmf.noarch 0:2007-39.el6_7
  texlive-texmf-dvips.noarch 0:2007-39.el6_7             texlive-texmf-errata.noarch 0:2007-7.1.el6
  texlive-texmf-errata-dvips.noarch 0:2007-7.1.el6       texlive-texmf-errata-fonts.noarch 0:2007-7.1.el6
  texlive-texmf-errata-latex.noarch 0:2007-7.1.el6       texlive-texmf-fonts.noarch 0:2007-39.el6_7
  texlive-texmf-latex.noarch 0:2007-39.el6_7             texlive-utils.x86_64 0:2007-60.el6_7
  tk.x86_64 1:8.5.7-5.el6                                tk-devel.x86_64 1:8.5.7-5.el6
  xdg-utils.noarch 0:1.0.2-17.20091016cvs.el6            xorg-x11-proto-devel.noarch 0:7.7-9.el6
  xz-devel.x86_64 0:4.999.9-0.5.beta.20091007git.el6

Complete!
```

4. Now we can download and install a 64-bit version of RStudio Server for RedHat/CentOS 6. Remember to check which exact Linux version your virtual machine is running on and the actual version of RStudio Server available at the time of writing, the current version of RStudio Server was 0.99.891:

```
$ sudo wget
https://download2.rstudio.org/rstudio-server-rhel-0.99.891-x86_64.rpm
...
$ sudo yum install --nogpgcheck rstudio-server-rhel-0.99.891-x86_64.rpm
```

5. Once the libraries and dependencies for RStudio Server are resolved you will be asked to confirm the installation of RStudio Server. Type y and press Enter to proceed:

```
Dependencies Resolved

===============================================================================
 Package             Arch        Version         Repository                Size
===============================================================================
Installing:
 rstudio-server      x86_64      0.99.891-1      /rstudio-server-rhel-0.99.891-x86_64    280 M

Transaction Summary
===============================================================================
Install       1 Package(s)

Total size: 280 M
Installed size: 280 M
Is this ok [y/N]:
```

After a minute or two a Terminal window should return the message that the job has been completed.

6. Before we can make the most of our freshly installed RStudio Server and connect it through a browser we need to make sure to add port `8787` to our existing network security rules for this virtual machine. Go to the main Azure Portal by clicking on the large Microsoft Azure icon in the top-left corner of the screen and select **Resource groups** from the left-hand menu. Choose a Resource Group that contains the virtual machine with Hortonworks Hadoop and RStudio Server you have just installed in our case, it will be **hadoopsw**:

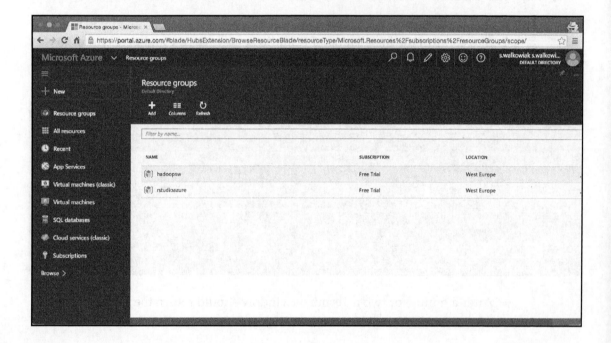

7. From the list of available resources, select the one with the icon in the shape of a shield, as shown in the following screenshot:

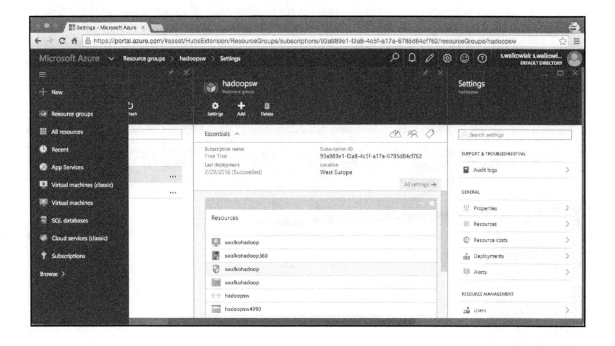

Hadoop and MapReduce Framework for R

8. Two new panels with **Essentials** and **Settings** should now appear on your screen. In the **Essentials** view you can probably see that there are already 34 inbound security rules defined in our virtual machine, due to the Hadoop distribution already pre-installed with all its data management and processing tools (in the preceding chapter there were none as we created our first virtual machine from scratch with only a bare minimum of in-built tools). To inspect the current inbound security rules and to add new ones go to the **Settings** view and choose **Inbound security rules** from the list of options:

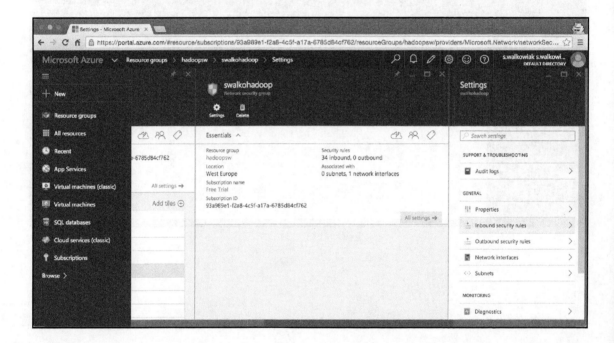

Chapter 4

9. Press the **Add** button in the top menu to add a new rule. A new side panel will show up. Give it a **Name**, for example, `RStudioServer`, keep the **Priority** as it is, select **TCPProtocol,** and set the **Destination port range** to `8787`. Keep the other settings unchanged. Click the **OK** button to confirm:

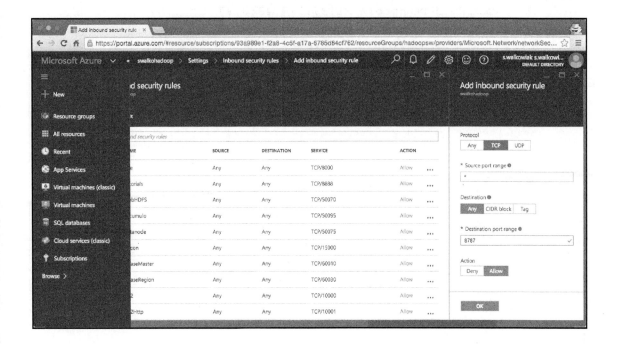

This will create a new inbound security rule which will allow us to connect to RStudio Server through a browser. To check whether the above worked, point your browser (Google Chrome, Mozilla Firefox, or Safari are recommended) to the URL: `http://<IPAddress>:8787`, where `<IPAddress>` is the IP address of your virtual machine; for example, in our case, it will be: `http://23.97.179.42:8787`:

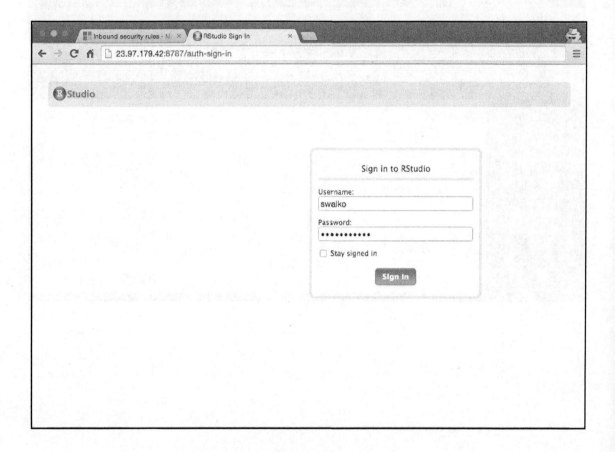

You can now sign in to RStudio using your existing credentials.

Unfortunately, this is not all that we have to do in order to connect R with Hadoop. This connectivity will be facilitated by a collection of R packages named `RHadoop` that support MapReduce and HDFS management directly from the console of the R language. The following section will introduce you to `RHadoop` packages and will explain their installation procedures and typical use cases.

Installing and configuring RHadoop packages

The `RHadoop` collection of packages contains five separate R packages that can be used to process, manage, and analyze large datasets making the most of the installed Hadoop infrastructure and its specific tools and frameworks such as HDFS, MapReduce, and HBase database. The packages have been developed by Revolution Analytics, but due to the acquisition of Revolution Analytics by Microsoft, the latter has recently become the lead maintainer of the packages. All five R packages, their binary files, documentation, and tutorials, are available at a GitHub repository at https://github.com/RevolutionAnalytics/RHadoop/wiki.

In this section, we will be using three of the RHadoop packages, namely:

- `rhdfs`: This provides connectivity to and support for operations on the HDFS
- `rmr2`: A package for carrying out MapReduce jobs on a Hadoop cluster
- `plyrmr`: This allows user-friendly data manipulations and transformations similar to the ones available in the `plyr` and `reshape2` packages

As our simple example will operate on a single-node Hadoop cluster, we don't have to worry about which specific packages should be installed on which particular nodes. However, note that for multi-node clusters, the `rmr2` and `plyrmr` packages should be installed on every node in the cluster, whereas the `rhdfs` package must only be installed on a node that runs the R client. In Chapter 6, *R with Non-Relational (NoSQL) Databases*, we will return to RHadoop packages and will explore some of the functionalities offered by the `rhbase` package that provides users with connectivity between R and HBase distributed databases.

However, for the time being, let's prepare our R to include fully configured and ready-to-use RHadoop packages:

1. In the initial stage, we will install essential `rmr2` dependencies (including a vital `rJava` package) and some additional R packages that may prove useful. As we want to install the packages for all potential users of the virtual machine, we will use the `sudo Rscript` command from the shell/Terminal window. The use of the `Rscript` command will ensure that the packages will be installed as from within the R environment without explicitly opening the R client. When connected using `ssh` to the virtual machine, type the following:

```
$ sudo Rscript -e 'install.packages(c("rJava", "Rcpp", "RJSONIO", "bitops", "digest", "functional", "stringr", "plyr", "reshape2", "caTools"), repos = "http://cran.r-project.org/")'
```

Hadoop and MapReduce Framework for R

This operation may take several minutes. You may notice several warning messages, but there is generally nothing to worry about, so we may ignore them.

2. Download the current version of the `rmr2` package (`tar` file) from the Revolution Analytics's GitHub account at https://github.com/RevolutionAnalytics/RHadoop/wiki/Downloads:

```
$ sudo wget
https://github.com/RevolutionAnalytics/rmr2/releases/download/3.3.1/rmr2_3.3.1.tar.gz
```

Once downloaded, install `rmr2` using the following line:

```
$ sudo R CMD INSTALL rmr2_3.3.1.tar.gz
```

3. We can now download and install the `plyrmr` package, but before that we should install some `plyrmr` dependencies in the same way as we did in the case of `rmr2`. Again, this operation may take several minutes to complete:

```
$ sudo Rscript -e 'install.packages(c("dplyr", "R.methodsS3", "Hmisc", "memoise", "lazyeval", "rjson"), repos = "http://cran.r-project.org/")'
```

4. We can now download and install the `plyrmr` package:

```
$ sudo wget
https://github.com/RevolutionAnalytics/plyrmr/releases/download/0.6.0/plyrmr_0.6.0.tar.gz
...
$ sudo R CMD INSTALL plyrmr_0.6.0.tar.gz
```

5. In order to install the `rhdfs` package we firstly need to set the `HADOOP_CMD` and `HADOOP_STREAMING` environment variables from within R. For that reason we can initialize the R client directly from the command line as a super-user:

```
$ sudo R
```

6. This will start the R console in the Terminal window, in which we can now execute the following code to find the locations of Hadoop binaries and Hadoop Streaming `jar` files:

```
> hcmd <-system("which hadoop", intern = TRUE)
> hcmd
[1] "/usr/bin/hadoop"
```

Set the `HADOOP_CMD` environment variable to the path from the `hcmd` object:

```
> Sys.setenv(HADOOP_CMD=hcmd)
```

Export the path to the Hadoop Streaming `jar` file to the `hstreaming` object:

```
> hstreaming <- system("find /usr -name hadoop-streaming*jar", intern=TRUE)
> hstreaming
[1] "/usr/hdp/2.3.2.0-2950/hadoop-mapreduce/hadoop-streaming.jar"
[2] "/usr/hdp/2.3.2.0-2950/hadoop-mapreduce/hadoop-streaming-2.7.1.2.3.2.0-2950.jar"
[3] "/usr/hdp/2.3.2.0-2950/oozie/share/lib/mapreduce-streaming/hadoop-streaming-2.7.1.2.3.2.0-2950.jar"
```

Select the first entry of the `hstreaming` object as your `HADOOP_STREAMING` environment variable:

```
> Sys.setenv(HADOOP_STREAMING = hstreaming[1])
```

7. Check whether both variables have been set correctly:

```
> Sys.getenv("HADOOP_CMD")
[1] "/usr/bin/hadoop"
> Sys.getenv("HADOOP_STREAMING")
[1] "/usr/hdp/2.3.2.0-2950/hadoop-mapreduce/hadoop-streaming.jar"
```

8. All looks fine, so we can now download and install the current version of the `rhdfs` package:

```
> system("wget --no-check-certificate http://github.com/RevolutionAnalytics/rhdfs/blob/master/build/rhdfs_1.0.8.tar.gz?raw=true")
...
> install.packages("rhdfs_1.0.8.tar.gz?raw=true", repos = NULL, type="source")
```

Once the `rhdfs` installation finishes, you may quit R environment and return to Linux shell/Terminal. You don't need to save the workspace image this time:

```
> q()
```

The last step completes the installation of the three essential `RHadoop` packages that we will be using during the R language implementation of our simple word count MapReduce example.

HDFS management and MapReduce in R – a word count example

As previously mentioned, in this section we will attempt to replicate the word count MapReduce task on Mark Twain's dataset which contains all his works available through Project Gutenberg, but this time we will do it in the R language.

We will begin our tutorial by logging in to RStudio Server. By now you should know that you need to direct your browser to the URL with your virtual machine's public IP address and port `8787`. For our machine, it will be `http://23.97.179.42:8787`. Sign in using your own credentials. Once logged in you should be able to see a standard RStudio Server layout with the main R console on the left-hand side and the remaining panes to the right, as shown in the following screenshot:

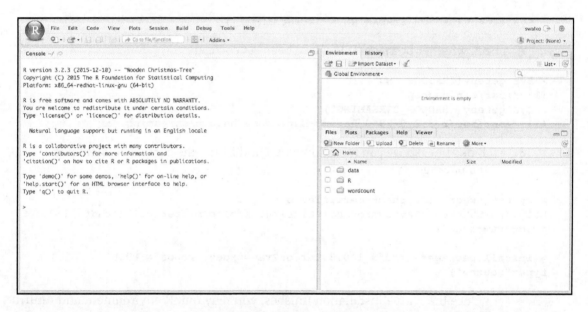

You can probably notice that the **Files** tab already includes a number of directories which we had created on the user's local area earlier when copying data and Hadoop Java files across. The `data` folder contains our original `Twain_data.txt` file, so we can now make this folder our current working directory:

```
> setwd("/home/swalko/data")
```

As this is a new R session, we also need to make sure that both the HADOOP_CMD and HADOOP_STREAMING variables are set to the correct paths before we start using RHadoop packages:

```
> cmd <- system("which hadoop", intern=TRUE)
> cmd
[1] "/usr/bin/hadoop"
> Sys.setenv(HADOOP_CMD=cmd) #setting HADOOP_CMD
> stream <- system("find /usr -name hadoop-streaming*jar", intern=TRUE)
find: `/usr/lib64/audit': Permission denied
Warning message:
running command 'find /usr -name hadoop-streaming*jar' had status 1
> stream
[1] "/usr/hdp/2.3.2.0-2950/hadoop-mapreduce/hadoop-streaming.jar"
[2] "/usr/hdp/2.3.2.0-2950/hadoop-mapreduce/hadoop-
streaming-2.7.1.2.3.2.0-2950.jar"
[3] "/usr/hdp/2.3.2.0-2950/oozie/share/lib/mapreduce-streaming/hadoop-
streaming-2.7.1.2.3.2.0-2950.jar"
attr(,"status")
[1] 1
> Sys.setenv(HADOOP_STREAMING=stream[1]) #setting HADOOP_STREAMING
```

When extracting paths for Hadoop Streaming `jar` files you may have come across a *permission denied* warning message; however, this in no way affects the execution of the command.

As always, it is a good habit to check whether the Hadoop environment variables are set correctly:

```
> Sys.getenv("HADOOP_CMD")
[1] "/usr/bin/hadoop"
> Sys.getenv("HADOOP_STREAMING")
[1] "/usr/hdp/2.3.2.0-2950/hadoop-mapreduce/hadoop-streaming.jar"
```

We can now load all three previously installed `RHadoop` packages: `rmr2`, `rhdfs`, and `plyrmr`:

```
> library(rmr2)
... #output truncated
> library(rhdfs)
... #output truncated
> library(plyrmr)
... #output truncated
```

Each of the preceding calls generated a number of messages informing users about additional dependencies being loaded. Some of them can be viewed in the following screenshot:

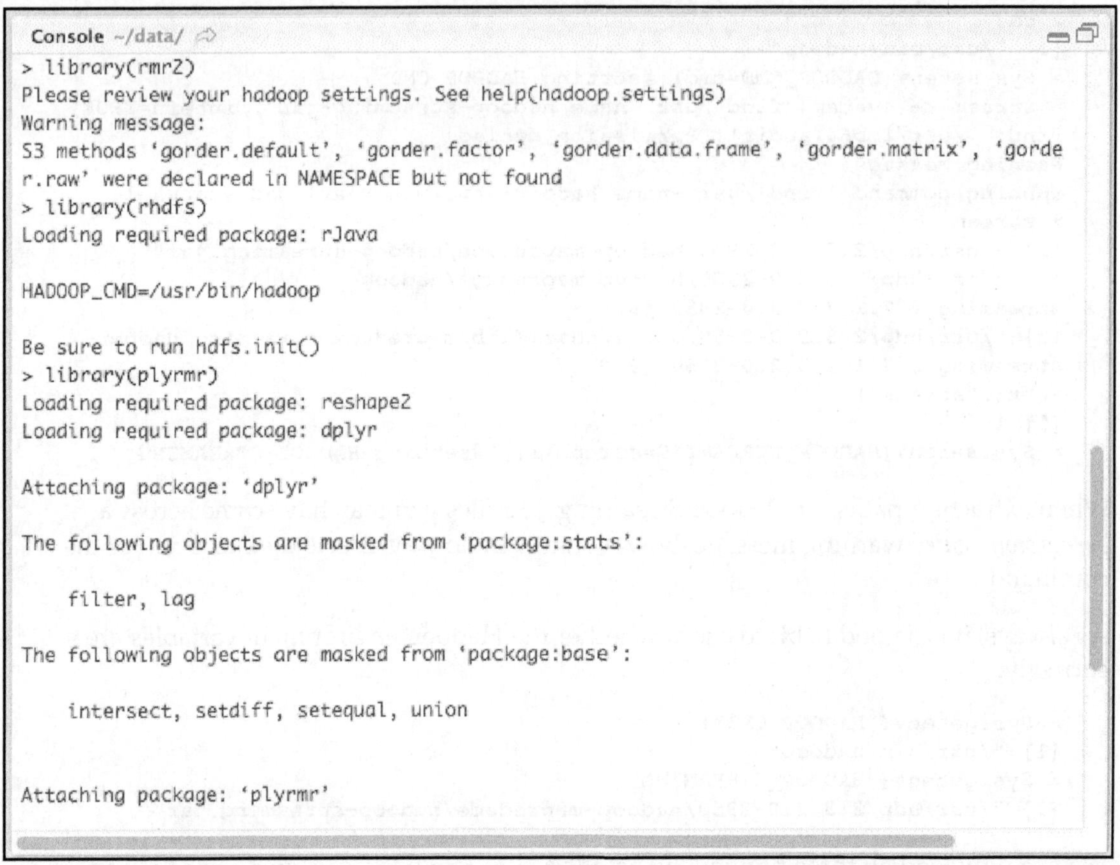

```
Console ~/data/
> library(rmr2)
Please review your hadoop settings. See help(hadoop.settings)
Warning message:
S3 methods 'gorder.default', 'gorder.factor', 'gorder.data.frame', 'gorder.matrix', 'gorde
r.raw' were declared in NAMESPACE but not found
> library(rhdfs)
Loading required package: rJava

HADOOP_CMD=/usr/bin/hadoop

Be sure to run hdfs.init()
> library(plyrmr)
Loading required package: reshape2
Loading required package: dplyr

Attaching package: 'dplyr'

The following objects are masked from 'package:stats':

    filter, lag

The following objects are masked from 'package:base':

    intersect, setdiff, setequal, union


Attaching package: 'plyrmr'
```

The most important information, however, is that all three `RHadoop` packages and their dependencies (most notably the `rJava` package) have been loaded without any errors. It means that all installation and configuration procedures that we carried out in the previous sections have succeeded.

At this point we can start the `rhdfs` package and initialize its connection with HDFS:

```
> hdfs.init()
...
```

The initialization may take up to several seconds and once finished you should see a standard R command line > prompt back in console.

The following short R snippet can be used to extract the path to a data file with a name matching the string specified in the `pattern` argument. We will use this statement to extract the actual full path to our `twain_data.txt` file in the local file system:

```
> file <- dir(getwd(), pattern = "_data.txt", full.names = TRUE)
> file
[1] "/home/swalko/data/twain_data.txt"
```

Knowing where our data is located on the user's local area, we can now copy the file to HDFS. Firstly, however, let's create a new directory called `twain` which will be used to store the data on HDFS. We will achieve this with the `hdfs.mkdir()` function:

```
> hdfs.mkdir("twain")
[1] TRUE
```

The returned Boolean `TRUE` output means that the `twain` directory has been successfully created on HDFS. You can inspect its content using the `hdfs.ls()` function:

```
> hdfs.ls("twain")
NULL
```

Of course, the directory is empty (`NULL`) as it doesn't store any data at the moment. In order to copy our `twain_data.txt` file to this newly created `twain` directory on HDFS, we will use the `hdfs.put()` function:

```
> hdfs.put(file, "twain")
[1] TRUE
```

In the `hdfs.put()` function above, we have used (as the first argument) the `file` object that stores the full path to the data on a local file system. The second argument is the name of the directory on HDFS where we want to move a copy of the data file. We can now check whether the data file has in fact been copied to the designated location on HDFS:

```
> hdfs.ls("twain")
   permission  owner  group      size           modtime
1  -rw-r--r-- swalko swalko  16013935  2016-03-02 22:23
                                          file
1  /user/swalko/twain/twain_data.txt
```

Using the following command you can easily retrieve the filename and the full path on HDFS:

```
> hdfs.ls("twain")$file
[1] "/user/swalko/twain/twain_data.txt"
```

In fact, we should keep the path to the data stored on HDFS assigned to a separate R object, so we can use it later:

```
> twain.path <- hdfs.ls("twain")$file
```

Similar to input and output formats defined by Java classes in the standard Java-based MapReduce job, R allows users to specify their own custom formats depending on required input and output configurations. The `make.input.format()` and `make.output.format()` functions provide R users with the ability to create combinations of input/output settings based on pre-defined formats (`format` argument), modes, for example, `text` or `binary` (`mode` argument), or an R function. Each format can further include additional argument characteristics for this specific format; for example, if the input data is in `csv` format, this enables users to specify most of the standard arguments used in `read.table()` and `write.table()` functions. We will create a more advanced input format when dealing with multivariate electricity meter data in the next part of this chapter dedicated to the multi-node Hadoop cluster service on Microsoft Azure called **HDInsight**. Mark Twain's data does not require us to create any custom input formats. As the file consists of free text only, we can simply select `"text"` as both the pre-defined `format` and `mode` for our data:

```
> twain.format <- make.input.format(format = "text", mode = "text")
```

At this point of the tutorial we can perform a simplified MapReduce job that will only contain a Mapper that will be responsible for merely splitting each line of the text into separate words. Sometimes it is a good idea to test our setup, input formats, and HDFS through this type of basic MapReduce task. This will also be your first time experiencing how easy it is to deploy a MapReduce job in the R language.

Our Mapper is in fact an R function that can be named any way you want it will be called `twain.map` in our example. The mapping function takes two arguments, multiple key-value pairs, and it emits the value of `keyval` a key-value object with a new pair of keys and their corresponding values. In our case, the Mapper simply takes each line of the text according to its input format (which is `text`) and splits it into individual words. It emits a list of words (keys) with each word on a new line and the numeric value of `1` for each extracted word. The `twain.map` Mapper can be written in R, as follows:

```
> twain.map <- function(k, v) {
+   words <- unlist(strsplit(v, " "))
+   keyval(words, 1)
+ }
```

The actual path to the data stored on HDFS (`twain.path` object), previously defined input format (`twain.format`) and the Mapper (`twain.map`), can now be used as arguments in the `mapreduce()` function that will perform a MapReduce task from R. The execution of the following line will initialize MapReduce:

```
> mr <- mapreduce(twain.path, input.format = twain.format, map = twain.map)
```

As in the standard Java-based MapReduce, Hadoop assigns application ID to the submitted job:

```
> mr <- mapreduce(twain.path, input.format = twain.format, map = twain.map)
WARNING: Use "yarn jar" to launch YARN applications.
packageJobJar: [] [/usr/hdp/2.3.2.0-2950/hadoop-mapreduce/hadoop-streaming-2.7.
1.2.3.2.0-2950.jar] /tmp/streamjob7729124526021459643.jar tmpDir=null
16/03/02 23:16:35 INFO impl.TimelineClientImpl: Timeline service address: htt
p://sandbox.hortonworks.com:8188/ws/v1/timeline/
16/03/02 23:16:35 INFO client.RMProxy: Connecting to ResourceManager at sandbox.
hortonworks.com/10.1.0.4:8050
16/03/02 23:16:36 INFO impl.TimelineClientImpl: Timeline service address: htt
p://sandbox.hortonworks.com:8188/ws/v1/timeline/
16/03/02 23:16:36 INFO client.RMProxy: Connecting to ResourceManager at sandbox.
hortonworks.com/10.1.0.4:8050
16/03/02 23:16:38 INFO mapred.FileInputFormat: Total input paths to process : 1
16/03/02 23:16:38 INFO mapreduce.JobSubmitter: number of splits:2
16/03/02 23:16:39 INFO mapreduce.JobSubmitter: Submitting tokens for job: job_14
56953689954_0001
16/03/02 23:16:40 INFO impl.YarnClientImpl: Submitted application application_14
56953689954_0001
16/03/02 23:16:40 INFO mapreduce.Job: The url to track the job: http://sandbox.h
ortonworks.com:8088/proxy/application_1456953689954_0001/
```

Hadoop and MapReduce Framework for R

However, this time, users do not need to worry about Java classes and creating `jar` files for MapReduce execution. This is done automatically, and the evidence of that can be viewed in the job tracker:

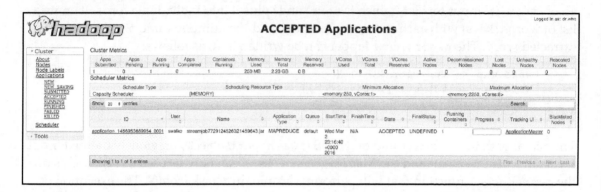

As before, the R console returns a real-time output with the progress information of the MapReduce job and confirms whether the job has been completed:

```
Console ~/data/
16/03/02 23:16:40 INFO mapreduce.Job: Running job: job_1456953689954_0001
16/03/02 23:17:22 INFO mapreduce.Job: Job job_1456953689954_0001 running in uber
mode : false
16/03/02 23:17:22 INFO mapreduce.Job:  map 0% reduce 0%
16/03/02 23:17:44 INFO mapreduce.Job:  map 8% reduce 0%
16/03/02 23:17:47 INFO mapreduce.Job:  map 16% reduce 0%
16/03/02 23:17:50 INFO mapreduce.Job:  map 22% reduce 0%
16/03/02 23:17:51 INFO mapreduce.Job:  map 25% reduce 0%
16/03/02 23:17:53 INFO mapreduce.Job:  map 29% reduce 0%
16/03/02 23:17:54 INFO mapreduce.Job:  map 33% reduce 0%
16/03/02 23:17:57 INFO mapreduce.Job:  map 39% reduce 0%
16/03/02 23:18:00 INFO mapreduce.Job:  map 47% reduce 0%
16/03/02 23:18:03 INFO mapreduce.Job:  map 53% reduce 0%
16/03/02 23:18:06 INFO mapreduce.Job:  map 56% reduce 0%
16/03/02 23:18:07 INFO mapreduce.Job:  map 59% reduce 0%
16/03/02 23:18:09 INFO mapreduce.Job:  map 62% reduce 0%
16/03/02 23:18:17 INFO mapreduce.Job:  map 68% reduce 0%
16/03/02 23:18:20 INFO mapreduce.Job:  map 78% reduce 0%
16/03/02 23:18:23 INFO mapreduce.Job:  map 88% reduce 0%
16/03/02 23:18:26 INFO mapreduce.Job:  map 91% reduce 0%
16/03/02 23:18:29 INFO mapreduce.Job:  map 97% reduce 0%
16/03/02 23:18:30 INFO mapreduce.Job:  map 100% reduce 0%
16/03/02 23:18:33 INFO mapreduce.Job: Job job_1456953689954_0001 completed succe
ssfully
```

The standard MapReduce metrics follow and the confirmation of completion is also viewable through the Resource Manager's job tracker in the normal way:

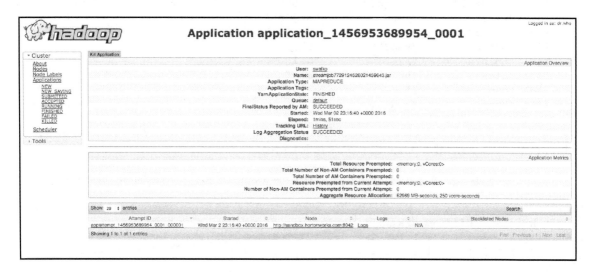

The data output of MapReduce processing is stored in a temporary file of which the location is easily retrievable by calling the name of the R object to which we have assigned the `mapreduce()` function:

```
> mr()
[1] "/tmp/file195d78f77ca6"
```

The structure of the `mr` object implies that our Mapper achieved the expected results as it contains both keys (individual words) and values (value `1` assigned to each word):

```
> str(from.dfs(mr))
List of 2
 $ key: chr [1:3043767] "was" "was" "was" "was" ...
 $ val: num [1:3043767] 1 1 1 1 1 1 1 1 1 1 ...
```

We may list an arbitrary number (for example, 50) of unique keys (individual words) returned by the Mapper, using the following line:

```
> head(unique(keys(from.dfs(mr))), n=50)
 [1] "was"       "introduced:" ""How"
 [4] "many"      "does"        "it"
 [7] "take"      "to"          "make"
[10] "a"         "pair?""       ""Well,"
[13] "two"       "generally"   "makes"
[16] "pair,"     "but"         "sometimes"
```

```
[19] "there"       "ain't"         "stuff"
[22] "enough"      "in"            "them"
[25] "whole"       "pair.""        ""
[28] "General"     "laugh."        """What"
[31] "were"        "you"           "saying"
[34] "about"       "the"           "English"
[37] "while"       "ago?"""        """Oh,"
[40] "nothing,"    "are"           "all"
[43] "right,"      "only--I--"""   "said"
[46] "them?"""     "I"             "only"
[49] "they"        "swallow"
```

As you can see, the returned keys are still quite messy as we haven't removed any special characters. We could also print all words in lowercase.

Similarly, we can extract a list of the top 50 values assigned to the words. We should expect here a value of 1 for each word:

```
> head(values(from.dfs(mr)), n=50)
 [1] 1 1 1 1 1 1 1 1 1 1 1 1 1 1 1 1 1 1 1 1 1 1 1 1
[25] 1 1 1 1 1 1 1 1 1 1 1 1 1 1 1 1 1 1 1 1 1 1 1 1
[49] 1 1
```

Let's now implement a full version of a MapReduce job with an improved Mapper and a Reducer function whose task will be to sum all counts of specific words and present them as a contingency table so exactly like in the Java-based example in the preceding part of this chapter. In the Mapper, we will add two lines of code which will let us do the following:

- Remove special characters from extracted words, and
- Transform all words to lowercase

Therefore, our new Mapper will look as follows:

```
> twain.map <- function(k, v) {
+   words <- unlist(strsplit(v, " "))
+   words <- gsub("[[:punct:]]", "", words, perl = TRUE)
+   words <- tolower(words)
+   keyval(words, 1)
+ }
```

Just like in Java, the Reducer function (`twain.reduce`) written in the R language will be very simple it will emit a key-value pair with specific words as keys and the total number of their occurrences in the text as their corresponding values:

```
> twain.reduce <- function(k, v) {
+   keyval(k, sum(v))
+ }
```

As this MapReduce job contains a Reducer, our new `mapreduce()` function will include an additional argument (`reduce`), enabling us to define the Reducer function, `twain.reduce`:

```
> mr <- mapreduce(twain.path, input.format = twain.format, map = twain.map,
  reduce = twain.reduce)
```

Once again, the execution of the preceding line of code will initialize the MapReduce job, and just like in the previous trials, users can track the application either through the output messages in the R console or in the browser (as shown in the following screenshot):

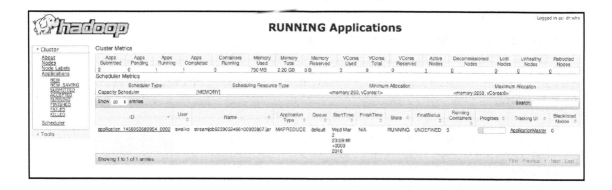

In addition to standard MapReduce performance metrics, the job carried out through `RHadoop` packages displays the number of reduce calls made:

```
Console ~/data/
                WRONG_MAP=0
                WRONG_REDUCE=0
        File Input Format Counters
                Bytes Read=16133432
        File Output Format Counters
                Bytes Written=4189830
    rmr
                reduce calls=77720
16/03/03 00:02:47 INFO streaming.StreamJob: Output directory: /tmp/file195d
115ae2c5
```

This number is consistent with the total number of unique keys (words) returned by the MapReduce task:

```
> length(unique(keys(from.dfs(mr))))
[1] 77720
```

We can now compare the achieved results by firstly extracting 50 unique keys (words) from the data output:

```
> head(unique(keys(from.dfs(mr))), n=50)
 [1] ""   "0"  "1"  "2"  "3"  "4"  "5"  "6"  "7"
[10] "8"  "9"  "a"  "b"  "c"  "d"  "e"  "f"  "g"
[19] "h"  "i"  "j"  "k"  "l"  "m"  "n"  "o"  "p"
[28] "q"  "r"  "s"  "t"  "u"  "v"  "w"  "x"  "y"
[37] "z"  "00" "01" "02" "03" "04" "05" "06" "07"
[46] "08" "09" "0d" "0s" "10"
```

And also we can obtain values for the first 50 keys:

```
> head(values(from.dfs(mr)), n=50)
 [1] 253713      5    361    305    312    231    157
 [8]    167    122    133    124  71432    135    247
[15]    205    142     85     82    244  44161    135
[22]     23    339    281    181    380    158     63
[29]     67    368    133    113     57    304     61
[36]     84     13      3      3      5      6     10
[43]      6      5      3     10      6      1      1
[50]    141
```

We have previously explained that the output data is stored on HDFS in a temporary directory. We can, however, very easily copy the output from HDFS to the local file system or store it in an R object. Just a word of caution, remember that when processing huge datasets your MapReduce data outputs may still be of very considerable sizes, so creating R objects from these outputs may easily use up your RAM resources. In our case, the output is small; therefore, we can extract it as a separate R object. You can achieve this with the `from.dfs()` function:

```
> output <- from.dfs(mr, format = "native")
> str(output)
List of 2
 $ key: chr [1:77720] "" "0" "1" "2" ...
 $ val: num [1:77720] 253713 5 361 305 312 ...
```

In the `from.dfs()` function, we have set the `format` argument to `"native"`, which forces the created R object to adopt the current format of the MapReduce output in this case it was a `list` with two components: one character `vector` with keys and one numeric `vector` with values.

We could have also defined a different output format and a separate directory for output files on HDFS and passed this information to the `mapreduce()` function. In the following example, we will set an output format to `csv` with a standard comma separator included in the `make.output.format()` function:

```
> out.form <- make.output.format(format = "csv", sep = ",")
```

The created output format named `out.form` can now be used along with the `output` argument pointing to the HDFS directory for output files in the new MapReduce job:

```
> mr <- mapreduce(twain.path, output = "/user/swalko/out1",
+          input.format = twain.format,
+          output.format = out.form,
+          map = twain.map, reduce = twain.reduce)
```

This will obviously initialize a new MapReduce application, but this time instead of the temporary file, the output file will be saved to the `/user/swalko/out1` directory on HDFS as indicated in the `output` argument. The output directory can be inspected using a standard `hdfs.ls()` command:

```
> hdfs.ls("out1")
  permission  owner   group     size             modtime
1 -rw-r--r--  swalko  swalko       0 2016-03-03 00:44
2 -rw-r--r--  swalko  swalko 1233188 2016-03-03 00:44
                            file
1    /user/swalko/out1/_SUCCESS
2 /user/swalko/out1/part-00000
```

The data can now be easily copied from `/user/swalko/out1` on HDFS to any directory on the local file system (for example, it may be a current working directory defined in R) using the `hdfs.get()` function:

```
> mr
[1] "/user/swalko/out1"

> wd <- getwd()
> wd
[1] "/home/swalko/data"
> hdfs.get(mr, wd)
[1] TRUE
```

Or if you want to extract an individual output file from HDFS to a data file, for example, `*.txt` file in a local file system, use the following command (making sure to adjust the paths to suit your personal directories):

```
> hdfs.get("/user/swalko/out1/part-00000",
+          "/home/swalko/data/output.txt")
[1] TRUE
```

Both the `out1` directory and the `output.txt` file should be visible through the **Files** tab view in RStudio:

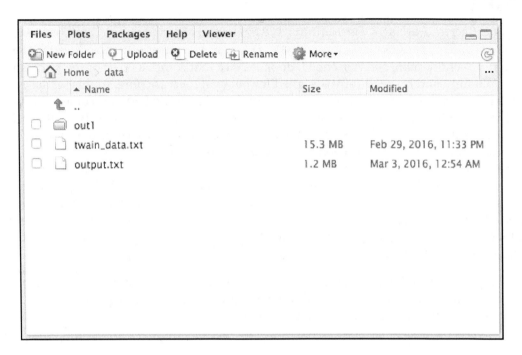

At the end, you can save your work and log out of RStudio. Don't forget to transfer all your important files back to your local personal computer out of the virtual machine. You should de-allocate your virtual machine by clicking on the **Stop** button in the **Virtual machine** view in the Azure Portal to avoid further charges.

In this section, we have guided you through a sequence of steps required to set up, configure, and deploy a Linux virtual machine with a single-node Hadoop installed and ready for data processing using the R language. We've carried out a number of HDFS management activities using a Linux Terminal line and straight from an RStudio console. Finally, we have run both Java and the R language implementations of MapReduce jobs on the same data.

In the next section, we are going to take a huge step forward, much greater than in the preceding sections of this book. You will learn how to create a multi-node Hadoop cluster that can potentially process terabytes or petabytes of data, and even better, you will be able to achieve it with the R language.

HDInsight – a multi-node Hadoop cluster on Azure

In *Online Chapter, Pushing R Further* (https://www.packtpub.com/sites/default/files/downloads/5396_6457OS_PushingRFurther.pdf), we briefly introduced you to **HDInsight-a** fully-managed Apache Hadoop service that comes as part of the Microsoft Azure platform and is specifically designed for heavy data crunching. In this section, we will deploy a multi-node HDInsight cluster with R and RStudio Server installed and will perform a number of MapReduce jobs on smart electricity meter readings (~414,000,000 cases, four variables, ~12 GB in size) of the **Energy Demand Research Project** available to download from UK Data Service's online **Discover** catalog at https://discover.ukdataservice.ac.uk/catalogue/?sn=7591. But before we can tap into the actual data crunching, we need to set up and prepare an HDInsight cluster to process our data. The configuration of HDInsight is not the simplest task to accomplish, especially if we need to install additional tools, for example, RStudio Server, but hopefully the following instruction guidelines presented will make your work in HDInsight with R much easier and more enjoyable.

Also note that depending on your subscription level of Microsoft Azure, the deployment of this HDInsight cluster may cost you additional charges. How much depends on your current subscription, support plans, and many other factors.

Creating your first HDInsight cluster

Deployment procedures of HDInsight clusters on Microsoft Azure with R and RStudio servers involve taking the following steps:

1. Creating a new Resource Group.
2. Deploying a Virtual Network.
3. Creating a Network Security Group.
4. Setting up and configuring HDInsight clusters (for example, adding storage containers and a custom installation script for core R).
5. Starting HDInsight clusters and exploring Ambari.
6. Connecting to the HDInsight cluster and installing RStudio Server.
7. Adding a new Inbound Security Rule for port 8787.
8. Editing the Virtual Network's public IP address for the head node.

It is possible to create an HDInsight cluster with just core R installed and available to users from the command line by completing just two steps from the above list namely steps 4 and 5. For a much better user experience and much greater data analytics capabilities, it is recommended to include RStudio Server integration. This, however, requires a number of additional stages to complete.

Creating a new Resource Group

We will begin by manually creating a new **Resource Group** that will contain all services and vital components of our HDInsight cluster:

1. Go to the main Azure Portal Dashboard view and choose **Resource groups** from the left-hand menu:

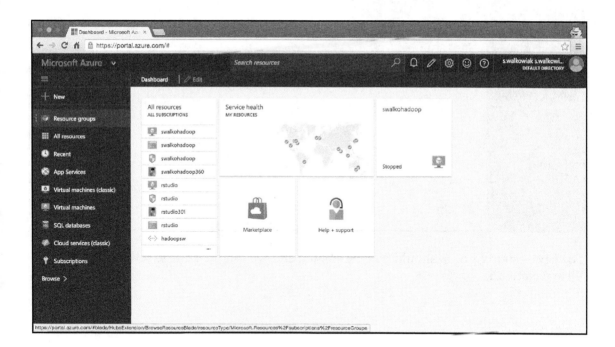

2. In the **Resource groups** view click on the **Add** button from the top menu to start creating a new group. A new **Resource group** pane will appear. Complete the form by setting the **Resource group name** to, for example, `testcluster` and choosing the correct **Resource group location** in our case, it will be `West Europe`. If you have a paid or a free subscription you can select its appropriate type in the **Subscription** drop-down menu. When you complete all fields, click the **Create** button below the form:

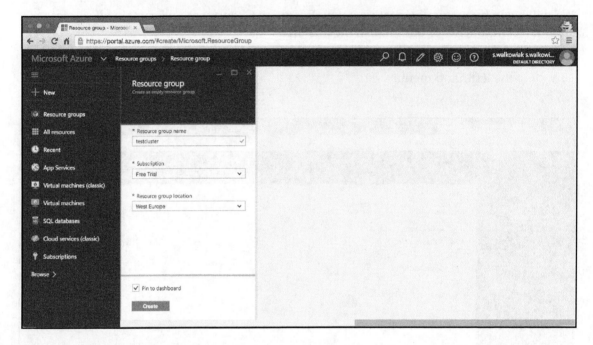

The new resource group should now be created and we can progress to setting up the Virtual Network.

Deploying a Virtual Network

Virtual Network will allow us to connect remotely to HDInsight services and build distributed applications:

1. Go back to the Azure Portal Dashboard and click on the **New** button with a + icon located at the very top of the left-hand menu. In the search bar of a new pane type *virtual network* and select the top **Virtual network**:

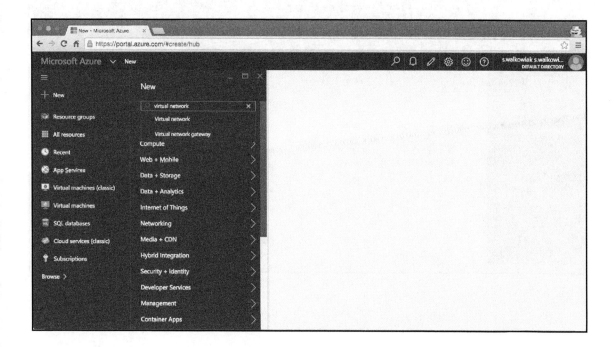

2. A new view called **Everything** will appear. Choose **Virtual network** from the list and a new pane describing **Virtual network** service will be shown. Keep the **Select a deployment model** set to **Resource Manager** and click **Create** to proceed:

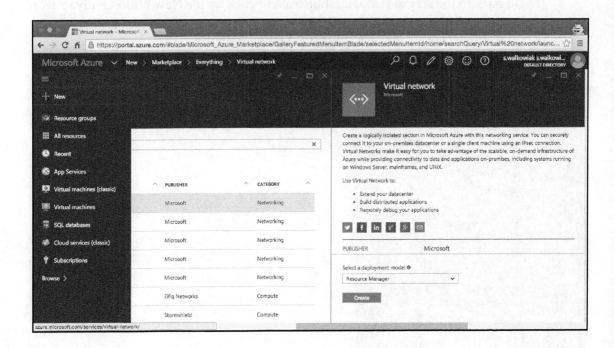

3. In the new **Create virtual network** view we can define a **Name** for our Virtual Network here set to `clusterVN`. Keep **Address space**, **Subnet name**, and **Subnet address range** unchanged. In the **Resource group** option, choose an existing group that we created in the previous section in our case, it was named `testcluster`. Make sure the **Location** is set accordingly for our purposes we chose **West Europe**. Click **Create** to accept the changes and deploy a new Virtual Network:

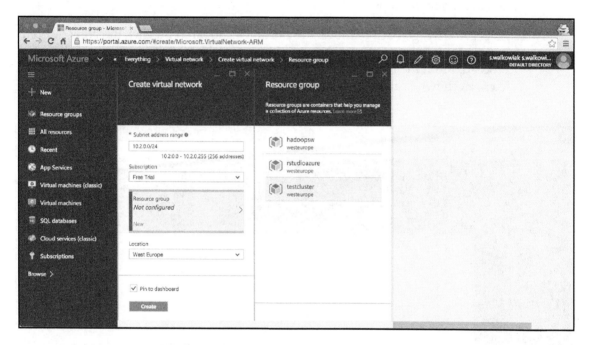

It may take several seconds to create the Virtual Network. Once it's done, you may move on to starting a new **Network Security Group**.

Creating a Network Security Group

Network Security Groups allow users to control in and out traffic from and to the cluster nodes by defining inbound and outbound security rules. By default, security groups created automatically during the deployment of HDInsight clusters block all incoming and outgoing connections apart from a few pre-defined. As we wish to access RStudio Server through a browser, we need to override these default settings by adding one more inbound rule for destination port `8787`. Let's then create a new Network Security Group called `clusterns` in a few simple steps:

1. As previously, go back to the Azure Portal Dashboard and click the **New** button. In the search bar, type *Network security group* and choose the matching option to proceed. From the **Everything** view select **Network security group** (with a shield icon), as shown in the following screenshot:

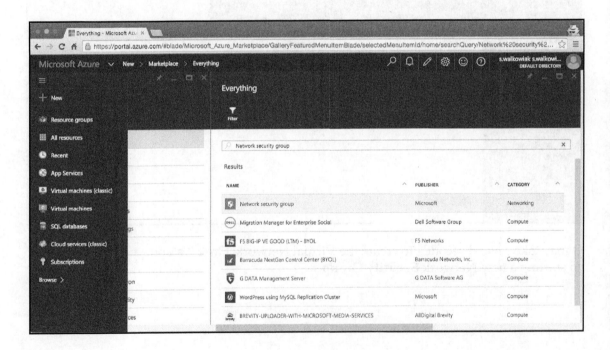

Chapter 4

2. In the **Network security group** view, choose the **Classic** option from the **Select a deployment model** drop-down menu and click **Create** to proceed:

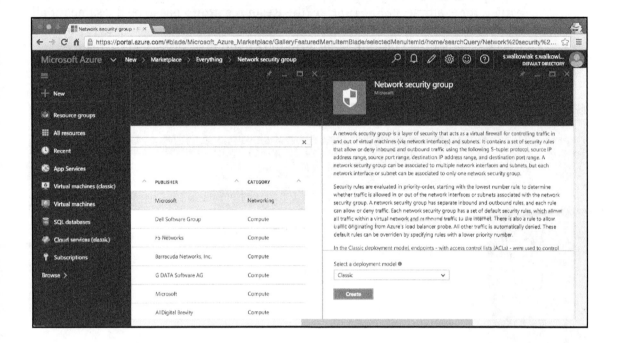

3. This will take you to a configuration pane where you can define a **Name** for your Network Security Group (we set it to `clusterns`) and will let you choose the appropriate **Resource group** make sure to select the existing one that you created earlier, for example, `testcluster` (the **Resource group** has to be the same for all the services connected to the HDInsight cluster which we are going to deploy). Check whether the **Location** is set as in the previous sections. Click **Create** to deploy the security group:

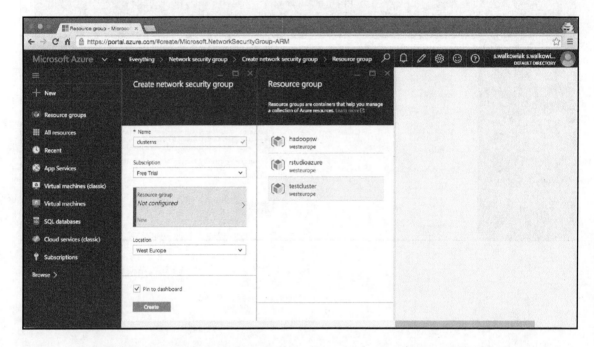

After a few seconds, the newly created security group should be up and running. At this stage, it should have no inbound or outbound security rules defined.

You should have by now started a Resource Group named `testcluster` with two services: one Virtual Network called `clusterVN` and one Network Security Group called `clusterns`. We may now proceed to the actual deployment of our HDInsight cluster.

Setting up and configuring an HDInsight cluster

In this section, we will explain all configuration options of our HDInsight cluster. We will also add storage containers and install all core R files using a custom script:

1. Click on the **New** button again and type *HDI* in the search bar. Select **HDInsight**. From the **Everything** pane, choose the **HDInsight** service (with an icon depicting a white elephant on a yellow background):

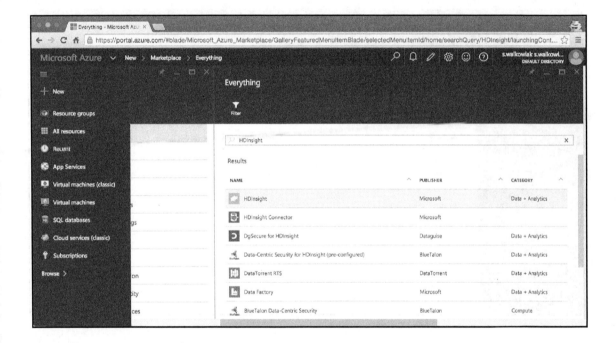

Hadoop and MapReduce Framework for R

2. You will now be able to review some general information about HDInsight. Once you're ready, click the **Create** button to proceed to the **New HDInsight Cluster** configuration view. Give a **Cluster Name** (we set it to `smallcluster`), choose **Hadoop** as a **Cluster Type** and **Linux** as a **Cluster Operating System**. We will select the most recent **Hadoop 2.7.0 (HDI 3.3)** as the preferred **Version** of Linux:

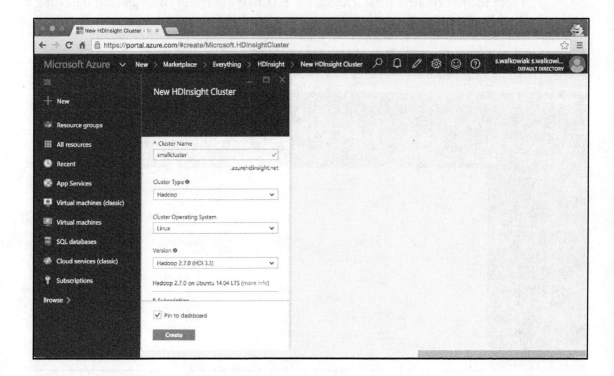

Chapter 4

3. Scroll down the form and click on the **Resource Group** option to select the correct **Resource Group** for the cluster. We will stick with `testcluster` that also contains our previously created Virtual Network and Network Security Group:

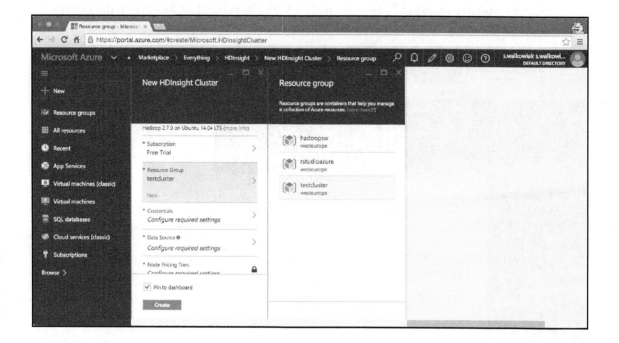

4. Click on the **Credentials** tab and complete the **Cluster Credentials** form. We will keep `admin` as **Cluster Login Username** and we will define and confirm its password. For the **SSH Username** we will use `swalko`, but you can choose any other preferred username. Keep the **SSH Authentication Type** as **PASSWORD** and specify the **SSH Password** and confirm it. When finished, click the **Select** button:

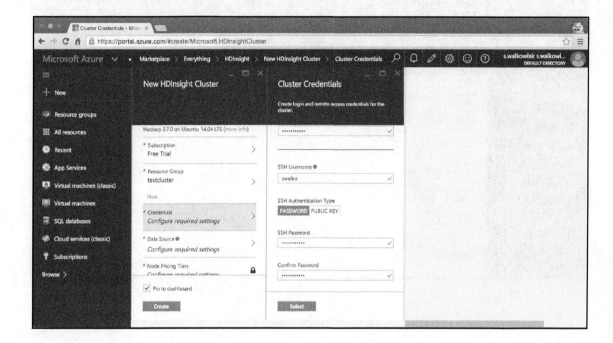

5. Select the **Data Source** tab and **Create a new storage account**. We will call it `clusterstore` and we will type `clusterstore1` in the **Choose Default Container** field. Make sure to set the **Location** of your storage container to the same value as the location of all other services provisioned for this HDInsight cluster. Click **Select** to accept changes:

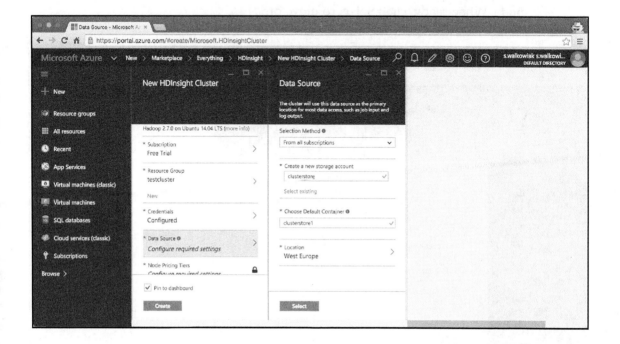

6. Click on the **Node Pricing Tiers** tab and adjust the number of workers. We will keep 4 as the default value. Next, you can configure **Worker Nodes** and the **Head Node Pricing Tier**. Select the appropriate specification based on your budget and/or subscription plan. As we are going to be crunching quite a large dataset of 12 GB, for both Head and Workers, we will choose `D13 V2 Optimized` virtual machines, with eight cores, 56 GB of RAM, and 400 GB of SSD drive for each node. When ready, click **Select** to move on:

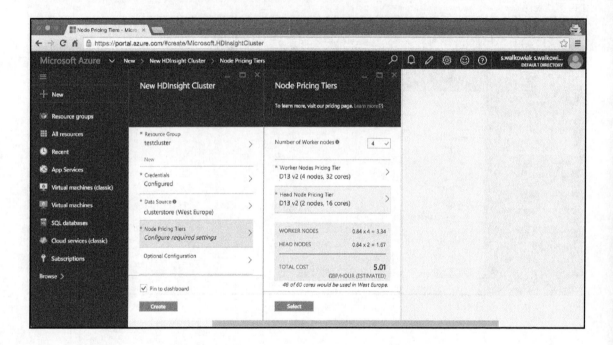

7. Finally, as we want to install core R files and make sure that our previously created Virtual Network is recognized by HDInsight cluster, we have to use the **Optional Configuration** tab to add certain settings. In the **Optional Configuration** view click on the **Virtual Network** tab; repeat the same action for the **Virtual Network** view, and in **Select Resource** choose the name of the deployed Virtual Network for this cluster in our case, it was `clusterVN`. In the **Subnet** tab, choose `default`. Press **Select** to continue:

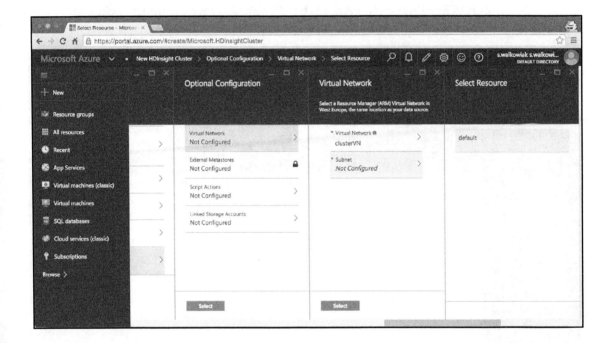

You should now be in the **Optional Configuration** menu. Choose the **Script Actions** tab. Using a custom shell script we will be able to install core R files on all selected nodes in our cluster in one go. In the **NAME** field, provide the name for the action we will call it `installR`. In the **SCRIPT URI** specify the link to a script which will be responsible for R installation. In our case, the script is located at `https://hdiconfigactions.blob.core.windows.net/linuxrconfigactionv01/r-installer-v01.sh`. Copy and paste this link into the **SCRIPT URI** field. Tick all fields for **HEAD**, **WORKER**, and **ZOOKEEPER**. Leave **PARAMETERS** blank. Once configured, click the **Select** button:

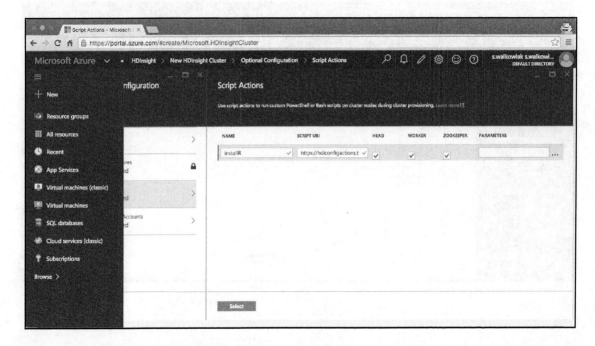

Click the **Select** button again to accept all changes and go back to the **New HDInsight Cluster** view.

8. We may now click on the **Create** button to start deploying the cluster. Depending on the location and specific configuration of the cluster it may take anything from 20 minutes up to an hour.

Chapter 4

Starting the cluster and exploring Ambari

Once deployed, you will be transferred to the HDInsight Cluster view, from which you can easily connect to your cluster, its dashboard and advanced settings, and administration controls:

1. In the **HDInsight Cluster** view click on the **All settings** link (or the **Settings** button located in top controls) and select the **Cluster Login** tab to double check your **Cluster Login Username**, which you set up earlier (it's `admin` in our case), and the **Remote address**:

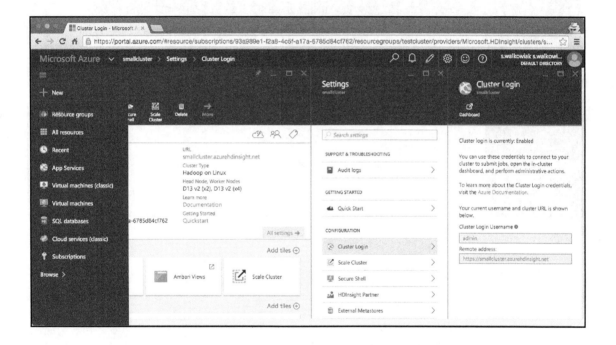

2. Go back to the **HDInsight Cluster** view and click on either the **Dashboard** button in the top controls or the **Cluster Dashboard** tile in the **Quick Links** section and then on the **HDInsight Cluster Dashboard** tile:

Chapter 4

This will take you to the **Ambari Dashboard** login page, where you can provide your credentials for **Cluster Username**:

In the Ambari Dashboard you will be able to inspect multiple performance metrics for all services connected to your cluster. Some of them are already prepared for you by default, as shown in the following screenshot:

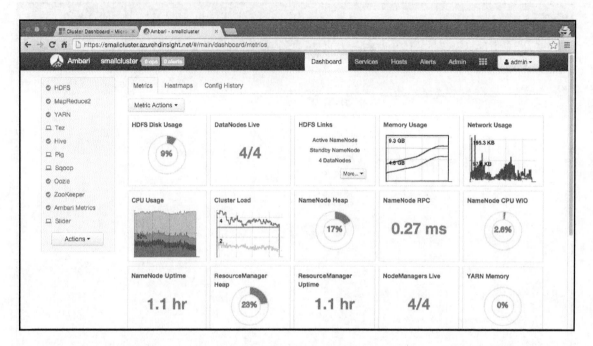

You can also edit and configure all settings and options related to individual tools and components of your Hadoop deployment. Unfortunately, it is beyond the subject matter of this book to include even a short description of several metrics and configuration options available in Ambari. However, please make sure to pause here for a moment and explore the Ambari Dashboard at your will.

Connecting to the HDInsight cluster and installing RStudio Server

At this stage, we may SSH to the cluster and install RStudio Server. As previously explained, our cluster has all current core R files already installed and ready to use. You could just use a standard base R straight from the command line, but of course it would be much easier for the user to be able to run R scripts from an R client with a user-friendly and interactive GUI, for example, RStudio Server. Also, this would allow users to not only run MapReduce jobs on a cluster straight from the R console, but also perform other types of analytics on R objects and create static and interactive data visualizations:

1. Connect to the HDInsight cluster using a standard `ssh` command in the new Terminal window. Before connecting through SSH to the cluster you need to obtain the hostname of the head node of the cluster. Note that this hostname is not the same as the URL available through the main **Essentials** window of the **HDInsight Cluster** view. For our cluster, we can obtain the hostname by clicking on the **All settings** link in the **HDInsight Cluster** and selecting the **Secure Shell** tab from the configuration options in the **Settings** pane, as shown in the following screenshot:

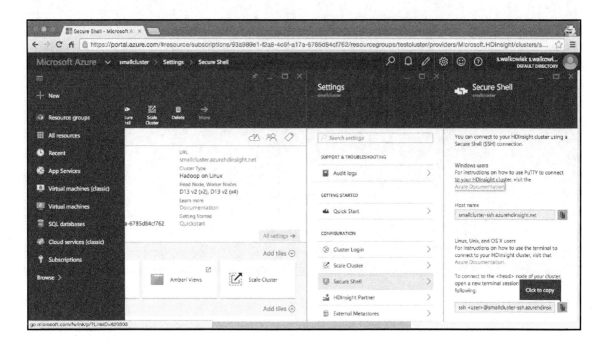

From the **Secure Shell** window we can see that, in our case, we should use `ssh <user>@smallcluster-ssh.azurehdinsight.net` to access the cluster from shell/Terminal:

```
$ ssh swalko@smallcluster-ssh.azurehdinsight.net
```

Type `yes` if prompted whether you are sure you want to continue and then provide your password for the username when asked. Upon verification, you should be connected to the head node of your cluster.

2. Before installing RStudio Server, check whether your core R is working. In order to do so, you can just simply invoke R GUI by typing the following command from shell/Terminal:

```
$ R
```

If you can see a welcome message and R information output similar to the one presented in the following screenshot it means that the installation of core R was successful:

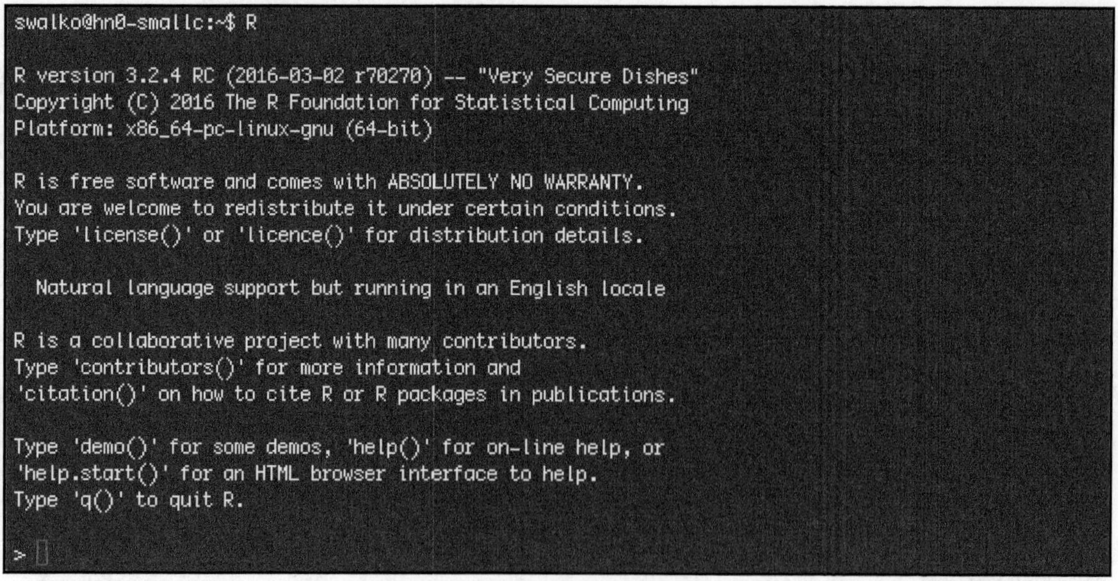

```
swalko@hn0-smallc:~$ R

R version 3.2.4 RC (2016-03-02 r70270) -- "Very Secure Dishes"
Copyright (C) 2016 The R Foundation for Statistical Computing
Platform: x86_64-pc-linux-gnu (64-bit)

R is free software and comes with ABSOLUTELY NO WARRANTY.
You are welcome to redistribute it under certain conditions.
Type 'license()' or 'licence()' for distribution details.

  Natural language support but running in an English locale

R is a collaborative project with many contributors.
Type 'contributors()' for more information and
'citation()' on how to cite R or R packages in publications.

Type 'demo()' for some demos, 'help()' for on-line help, or
'help.start()' for an HTML browser interface to help.
Type 'q()' to quit R.

>
```

You can quit R (using the `q()` function) for the time being and go back to the shell/Terminal.

3. Once logged to your cluster, you may now install preferred R packages if you wish so. Note that when we were installing core R using a custom script for the HDInsight cluster, some packages that support data management and processing in Hadoop ecosystem, for example, `rmr2` and `rhdfs`, and their dependencies, have already been installed for us. There is therefore no need for us to worry about the `rJava` package and others these have been taken care of automatically. We can then just proceed to the installation of RStudio Server. As our operating system is Linux Ubuntu 14.04, visit the RStudio Server download page, https://www.rstudio.com/products/rstudio/download-server/, and execute a number of lines of Linux commands to download and install it:

```
$ sudo apt-get install gdebi-core
...
$ wget https://download2.rstudio.org/rstudio-server-0.99.891-amd64.deb
...
$ sudo gdebi rstudio-server-0.99.891-amd64.deb
...
```

Hadoop and MapReduce Framework for R

Adding a new inbound security rule for port 8787

We have now installed RStudio Server, but we still can't access it through the browser. Therefore, we need to add a new inbound security rule for port `8787`. In order to accomplish it, carry out the following instructions:

1. From the left-hand menu choose the **Resource groups** option and select the Resource Group with the HDInsight cluster for us, it will be `testcluster`. A new **testcluster** pane will appear with a list of all resources assigned to the group. Choose the Network Security Group (a resource with an icon in the shape of a shield) named `clusterns`, as depicted in the following screenshot:

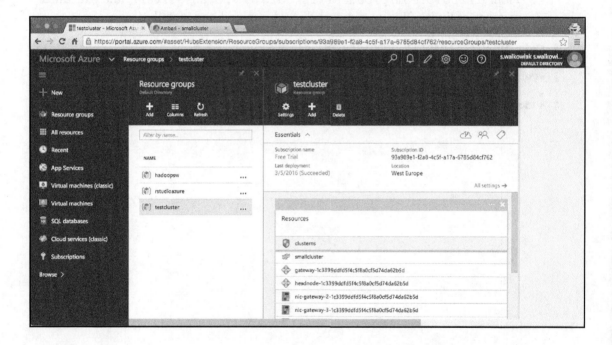

Chapter 4

2. From the **Essentials** view of the chosen Network Security Group, click on the **All settings** link and then on the **Inbound security rules** in the **Settings** view:

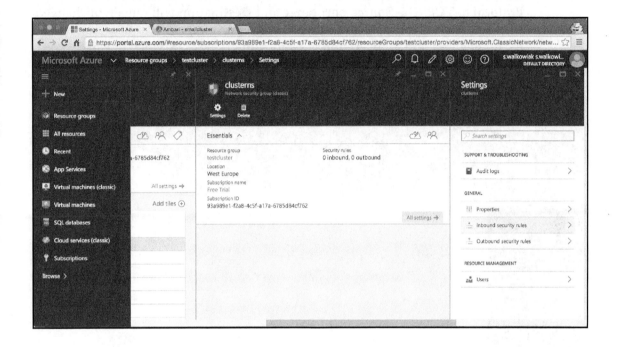

3. In the **Inbound security rules** click on the **Add** button in the top control panel and fill in the form to add a new inbound security rule. Give it a **Name**, for example, `rstudio`, keep the default **Priority** level and **Source**, change the **Protocol** to **TCP**. The **Source port range** and **Destination** will remain unchanged; however, please type `8787` in the **Destination port range** field. Make sure that the highlighted option for **Action** is set to **Allow**. Click **OK** when finished:

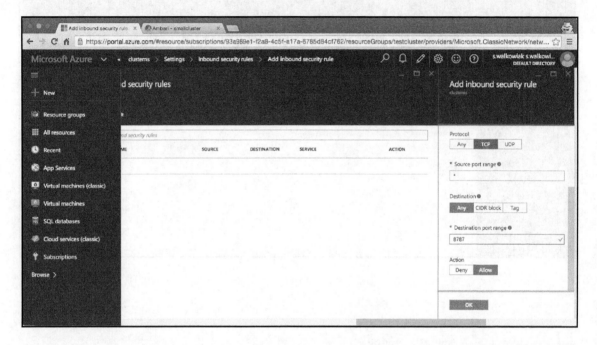

This will create a new security rule and update the current Network Security Group.

Editing the Virtual Network's public IP address for the head node

Despite configuring inbound security rules, we are still unable to connect to RStudio Server through a browser. In order to make it happen, we need to obtain a public IP address for the node (a head node to be precise) where we installed RStudio Server:

1. Click on the **Resource groups** (a menu panel on the left), select the Resource Group with the HDInsight cluster for us it will be `testcluster`. Click on the tile with the list of **Resources** to show all resources and services related to this Resource Group:

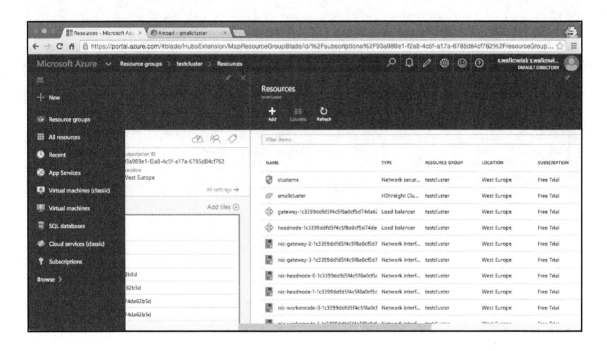

2. As the list is pretty long, we need to know which specific service we have to edit the public IP address for. The way of finding it out is to type the `ifconfig` command (the equivalent of `ipconfig` for Windows) in the shell/Terminal window and check the private IP address of the head node to which we are currently connected using SSH:

```
$ ifconfig
```

```
swalko@hn0-smallc:~$ ifconfig
eth0      Link encap:Ethernet  HWaddr 00:0d:3a:22:91:1d
          inet addr:10.2.0.18  Bcast:10.2.0.255  Mask:255.255.255.0
          inet6 addr: fe80::20d:3aff:fe22:911d/64 Scope:Link
          UP BROADCAST RUNNING MULTICAST  MTU:1500  Metric:1
          RX packets:1375224 errors:0 dropped:0 overruns:0 frame:0
          TX packets:806024 errors:0 dropped:0 overruns:0 carrier:0
          collisions:0 txqueuelen:1000
          RX bytes:800727209 (800.7 MB)  TX bytes:1075674669 (1.0 GB)

lo        Link encap:Local Loopback
          inet addr:127.0.0.1  Mask:255.0.0.0
          inet6 addr: ::1/128 Scope:Host
          UP LOOPBACK RUNNING  MTU:65536  Metric:1
          RX packets:2490610 errors:0 dropped:0 overruns:0 frame:0
          TX packets:2490610 errors:0 dropped:0 overruns:0 carrier:0
          collisions:0 txqueuelen:0
          RX bytes:1607036428 (1.6 GB)  TX bytes:1607036428 (1.6 GB)
```

Chapter 4

3. From the preceding output, we gather that our head node has the following private IP address: `10.2.0.18`. Click on the name of the head node with the matching IP address (usually it's the head node 0):

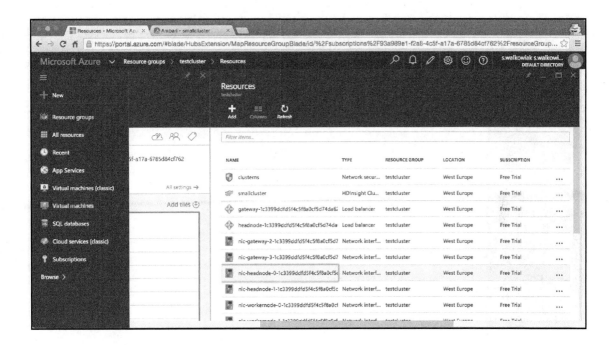

4. The head node's **Essentials** will appear, and we can see that the head node does not have a **Public IP address** set up. Click on the **All settings** link and select **IP addresses** from a menu of **General** configuration options:

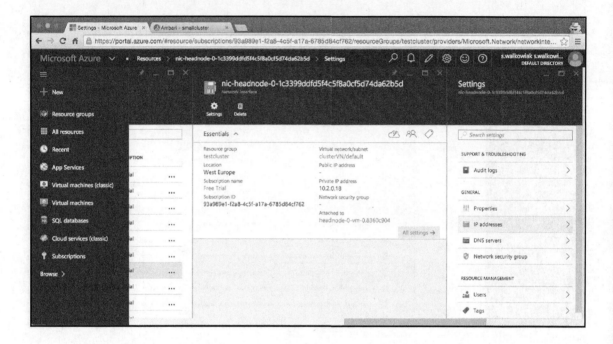

Chapter 4

5. In the **IP addresses** view, make sure to select **Enabled** for the **Public IP address settings**. Click on the **IP address** tab and then select **Create new**, as shown in the following screenshot:

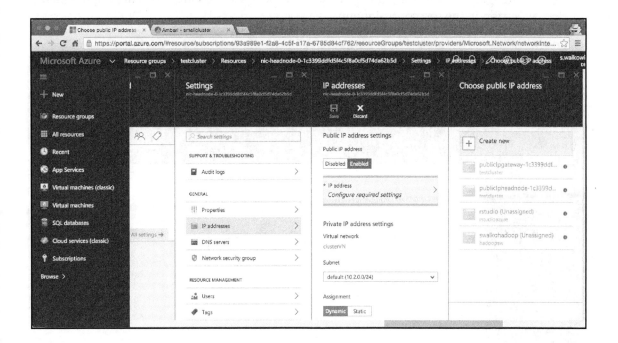

Hadoop and MapReduce Framework for R

6. In the **Create public IP address** view, give your IP address a **Name**, for example, `rstudio`, and click the **OK** button below to proceed.
7. You will be taken back to the **IP addresses** view, which should include all changes made by you, and you can now accept the changes by clicking the **Save** button in the top control menu:

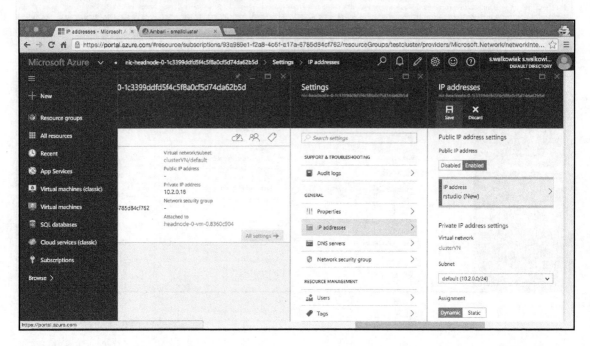

This will create a new public IP address for the head node and will update the network interface.

[226]

Chapter 4

8. To check the public IP address assigned to the head node, go back to its **Essentials** view by closing all other consecutive views and click on the **All settings** link and the **IP addresses** tab. In the **IP addresses** view, you should see the public IP address assigned to the head node:

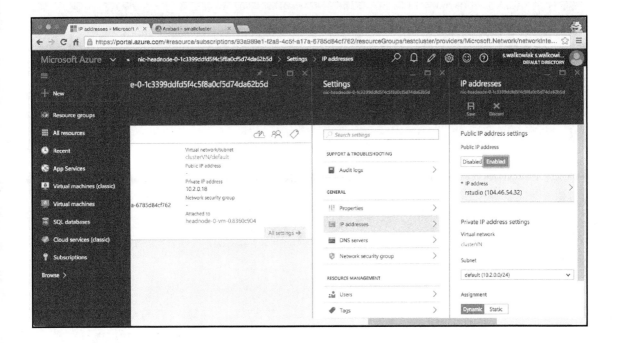

Hadoop and MapReduce Framework for R

9. To check whether you can connect to RStudio Server through a browser, point your browser to `http://<headnode_public_IP_address>:8787`. In our case, it will be `http://104.46.54.32:8787`:

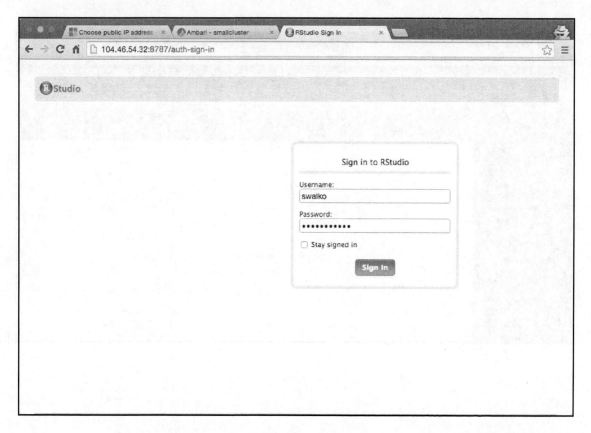

It worked! You can now type your standard credentials used for connecting through SSH to the cluster to log in to a new RStudio Server session.

This concludes the section on HDInsight setup and configuration with RStudio Server installation. In the next section, we will use RStudio to perform a number of MapReduce tasks on a dataset with smart electricity meter readings.

Smart energy meter readings analysis example – using R on HDInsight cluster

Upon logging into RStudio Server on HDInsinght you should be able to see a standard RStudio GUI with a console and other windows ready for data processing.

In this section, we will perform a number of data aggregations, cross-tabulations, and other analytics using MapReduce frameworks on a large dataset of approximately 414,000,000 rows and four variables (~12 GB) of the **Energy Demand Research Project,** which we briefly introduced you to earlier in this chapter. As the data is quite big and it's not open-access, we can't share it through the Packt Publishing website, but you can obtain the full dataset by registering at **UK Data Service** and downloading the file from `https://discover.ukdataservice.ac.uk/catalogue/?sn=7591`. The dataset contains electricity meter readings taken at 30-minute intervals from a representative sample of over 14,000 households in the UK between 2007 and 2010. Before you use the file, however, you should prepare the data so that it does not include the header with the names of variables. For the purpose of our analysis, we have also added a fifth variable with the `HOUR` of meter reading. We have simply extracted the hours from the `ADVANCEDATETIME` variable using the `lubridate` package for R. The whole `data.frame` should then be exported to a separate file (named, for example, `elec_noheader.csv`), which can be used for Hadoop processing. You can replicate all these data preparation procedures by following the R script available to download from the Packt Publishing website created for this book. Make sure you carry out all activities on a server or a virtual machine with at least 64 GB of RAM and 100 GB of free hard-drive space. As a result, you should end up with a single `csv` file with ~414,000,000 rows and five columns of data, roughly ~13.5 GB in size.

Hadoop and MapReduce Framework for R

Once we are logged to RStudio, we have to make sure that the `HADOOP_CMD` and `HADOOP_STREAMING` variables are set properly:

```
> cmd <- system("which hadoop", intern=TRUE)
> cmd
[1] "/usr/bin/hadoop"
> Sys.setenv(HADOOP_CMD=cmd)
> stream <- system("find /usr -name hadoop-streaming*jar", intern=TRUE)
...#output truncated
> stream
```

```
> stream
[1] "/usr/hdp/2.3.3.1-7/hadoop-mapreduce/hadoop-streaming.jar"
[2] "/usr/hdp/2.3.3.1-7/hadoop-mapreduce/hadoop-streaming-2.7.1.2.3.3.1-7.jar"
[3] "/usr/hdp/2.3.3.1-7/oozie/share/lib/mapreduce-streaming/hadoop-streaming-2.7.1.2.3.3.1-7.jar"
attr(,"status")
[1] 1
```

```
> Sys.setenv(HADOOP_STREAMING=stream[1])
```

We should quickly check whether the assignments were completed successfully:

```
> Sys.getenv("HADOOP_CMD")
[1] "/usr/bin/hadoop"

> Sys.getenv("HADOOP_STREAMING")
[1] "/usr/hdp/2.3.3.1-7/hadoop-mapreduce/hadoop-streaming.jar"
```

Let's load `rmr2` and `rhdfs` packages:

```
> library(rmr2)
...#output truncated
> library(rhdfs)
...#output truncated
```

And let's start the HDFS connection:

```
> hdfs.init()
...#output truncated
```

Some of the preceding commands may produce shorter or slightly longer outputs with details about loaded dependencies or may also include some warning messages, but generally they are not a reason for any concern, so we may as well disregard them at this time.

In order to get the data in, open a new Terminal window and use a previously presented script for data transfer between local and virtual machines. In our case, provided that the `data` folder is located on a Desktop of a local machine, the shell/Terminal command will look as follows:

```
$ scp -r ~/Desktop/data/ swalko@smallcluster-ssh.azurehdinsight.net:~/
```

As a result, you should see a data folder on your local area on the head node. The folder includes two files: a data `csv` file with no header (no variables labels/names) and a small `csv` file containing input formats variable names and types of variable only.

You should now set your R working directory to the `data` folder with both the files we transferred to the head node:

```
> getwd()
[1] "/home/swalko"
> setwd("/home/swalko/data")
```

Extract the full path to the data file in the working directory:

```
> file <- dir(getwd(), pattern = "_noheader.csv", full.names = TRUE)
> file
[1] "/home/swalko/data/elec_noheader.csv"
```

We can now put the file into the HDFS. Before we do so though, we should firstly create a new directory on HDFS:

```
> hdfs.mkdir("elec/data")
[1] TRUE
```

Transfer the data to the `elec/data` directory on HDFS. The process may take several seconds, so wait patiently until you see the TRUE output:

```
> hdfs.put(file, "elec/data")
[1] TRUE
```

We can now check whether the file has been copied properly to HDFS:

```
> hdfs.ls("elec/data")
  permission  owner      group         size              modtime
1 -rw-r--r--  swalko  supergroup  14443144092  2016-03-06 19:09
                                         file
1 /user/swalko/elec/data/elec_noheader.csv
```

It is also recommended to create a *shortcut* to the file by extracting its full path to HDFS:

```
> elec.data <- hdfs.ls("elec/data")$file
> elec.data
[1] "/user/swalko/elec/data/elec_noheader.csv"
```

We can now start preparing an input format for the data file. Note that we need to create an input format for our numerical dataset as it comes without any header. It is important to remember that standard datasets with rows and columns should be uploaded to HDFS without any names for the variables. The shuffling process on nodes may result in some variable names being wrongly assigned to different columns; therefore, we need to create an input format independently from the main data file. Firstly, we will retrieve the variable information from the supplementary input_format.csv file:

```
> elec.format <- read.csv("input_format.csv", sep = ",", header=TRUE, stringsAsFactors = FALSE)
> str(elec.format)
'data.frame':   1 obs. of  5 variables:
 $ ANON_ID       : chr "character"
 $ ADVANCEDATETIME: chr "character"
 $ HH            : chr "integer"
 $ ELECKWH       : chr "numeric"
 $ HOUR          : chr "integer"
```

From the preceding output, you can see that our input format is very simple it is, in fact, a data.frame with variable names and types of variables only.

Let's extract column classes for each variable we will need this information in the next step:

```
> colClasses <- as.character(as.vector(elec.format[1, ]))
> colClasses
[1] "character" "character" "integer"   "numeric"   "integer"
```

Finally, we can create an input format for Hadoop processing using the make.input.format() function, which you should already be familiar with from the previous parts of this chapter. As our data comes in csv format we will include it in the format argument of the function and we will also define the separator (sep). Additionally, we will make use of the column classes object (colClasses) that we created in the previous step and we will extract the names of all variables and pass them to the col.names argument:

```
> data.format <- make.input.format(format = "csv", sep = ",",
                                   col.names = names(elec.format),
                                   colClasses = colClasses,
                                   stringsAsFactors = FALSE)
```

We can now proceed to our first MapReduce job, but initially we will just run a Mapper function. In this job, we would only extract the timestamp in the form of a weekday, for example, Monday, Tuesday, and so on, when the electricity meter reading was collected. Each row of the output will be assigned the value 1. This simple Mapper will convert the date extracted from the second column of our data into a weekday through the `weekdays()` function available in base R:

```
> elec.map <- function(k, v) {
+   timestamp <- v[[2]]
+   wkday <- weekdays(as.Date(timestamp, format = "%d%b%y"))
+   keyval(wkday, 1)
+ }
```

To initialize the MapReduce job we can use the familiar `mapreduce()` function in which we specify the path to data on HDFS (`elec.data`), its input format (`data.format`), and the mapper function (`elec.map`):

```
> mr <- mapreduce(elec.data, input.format = data.format, map = elec.map)
```

From the R console you can easily observe the progress of the task. YARM submits the application and starts the MapReduce job:

```
WARNING: Use "yarn jar" to launch YARN applications.
packageJobJar: [] [/usr/hdp/2.3.3.1-7/hadoop-mapreduce/hadoop-streaming-2.7.1.2.3.3.1-7.jar]
/tmp/streamjob7498273722666106155.jar tmpDir=null
16/03/06 19:29:37 INFO impl.TimelineClientImpl: Timeline service address: http://hn0-smallc.
bauchabapwtuji1wpnn2paxlfg.ax.internal.cloudapp.net:8188/ws/v1/timeline/
16/03/06 19:29:38 INFO impl.TimelineClientImpl: Timeline service address: http://hn0-smallc.
bauchabapwtuji1wpnn2paxlfg.ax.internal.cloudapp.net:8188/ws/v1/timeline/
16/03/06 19:29:39 INFO mapred.FileInputFormat: Total input paths to process : 1
16/03/06 19:29:39 INFO mapreduce.JobSubmitter: number of splits:27
16/03/06 19:29:39 INFO mapreduce.JobSubmitter: Submitting tokens for job: job_1457287599108_
0002
16/03/06 19:29:40 INFO impl.YarnClientImpl: Submitted application application_1457287599108_
0002
16/03/06 19:29:40 INFO mapreduce.Job: The url to track the job: http://hn0-smallc.bauchabapw
tuji1wpnn2paxlfg.ax.internal.cloudapp.net:8088/proxy/application_1457287599108_0002/
16/03/06 19:29:40 INFO mapreduce.Job: Running job: job_1457287599108_0002
16/03/06 19:29:57 INFO mapreduce.Job: Job job_1457287599108_0002 running in uber mode : fals
e
16/03/06 19:29:57 INFO mapreduce.Job:  map 0% reduce 0%
16/03/06 19:30:21 INFO mapreduce.Job:  map 1% reduce 0%
16/03/06 19:30:24 INFO mapreduce.Job:  map 2% reduce 0%
16/03/06 19:30:25 INFO mapreduce.Job:  map 3% reduce 0%
16/03/06 19:30:27 INFO mapreduce.Job:  map 4% reduce 0%
16/03/06 19:30:28 INFO mapreduce.Job:  map 6% reduce 0%
16/03/06 19:30:31 INFO mapreduce.Job:  map 8% reduce 0%
```

At the same time, you can control the application from the Resource Manager using your browser at `https://<cluster_name>.azurehdinsight.net/yarnui/hn/cluster/app/RUNNING`. In our case, it was accessible at `https://smallcluster.azurehdinsight.net/yarnui/hn/cluster/apps/RUNNING`:

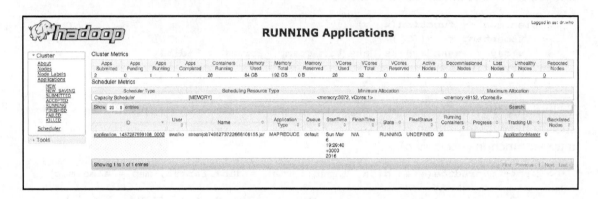

When the jobs have been completed, MapReduce will issue a message in the R console to confirm successful completion and will produce a list of performance-related metrics, for example:

```
16/03/06 19:33:00 INFO mapreduce.Job:  map 100% reduce 0%
16/03/06 19:33:04 INFO mapreduce.Job: Job job_1457287599108_0002 completed successfully
16/03/06 19:33:04 INFO mapreduce.Job: Counters: 30
        File System Counters
                FILE: Number of bytes read=0
                FILE: Number of bytes written=3744971
                FILE: Number of read operations=0
                FILE: Number of large read operations=0
                FILE: Number of write operations=0
                WASB: Number of bytes read=14446607529
                WASB: Number of bytes written=3324939256
                WASB: Number of read operations=0
                WASB: Number of large read operations=0
                WASB: Number of write operations=0
```

Finally, it will also present the final counters for input and output formats, which refer to read and written bytes of data, respectively. It will also inform you about the location of the output file:

```
       File Input Format Counters
               Bytes Read=14446551963
       File Output Format Counters
               Bytes Written=3324939256
16/03/06 19:33:04 INFO streaming.StreamJob: Output directory: /tmp/filea4c333ead425
```

For your information it took around 3 minutes and 22 seconds to run this MapReduce on ~13.5 GB of heavy data. It's a pretty good result considering the size of the individual file and that we only had two head and four worker nodes powering our task.

If you wish to inspect the top 50 keys and values of the output, you may run the following line, but remember that the returned output from the MapReduce job is quite large so you may have to wait for a while to obtain the results:

```
> head(keys(from.dfs(mr)), n=50)
> head(values(from.dfs(mr)), n=50)
```

We will now run a simple MapReduce job with the same Mapper as done previously, but this time we will also add a Reducer function. The Reducer will simply sum all occurrences of weekdays the task will return the number of each weekday for which the electricity meter readings were taken. As we will use the same Mapper, we will only show an example of the Reducer function:

```
> elec.reduce <- function(k, v) {
+   keyval(k, sum(v))
+ }
```

Unfortunately, on HDInsight there is a known `Java heap space` error when running a MapReduce task with both a Mapper and a Reducer. To avoid this issue, adjust the `rmr.options()` by setting additional Hadoop parameters. In our case, we will change the value of memory assigned to the Mapper task. Unfortunately, for the time being there is no clear explanation as to what causes these issues and users are advised to try to configure this value based on specific tasks:

```
> rmr.options(backend = "hadoop",
+             backend.parameters = list(hadoop = list(D =
+             "mapreduce.map.memory.mb=1024")))
```

Once setup we can initialize the MapReduce application:

```
> mr <- mapreduce(elec.data, input.format = data.format, map = elec.map,
reduce = elec.reduce)
```

As the Mapper returns quite a lot of data, it slows down the Reducer. Therefore, it may take up to an hour for the Reducer to complete. The MapReduce job reduced the data from 14,446,551,963 to only 2,434 bytes:

```
        File Input Format Counters
                Bytes Read=14446551963
        File Output Format Counters
                Bytes Written=2434
        rmr
                reduce calls=7
16/03/06 22:31:00 INFO streaming.StreamJob: Output directory: /tmp/fileacb12fdf4c68
```

As we expect the output to include only seven keys, one for each weekday, we can easily extract their values to the R console directly from HDFS:

```
> keys(from.dfs(mr))
[1] "Friday"    "Monday"    "Sunday"    "Tuesday"   "Saturday"
[6] "Thursday"  "Wednesday"
```

The same refers to the actual values of the key-value pairs. The values return the number electricity meter readings for each weekday:

```
> values(from.dfs(mr))
[1] 59178172 58002880 58107070 59482118 58978382 60111058
[7] 59976358
```

We can now perform a slightly more complex, but also more interesting, MapReduce task. We will calculate the average electricity consumption per hour across all data points. Therefore, our Mapper will gather values of hours (stored in the fifth column of our data) and half-hour electricity consumption (from column four). The Reducer, on the other hand, will return a `data.frame` as an output, which will contain a specific single value of an hour and the arithmetic mean of electricity consumption for each specific hour. Both functions can be written as follows:

```
> elec.map <- function(k, v) {
+   keyval(v[[5]], v[[4]])
+ }
> elec.reduce <- function(k, v) {
+   data.frame(hour=k, electricity=mean(v), row.names = k)
+ }
```

We can start this MapReduce task in a standard way through the `mapreduce()` function:

```
> mr <- mapreduce(elec.data, input.format = data.format, map = elec.map,
reduce = elec.reduce)
```

This time, the job was much faster and it has completed in just over 23 minutes:

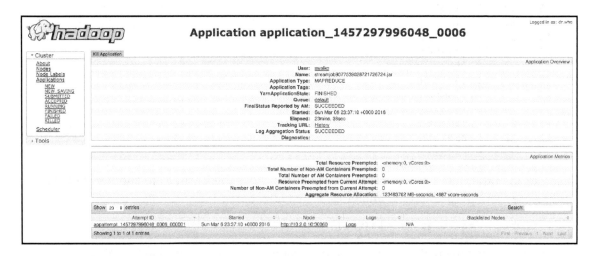

The data has been reduced to 13,768 bytes, as shown by the following output:

```
        File Input Format Counters
                Bytes Read=14446551963
        File Output Format Counters
                Bytes Written=13768
    rmr
                reduce calls=24
16/03/07 00:00:50 INFO streaming.StreamJob: Output directory: /tmp/file7f7b22bf699d
```

As we wanted to have a `data.frame` returned as the output we can find it in the values of the key-value pair given by the Reducer:

```
> values(from.dfs(mr))
  hour electricity
0    0   0.1644723
1    1   0.1625018
2    2   0.1550308
3    3   0.1477970
4    4   0.1490879
5    5   0.1760863
```

```
6    6   0.2290345
7    7   0.2625266
8    8   0.2621741
9    9   0.2559816
10   10  0.2522044
11   11  0.2561508
12   12  0.2537384
13   13  0.2428750
14   14  0.2429057
15   15  0.2723253
16   16  0.3349504
17   17  0.3817152
18   18  0.3866054
19   19  0.3724855
20   20  0.3535727
21   21  0.3200561
22   22  0.2552215
23   23  0.1889879
```

Just like any other data.frame, we can, of course, store these values to another R object and re-use them in further data analytics or visualizations:

```
> plot1 <- values(from.dfs(mr))
```

For example, we can create a simple line plot using the ggplot2 package, which will display the mean electricity consumption for each specific hour:

```
> install.packages("ggplot2")
... #output truncated - installing dependencies
> library(ggplot2)
> ggplot(plot1, aes(x=factor(hour), y=electricity, group=24)) +
+   geom_line(colour="blue", linetype="longdash", size=1.5) +
+   geom_point(colour="blue", size=4, shape=21, fill="white") +
+   xlab("Hour of measurement") +
+   ylab("Units of kilowatt-hours consumed") +
+   ggtitle("A line graph of kilowatt-hour consumed per Hour") +
+   theme_bw()
```

The preceding code snippet will produce the following plot:

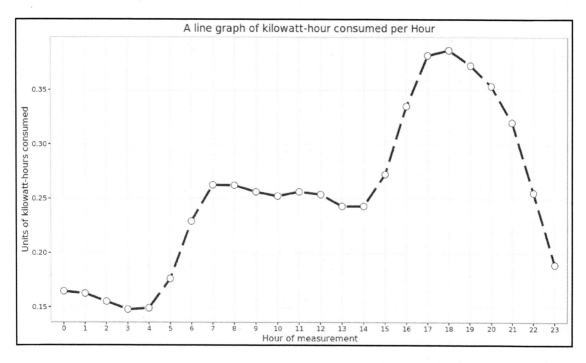

As in the word count example, you can redirect the output of your MapReduce job to another arbitrary destination folder on HDFS. In order to do so, we need to create an output format through the `make.output.format()` function and include it as well as the path to the output directory in the `mapreduce()` function call:

```
> out.form <- make.output.format(format = "csv", sep = ",")
> mr <- mapreduce(elec.data, output = "/user/swalko/output",
+                 input.format = data.format, output.format = out.form,
+                 map = elec.map, reduce = elec.reduce)
```

Once the job completes, you can find the output files in the `output` directory on HDFS created with the `mapreduce()` function. This is also confirmed by the job information message:

```
        File Input Format Counters
                Bytes Read=14446551963
        File Output Format Counters
                Bytes Written=491
        rmr
                reduce calls=24
16/03/07 01:09:03 INFO streaming.StreamJob: Output directory: /user/swalko/output
```

```
> hdfs.ls("output")
  permission  owner    group      size          modtime
1 -rw-r--r--  swalko   supergroup    0 2016-03-07 01:09
2 -rw-r--r--  swalko   supergroup  491 2016-03-07 01:09
                                  file
1   /user/swalko/output/_SUCCESS
2 /user/swalko/output/part-00000

> hdfs.file <- hdfs.ls("output")$file[2]
```

In the last line in the preceding code snippet, we have stored the path to the output files on HDFS as a new R object. We can now use it to transfer the returned values to a new `output.txt` file in the local file system:

```
> hdfs.get(hdfs.file, "/home/swalko/data/output.txt")
[1] TRUE
```

From there, we can simply download the file to a personal computer using the `scp` command in the shell/Terminal window.

This brings us to the end of the chapter. We have guided you through some more or less challenging Hadoop and R related tasks and tutorials.

Summary

Although the main goal of this chapter was to focus on data processing in Hadoop using the R language, throughout its various parts and sections you've been exposed to numerous different techniques and approaches used in Big Data analytics. We only hope that it wasn't too overwhelming!

We kicked off by introducing you to the diversity of Hadoop ecosystem, its tools and applications available to users, HDFS, and MapReduce frameworks.

We then created a single-node Hadoop cluster in which we carried out a simple word count MapReduce exercise in Java and the R languages, and we also showed you how to manage HDFS from the Linux command line and RStudio Server.

Finally, we achieved something that you probably won't be able to find in many (if any!) R books currently available on the market. We setup and configured a fully operational multi-node Hadoop cluster with R and RStudio Server installed and we crunched some real Big Data around 414,000,000 rows of electricity smart meter readings.

In the next two chapters, we will be working with data stored in databases. Firstly, in `Chapter 5`, *R with Relational Database Management Systems (RDBMSs)*, we will explore the R language connectivity with traditional **Relational Database Management Systems (RDBMSs)**, and in `Chapter 6`, *R with Non-Relational (NoSQL) Databases*, we will spend some time with more flexible, schema-free **NoSQL** databases such as **MongoDB**. Of course, we will be able to do all these things using our favorite language R.

5
R with Relational Database Management Systems (RDBMSs)

We have already done quite a lot of Big Data analytics using the R language in the preceding chapters, but this book would not be complete if we didn't touch on the subject of databases. To be precise, in this chapter we will explore connectivity between R and very popular **Relational Database Management Systems (RDBMSs)**, more commonly known as SQL databases. After reading the contents of this chapter, you will know how:

- To set up a number of local and/or remote SQL databases, for example SQLite, PostgreSQL, and MariaDB/MySQL
- To query and manage SQL databases directly from R (both locally and remotely) using a selection of R packages
- To launch fully-managed, highly-scalable Amazon RDS database instances (of different types) and query their records using the R language

During the process of achieving the preceding goals, you will also be exposed to a variety of methods and techniques that will help you with the installation of specific components required by individual databases and the R environment. Also, as we are going to query RDBMS, this chapter may serve as a very brief introduction to **Structured Query Language (SQL)**. As you will soon see, knowledge of several basic SQL commands will come in extremely useful throughout this chapter, from the moment you start the newly installed database, through reading the data in and their processing it. But before we immerse ourselves with practical exercises, let's firstly explore some features of databases that we will be using in this section of the book.

Relational Database Management Systems (RDBMSs)

The abundance of RDBMSs currently available means that it's nearly impossible to describe all or at least a large majority of them in one single chapter. If you haven't worked with any such databases in your analytical or research career, now is the best time to explore how they can benefit your Big Data processing and management activities.

A short overview of used RDBMSs

In order to give you a taste of the variety of databases available to R users, we decided to present three of them, which can be launched and connected from R in three different scenarios:

- Locally on a personal computer
- Locally on a virtual machine
- Remotely with a database on a server and RStudio installed on a personal local machine

Our selection criteria also included the requirements that all databases are open-source or at least free to use, are well-maintained with an active community of users, and can operate on multiple platforms (at least on Mac OS X, Windows, and Linux).

In the first part of the chapter, we will connect to a local **SQLite** database (https://www.sqlite.org/). SQLite is probably the most widely deployed RDBMS as it has become the favorite database engine commonly used in web browsers or distributed by default as part of several operating systems (for example Mac OS X, or Windows 10). SQLite, which first appeared in 2000, owes its popularity and good reputation to its simplicity of configuration and the fact that it can be easily linked to proprietary applications, for example web browsers. SQLite connects well to other programming languages (for example R) making it a good choice as an embedded database in multi-platform applications. As SQLite is now shipped with popular operating systems, we will use it as a data source for our first tutorial later in this chapter. Even if you don't work on Windows 10 or Mac OS X for which SQLite comes pre-installed, you can download SQLite installation files and quickly install them on your machine. We will then create a new database in SQLite and connect it with a locally run RStudio, which will be used to query and process the data stored in a SQLite database.

In the second tutorial, we will introduce you to **MariaDB** (`https://mariadb.com/`) and its connectivity with R. MariaDB was created in 2009 to mimic the functionalities of MySQL – a powerful and extremely popular open-source relational database management system which was purchased by Oracle in 2008. This acquisition inspired the original MySQL founders to create a highly-compatible relational database, which was offered to the public free of charge under the GNU GPL. In fact, MariaDB uses the same framework as MySQL, but it also includes several other extensions. Since its origin, it's become very popular with a growing community of passionate users. This popularity is due to MariaDB being available as a cross-platform database, its high scalability, and its good performance. During the tutorial, we will guide you through the process of installing MariaDB on an Amazon EC2 virtual machine operating on a Linux Ubuntu system. We will then connect the database with RStudio Server installed on the same instance by providing you with useful details of how to set up all Ubuntu dependencies and specific R packages required to make a connection with the MariaDB driver.

Finally, in the third tutorial we will show you methods to quickly deploy a PostgreSQL database (`https://www.postgresql.org/`) on the **Amazon Relational Database Service (RDS)** – a user-friendly, fully-managed, and highly scalable AWS solution for relational databases. PostgreSQL is another well-known and frequently used open-source, cross-platform, SQL-based database. It allows users to connect to a large number of other programming languages with ease (including R) and is usually praised for its security and reliability. In our tutorial, we will connect to PostgreSQL remotely from RStudio installed locally on a personal computer. We will also introduce you to other tools (for example RazorSQL) that facilitate cross-platform connectivity with a PostgreSQL server operated through Amazon RDS.

Structured Query Language (SQL)

Most Relational Database Management Systems use **Structured Query Language (SQL)** for data management, processing, and querying. Although publicly released for the first time in the mid-1980s, SQL is still very popular amongst database engineers, computer scientists, and everyone else who interacts with data on a daily basis. In fact, according to many modern data analysts, a mixture of good skills in R, Python, and SQL can give you a pretty successful career in the data analytics industry. Add to this some Java, Scala, and a bit of the C family of languages, and you can quickly become an expert in the Big Data world.

The best thing about SQL is that it is quite easy to learn and understand, even by a person without any knowledge of SQL, as its elements are largely inspired by words used in standard, natural English, and its structure is easily interpretable. SQL is built of several components, which can be arranged into queries or larger statements. It also includes operators well-known from other programming languages, such as equal to, greater than, and so on.

Most of the queries are `SELECT` operations that retrieve data from a table held within a database. The following simple query extracts all records from a table named `table1`:

```
SELECT * FROM table1;
```

In the majority of cases, the queries are followed by a semi-colon (;) sign. Some exceptions include `USE` and `EXIT` commands, which are not followed by any sign.

Usually your SQL statements will include more than just one query. They may also contain expressions, clauses, predicates, and other elements that will be used to manage and crunch your data. An example of a more advanced statement that calculates an average value of an order is shown in the following query:

```
SELECT clientCity, clientCountry, AVG(orderValue) as avgValue
FROM orders
WHERE clientCountry = 'Germany'
GROUP BY clientCity, clientCountry
ORDER BY clientCity;
```

In the preceding statement, we have included several queries. Firstly, the `SELECT` command extracts two already present variables in the table called `orders`: `clientCity` and `clientCountry`. However, we have also created and retrieved a third variable called `avgValue`, which represents the arithmetic mean of all orders stored in the `orderValue` variable. The mean was calculated for each level of two variables, `clientCity` and `clientCountry`, as indicated by the `GROUP BY` command. As we explicitly wanted to show only the records of orders made by clients based in Germany (as defined by the `WHERE` query), only such records will be returned by the `SELECT` command. Finally, all orders will be sorted alphabetically by the `clientCity` variable (`ORDER BY` command).

The resulting set may look as follows:

```
clientCity clientCountry avgValue
  Augsburg       Germany  2320.21
    Berlin       Germany  3712.39
   Cologne       Germany  2688.90
   Hamburg       Germany  4101.15
    Munich       Germany  5093.13
 Stuttgart       Germany  4266.91
```

In order to get more experience with SQL queries, feel free to pause for a moment and go through some SQL language resources and tutorials available online, for example:

- http://beginner-sql-tutorial.com/sql.htm
- http://www.tutorialspoint.com/sql/index.htm
- http://www.w3schools.com/sql/

As a word of caution, note that some relational databases may also include their own implementations of database-specific SQL-like commands, which do not follow standardized SQL queries. For example, a SQLite database has a range of *dot* commands for example, .databases, .help, .exit, and several others that are exclusively used in SQLite only. Some other databases on the other hand, may not include general SQL queries, for example PostgreSQL does not contain a DATEDIFF() function, which in standard SQL is used to calculate the difference between two dates.

SQLite with R

In this part of the chapter, we will query a SQLite database installed on a local, personal computer directly from RStudio. But before we can do it, follow the next section to prepare a SQLite database and read the data in.

Preparing and importing data into a local SQLite database

We mentioned earlier that SQLite is, by default, included in some distributions of popular operating systems, for example Mac OS X (since version 10.4) and in Windows 10. You can easily check whether your machine has SQLite installed by starting it through a Terminal/shell window:

```
$ sqlite3
SQLite version 3.12.1 2016-04-08 15:09:49
Enter ".help" for usage hints.
Connected to a transient in-memory database.
Use ".open FILENAME" to reopen on a persistent database.
sqlite>
```

If the command produces the preceding output (or similar) your machine is already equipped with SQLite database. If for some reason your operating system does not contain SQLite, visit http://www.sqlite.org/download.html to download and install the binaries for your particular OS.

Once installed, in a Terminal/shell navigate to the directory with our data, need_puf_2014.csv for example:

```
$ cd ~/Desktop/B05396_Ch06_Code/
```

The data that we are going to use in this tutorial is the *National Energy Efficiency Data – Framework: anonymised data 2014 (NEED)* dataset provided by the **Department of Energy & Climate Change**. The NEED data includes household-level information on annual consumption of gas and electricity across different geographical locations in the United Kingdom, and covers the years from 2005 until 2012.

The supplied **Public Use File** (PUF) is available for download from the following website:
https://www.gov.uk/government/statistics/national-energy-efficiency-data-framework-need-anonymised-data-2014. The NEED PUF file comes in a comma-separated (CSV) format (7.48 MB) and it contains a small representative sample of 49,815 records drawn from a full Big Data dataset of 4,086,448 records (1.38 GB as a SPSS *.sav file or 719 MB as a tab-delimited file) available to download through End-User License from the already introduced **UK Data Archive** at:
https://discover.ukdataservice.ac.uk/catalogue/?sn=7518.

Once in the directory with the data file, start SQLite by creating a new database called `need_data`:

```
$ sqlite3 need_data
```

Type `.databases` to show all currently available databases:

```
sqlite> .databases
seq  name              file
---  ----------------  -------------------------------------------------------
0    main              /Users/simonwalkowiak/Desktop/B05396_Ch06_Code/need_data
```

At this stage, you may want to physically open the folder where your data is stored. You may now notice that a new empty file called `need_data` has been created.

Then set the column separator to comma and import the `need_puf_2014.csv` data file as a new table called `need`:

```
sqlite> .separator ","
sqlite> .import need_puf_2014.csv need
```

You can check the available tables by running the `.tables` command:

```
sqlite> .tables
need
```

We can now view the folder with our data again. The `need_data` file has now been populated with the imported data.

Also, we may inspect the structure of the table using a `PRAGMA` statement:

```
sqlite> PRAGMA table_info('need');
0,HH_ID,TEXT,0,,0
1,REGION,TEXT,0,,0
2,IMD_ENG,TEXT,0,,0
3,IMD_WALES,TEXT,0,,0
4,Gcons2005,TEXT,0,,0
5,Gcons2005Valid,TEXT,0,,0
6,Gcons2006,TEXT,0,,0
7,Gcons2006Valid,TEXT,0,,0
8,Gcons2007,TEXT,0,,0
9,Gcons2007Valid,TEXT,0,,0
10,Gcons2008,TEXT,0,,0
...#output truncated
```

The `.schema` function allows us to print the schema of the table. A schema is a structure of a database object, for example the names of variables, their classes, and other features of a table. In other words, the schema describes the design of the table. The following snippet creates a schema for our `need` table:

```
sqlite> .schema need
CREATE TABLE need(
   "HH_ID" TEXT,
   "REGION" TEXT,
   "IMD_ENG" TEXT,
   "IMD_WALES" TEXT,
   "Gcons2005" TEXT,
   "Gcons2005Valid" TEXT,
   "Gcons2006" TEXT,
   "Gcons2006Valid" TEXT,
   "Gcons2007" TEXT,
   "Gcons2007Valid" TEXT,
...#output truncated
);
```

Once the table is created and our data is in it, we can open the RStudio application and connect to the SQLite database.

Connecting to SQLite from RStudio

When in RStudio, make sure that your working directory is set to the directory with the `need_data` file. If you downloaded the data and script files for this chapter from the Packt Publishing website to your Desktop area, it is very likely that your R working directory should be set as follows:

```
> setwd("~/Desktop/B05396_Ch06_Code")
```

As R requires the `RSQLite` package to make a connection with the SQLite database using the `DBI` package, we have to download the new versions of `DBI` and its dependency `Rcpp` beforehand. Note that in order to install recent releases of these packages from GitHub repositories, you first need to install the `devtools` package – it allows connectivity with GitHub:

```
> install.packages("devtools")
...#output truncated
> devtools::install_github("RcppCore/Rcpp")
...#output truncated
> devtools::install_github("rstats-db/DBI")
...#output truncated
```

Then, install and load the RSQLite package:

```
> install.packages("RSQLite")
...#output truncated
> library(RSQLite)
```

Let's then create a connection with the need_data SQLite database:

```
> con <- dbConnect(RSQLite::SQLite(), "need_data")
> con
<SQLiteConnection>
```

The dbListTables() and dbListFields() functions provide information on the available tables in the connected database and columns within a specified table, respectively:

```
> dbListTables(con)
[1] "need"
> dbListFields(con, "need")
 [1] "HH_ID"           "REGION"            "IMD_ENG"
 [4] "IMD_WALES"       "Gcons2005"         "Gcons2005Valid"
 [7] "Gcons2006"       "Gcons2006Valid"    "Gcons2007"
[10] "Gcons2007Valid"  "Gcons2008"         "Gcons2008Valid"
[13] "Gcons2009"       "Gcons2009Valid"    "Gcons2010"
[16] "Gcons2010Valid"  "Gcons2011"         "Gcons2011Valid"
[19] "Gcons2012"       "Gcons2012Valid"    "Econs2005"
[22] "Econs2005Valid"  "Econs2006"         "Econs2006Valid"
[25] "Econs2007"       "Econs2007Valid"    "Econs2008"
[28] "Econs2008Valid"  "Econs2009"         "Econs2009Valid"
[31] "Econs2010"       "Econs2010Valid"    "Econs2011"
[34] "Econs2011Valid"  "Econs2012"         "Econs2012Valid"
[37] "E7Flag2012"      "MAIN_HEAT_FUEL"    "PROP_AGE"
[40] "PROP_TYPE"       "FLOOR_AREA_BAND"   "EE_BAND"
[43] "LOFT_DEPTH"      "WALL_CONS"         "CWI"
[46] "CWI_YEAR"        "LI"                "LI_YEAR"
[49] "BOILER"          "BOILER_YEAR"
```

We may now query the data using the dbSendQuery() function; for example, we can retrieve all records from the table for which the value of the FLOOR_AREA_BAND variable equals 1:

```
> query.1 <- dbSendQuery(con, "SELECT * FROM need WHERE FLOOR_AREA_BAND = 1")
> dbGetStatement(query.1)
[1] "SELECT * FROM need WHERE FLOOR_AREA_BAND = 1"
```

In case you want to extract the string representing the SQL query used, you may apply the `dbGetStatement()` function on the object created by the `dbSendQuery()` command, as shown in the preceding code.

The results set may now be easily pulled to R (note that all queries and data processing activities run directly on the database, thus saving valuable resources normally used by R processes):

```
> query.1.res <- fetch(query.1, n=50)
> str(query.1.res)
'data.frame':  50 obs. of  50 variables:
 $ HH_ID           : chr  "5" "6" "12" "27" ...
 $ REGION          : chr  "E12000003" "E12000007" "E12000007" "E12000004" ...
 $ IMD_ENG         : chr  "1" "2" "1" "1" ...
 $ IMD_WALES       : chr  "" "" "" "" ...
 $ Gcons2005       : chr  "" "" "" "5500" ...
 $ Gcons2005Valid  : chr  "M" "O" "M" "V" ...
...#output truncated
> query.1.res
  HH_ID    REGION IMD_ENG IMD_WALES Gcons2005 Gcons2005Valid
1     5 E12000003       1                                  M
2     6 E12000007       2                                  O
3    12 E12000007       1                                  M
4    27 E12000004       1                5500              V
5    44 W99999999                 1     18000              V
...#output truncated
```

After performing the query, we can obtain additional information, for example its full SQL statement, the structure of the results set, and how many rows it returned:

```
> info <- dbGetInfo(query.1)
> str(info)
List of 6
 $ statement    : chr "SELECT * FROM need WHERE FLOOR_AREA_BAND = 1"
 $ isSelect     : int 1
 $ rowsAffected : int -1
 $ rowCount     : int 50
 $ completed    : int 0
 $ fields       :'data.frame': 50 obs. of  4 variables:
  ..$ name  : chr [1:50] "HH_ID" "REGION" "IMD_ENG" "IMD_WALES" ...
  ..$ Sclass: chr [1:50] "character" "character" "character" "character" ...
  ..$ type  : chr [1:50] "TEXT" "TEXT" "TEXT" "TEXT" ...
  ..$ len   : int [1:50] NA NA NA NA NA NA NA NA NA NA ...
> info
$statement
```

```
[1] "SELECT * FROM need WHERE FLOOR_AREA_BAND = 1"
$isSelect
[1] 1
$rowsAffected
[1] -1
$rowCount
[1] 50
$completed
[1] 0
$fields
             name     Sclass type len
1           HH_ID  character TEXT  NA
2          REGION  character TEXT  NA
3         IMD_ENG  character TEXT  NA
4       IMD_WALES  character TEXT  NA
5        Gcons2005 character TEXT  NA
...#output truncated
```

Once we complete a particular query, it is recommended to free the resources by clearing the obtained results set:

```
> dbClearResult(query.1)
[1] TRUE
```

We may now run a second query on the need table contained within the need_data SQLite database. This time, we will calculate the average electricity consumption for 2012 grouped by the levels of categorical variables: electricity efficiency band (EE_BAND), property age (PROP_AGE), and property type (PROP_TYPE). The statement will also sort the results set in ascending order, based on the values of two variables, EE_BAND and PROP_TYPE:

```
> query.2 <- dbSendQuery(con, "SELECT EE_BAND, PROP_AGE, PROP_TYPE,
+                              AVG(Econs2012) AS 'AVERAGE_ELEC_2012'
+                              FROM need
+                              GROUP BY EE_BAND, PROP_AGE, PROP_TYPE
+                              ORDER BY EE_BAND, PROP_TYPE ASC")
```

By running the statement, we simply apply a set of queries on the database; we then need to get the results into R using the fetch() function, just like we did in the first query. If you want to fetch all records, set the n parameter to -1:

```
> query.2.res <- fetch(query.2, n=-1)
```

It's always a good idea to inspect the structure and the size of the results set created with the query:

```
> info2 <- dbGetInfo(query.2)
> info2
$statement
[1] "SELECT EE_BAND, PROP_AGE, PROP_TYPE, \n AVG(Econs2012) AS 'AVERAGE_ELEC_2012' \n FROM need \n GROUP BY EE_BAND, PROP_AGE, PROP_TYPE \n ORDER BY EE_BAND, PROP_TYPE ASC"
$isSelect
[1] 1
$rowsAffected
[1] -1
$rowCount
[1] 208
$completed
[1] 1
$fields
                name     Sclass type len
1            EE_BAND  character TEXT  NA
2           PROP_AGE  character TEXT  NA
3          PROP_TYPE  character TEXT  NA
4 AVERAGE_ELEC_2012     double REAL   8
```

From the preceding output, you can see that our results set consists of 208 rows of data with the variables as outlined in the `fields` attribute of the output. We can finally view the first six rows of data:

```
> head(query.2.res, n=6)
  EE_BAND PROP_AGE PROP_TYPE AVERAGE_ELEC_2012
1       1      101       101          2650.000
2       1      102       101         12162.500
3       1      103       101          3137.500
4       1      104       101          4200.000
5       1      105       101          3933.333
6       1      106       101          5246.774
```

Before disconnecting from the database, you can also export the results set into a new table within the database:

```
> dbWriteTable(con, "query_2_result", query.2.res)
[1] TRUE
```

The new table named `query_2_result` has now been created in the `need_data` database:

```
> dbListTables(con)
[1] "need"             "query_2_result"
```

Once you finish all the processing, make sure to clear the results of the most recent query and disconnect from the active connection:

```
> dbClearResult(query.2)
[1] TRUE
> dbDisconnect(con)
[1] TRUE
```

This completes the tutorial on SQLite database connectivity with the R language as a data source for locally run SQL queries. In the next section, we will explore how easily R can operate with a MariaDB database deployed on an Amazon EC2 instance.

MariaDB with R on a Amazon EC2 instance

In *Online Chapter, Pushing R Further* (https://www.packtpub.com/sites/default/files/downloads/5396_6457OS_PushingRFurther.pdf), apart from creating a Linux Ubuntu virtual machine with RStudio Server on Microsoft Azure, we have also launched an Amazon Linux EC2 instance. In this section we will deploy an Ubuntu instance with RStudio Server, but this time on Amazon EC2.

Preparing the EC2 instance and RStudio Server for use

The good news is that you should already know how to do this. Initially, simply follow steps 1 through 13 of the section on *Creating your first Amazon EC2 instance* from *Online Chapter, Pushing R Further* (https://www.packtpub.com/sites/default/files/downloads/5396_6457OS_PushingRFurther.pdf). Choose a Free Usage Tier Linux Ubuntu instance and (in step 6) create your new key pair with a distinct name, for example `rstudio_mariadb.pem`. Also in the same step (step 6), add another custom TCP rule for port `3306`. This will allow connectivity with the MariaDB database.

R with Relational Database Management Systems (RDBMSs)

Launch the instance and wait until Instance State is set to **running**, as shown in the following screenshot:

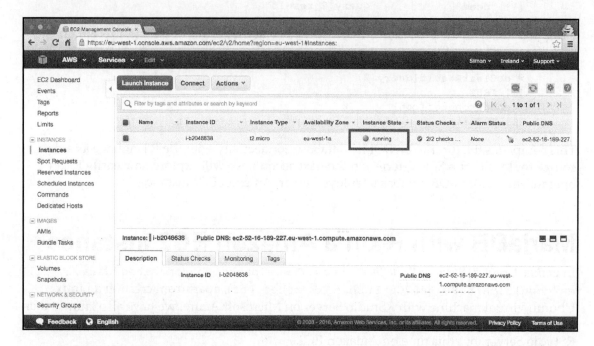

Once this happens, you may `ssh` to your new instance to check whether it's operational. Don't forget to navigate to the directory with the key pair and execute the `chmod` command:

```
$ cd Downloads/
$ chmod 400 rstudio_mariadb.pem
```

As we have created an Ubuntu instance, we need to follow the master user naming convention of Amazon and type `ubuntu` before the public DNS address of our instance:

```
$ ssh -i "rstudio_mariadb.pem" ubuntu@ec2-52-16-189-227.eu-west-1.compute.amazonaws.com
```

You should now be able to see the EC2 instance's standard welcome message and basic stats about the machine.

For RStudio Server installation, follow the instructions provided in *Online Chapter, Pushing R Further* (https://www.packtpub.com/sites/default/files/downloads/5396_64570S_PushingRFurther.pdf), in the section titled *Installing RStudio Server on a Microsoft Azure virtual machine*. We have already covered step 1, so you just need to apply steps 2 through 6. In the final stage, you may create a new user – in our case the username will be swalko:

```
$ sudo adduser swalko
```

When you complete all the preceding instructions, you may test the connection to RStudio Server by navigating your web browser to http://IP:8787, where IP is the unique IP address of your instance – in our case it would be http://52.16.189.227:8787. Input your credentials for the created username (in our case swalko) to log in to RStudio Server.

Preparing MariaDB and data for use

At this stage, we have our EC2 instance launched with RStudio Server installed. We have also created a new user, swalko, who can log in to RStudio Server in the web browser. Once we are logged in to the instance as the ubuntu user it's worth doing some extra organizational activities by granting swalko permissions to ssh directly to the instance using the same key pair, rstudio_mariadb.pem. This small maintenance work will prove very useful in the future, when we will need to manage the MariaDB database. In order to perform the preceding task, execute the following lines of code one by one (if you created a different username, make sure to substitute your chosen username for swalko in the following code):

```
$ sudo cp -r /home/ubuntu/.ssh /home/swalko/
$ cd /home/swalko/
$ sudo chown -R swalko:swalko .ssh
```

Log out from the instance and try to log back in as swalko:

```
$ logout
$ ssh -i "rstudio_mariadb.pem" swalko@ec2-52-16-189-227.eu-west-1.compute.amazonaws.com
```

It is very likely that everything went well and you were just greeted with the instance's standard welcome message. As user access has been configured correctly, you can now copy the data files from your local computer to the `/home/swalko/` directory on the EC2 instance. Note that we are not inputting data into a database at the moment – we haven't installed it yet. We simply move the data from a local computer to a virtual machine. Open a new terminal window and execute the following line:

```
$ scp -r -i "rstudio_mariadb.pem" ~/Desktop/data/need_puf_2014.csv swalko@ec2-52-16-189-227.eu-west-1.compute.amazonaws.com:~/
```

The preceding code requires that the `swalko` user has permissions to the key pair of the instance stored in the `rstudio_mariadb.pem` file (we set it up earlier). It also assumes that our data file is stored on a local computer in the `~/Desktop/data/` directory and that we want it to be moved to the `swalko` user home directory on the `/home/swalko/` instance. Make sure to adjust the directories accordingly.

The following instructions will guide you through the process of installing MariaDB and reading the data into the database.

Go back to your previous terminal window, which you used to `ssh` to the instance. Make sure to log in to the virtual machine again, as the `ubuntu` user:

```
$ ssh -i "rstudio_mariadb.pem" ubuntu@ec2-52-16-189-227.eu-west-1.compute.amazonaws.com
```

From the welcome message, note the version of Ubuntu (in our case it was 14.04.3 LTS). Alternatively, you can obtain this information by using the following command in the terminal window:

```
$ lsb_release -d
```

Or you can use this:

```
$ cat /etc/lsb-release
```

The Ubuntu release information is essential to obtain a correct version of MariaDB. In order to download the correct MariaDB installation packages, visit `https://downloads.mariadb.org/` and click on the **See our repository configuration tool** link. displayed in the top part of the **Downloads** page:

Chapter 5

On the next page, choose your Ubuntu distribution and its correct release. Then select the preferred version of MariaDB. We will go for the recent stable release, 10.1. A web page will now be populated with up-to-date commands to install MariaDB on Ubuntu:

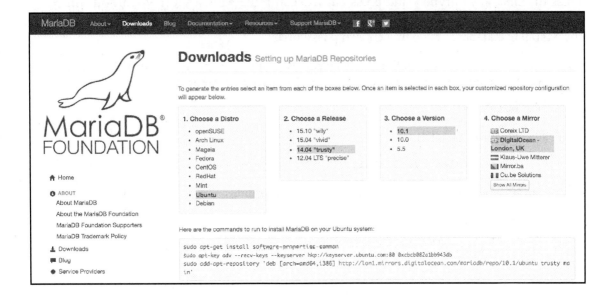

The following are the provided guidelines, we firstly need to install the `software-properties-common` packages. They are most likely already included in our instance, but it doesn't hurt to check again:

```
$ sudo apt-get install software-properties-common
```

Import the public key for the MariaDB installation packages:

```
$ sudo apt-key adv --recv-keys --keyserver hkp://keyserver.ubuntu.com:80 0xcbcb082a1bb943db
```

Add the MariaDB repository link:

```
$ sudo add-apt-repository 'deb [arch=amd64,i386] http://lon1.mirrors.digitalocean.com/mariadb/repo/10.1/ubuntu trusty main'
```

Once the preceding step is done, we can proceed to the actual installation of MariaDB:

```
$ sudo apt-get update
$ sudo apt-get install mariadb-server
```

Here you will be most likely prompted whether you want to continue with the installation. Press the *Enter* key or type `y` and press *Enter* to continue. After a short while, you will be asked to provide a new password for the `root` user, who will have all privileges to manage the MariaDB database. Type in a chosen password.

After several seconds the installation of MariaDB should complete. You can check the version of the installed MariaDB by typing the following command:

```
$ mysql -V
```

You can now start MariaDB:

```
$ sudo service mysql start
```

Once started, you can now log in to the MariaDB command client as `root` user:

```
$ mysql -uroot -p
```

Provide a previously specified password for the `root` user to authorize the login. If identified correctly, you should now see a welcome message similar to the following output:

```
Welcome to the MariaDB monitor.  Commands end with ; or \g.
Your MariaDB connection id is 2
Server version: 10.1.8-MariaDB mariadb.org binary distribution

Copyright (c) 2000, 2015, Oracle, MariaDB Corporation Ab and others.

Type 'help;' or '\h' for help. Type '\c' to clear the current input statement.

MariaDB [(none)]>
```

Congratulations, you have installed MariaDB on a Linux-Ubuntu Amazon EC2 instance with RStudio Server. We have also moved the data from your computer to the virtual machine. Therefore, there are only a few simple tasks ahead of us before we can start thinking of how to connect MariaDB with the R environment:

1. Create a database and a table that will hold our data.
2. Grant `swalko` user permissions to read/write and manage the database.
3. Move the data from the `/home/swalko/` directory to the database.

The preceding tasks have been named by us as *simple*, because frankly speaking there is not much difference between the MariaDB commands and the ones that you may be familiar with from your work with MySQL. As explained earlier, MariaDB is largely based on the MySQL framework and as such it operates in a very similar way. We therefore reiterate the following commands just as a revision and to provide a full explanation of activities users should undertake to achieve the remaining tasks listed above.

Upon logging in to MariaDB, you can view all available databases to the `root` user:

```
MariaDB [(none)]> SHOW databases;
+--------------------+
| Database           |
+--------------------+
| information_schema |
| mysql              |
| performance_schema |
+--------------------+
3 rows in set (0.00 sec)
```

We can create a new database named `data1` using the CREATE command:

```
MariaDB [(none)]> CREATE database data1;
Query OK, 1 row affected (0.00 sec)
MariaDB [(none)]> SHOW databases;
+--------------------+
| Database           |
+--------------------+
| data1              |
| information_schema |
| mysql              |
| performance_schema |
+--------------------+
4 rows in set (0.01 sec)
```

We will now select the `data1` database, in which we will create a new table called `need`:

```
MariaDB [(none)]> USE data1
Database changed
MariaDB [data1]>
```

As you can see, the MariaDB command prompt now includes (in the square brackets) the reference to the selected database `data1`. You can double-check the selection at any time using the following line:

```
MariaDB [data1]> SELECT database();
+------------+
| database() |
+------------+
| data1      |
+------------+
1 row in set (0.00 sec)
```

The `data1` database is currently empty, as confirmed by the output of the SHOW tables; command:

```
MariaDB [data1]> SHOW tables;
Empty set (0.00 sec)
```

We will now create a new table called `need`, which will store the data according to the defined schema:

```
MariaDB [data1]> CREATE TABLE need(
    -> hh_id INTEGER,
    -> region VARCHAR(25),
    -> imd_eng VARCHAR(25),
    -> imd_wales VARCHAR(25),
    -> gcons2005 VARCHAR(25),
```

```
    -> gcons2005valid VARCHAR(25),
    -> gcons2006 VARCHAR(25),
    -> gcons2006valid VARCHAR(25),
    -> gcons2007 VARCHAR(25),
    -> gcons2007valid VARCHAR(25),
    -> gcons2008 VARCHAR(25),
    -> gcons2008valid VARCHAR(25),
    -> gcons2009 VARCHAR(25),
    -> gcons2009valid VARCHAR(25),
    -> gcons2010 VARCHAR(25),
    -> gcons2010valid VARCHAR(25),
    -> gcons2011 VARCHAR(25),
    -> gcons2011valid VARCHAR(25),
    -> gcons2012 VARCHAR(25),
    -> gcons2012valid VARCHAR(25),
    -> econs2005 VARCHAR(25),
    -> econs2005valid VARCHAR(25),
    -> econs2006 VARCHAR(25),
    -> econs2006valid VARCHAR(25),
    -> econs2007 VARCHAR(25),
    -> econs2007valid VARCHAR(25),
    -> econs2008 VARCHAR(25),
    -> econs2008valid VARCHAR(25),
    -> econs2009 VARCHAR(25),
    -> econs2009valid VARCHAR(25),
    -> econs2010 VARCHAR(25),
    -> econs2010valid VARCHAR(25),
    -> econs2011 INTEGER,
    -> econs2011valid VARCHAR(25),
    -> econs2012 VARCHAR(25),
    -> econs2012valid VARCHAR(25),
    -> e7flag2012 VARCHAR(25),
    -> main_heat_fuel INTEGER,
    -> prop_age INTEGER,
    -> prop_type INTEGER,
    -> floor_area_band INTEGER,
    -> ee_band INTEGER,
    -> loft_depth INTEGER,
    -> wall_cons INTEGER,
    -> cwi VARCHAR(25),
    -> cwi_year VARCHAR(25),
    -> li VARCHAR(25),
    -> li_year VARCHAR(25),
    -> boiler VARCHAR(25),
    -> boiler_year VARCHAR(25));
Query OK, 0 rows affected (0.02 sec)
```

In the preceding call, we have indicated the data types for each variable in the data. The output informs us that zero rows were affected, as we have only created the structure (schema) for our data, with no data read in yet. You can inspect the schema of the `need` table within the `data1` database, using the `DESCRIBE` command:

```
MariaDB [data1]> DESCRIBE need;
... #output truncated
```

We can now upload the data to the `need` table we created. As the data is stored in the `/home/swalko/` directory, we will first have to create a new user for the database with all privileges for reading and writing from/to this database. To keep it simple and transparent, our new user will be called `swalko` and its access will be identified by a password, for example `Password1`:

```
MariaDB [data1]> CREATE USER 'swalko'@'localhost' IDENTIFIED BY 'Password1';
Query OK, 0 rows affected (0.00 sec)
```

We will then grant all privileges to `swalko` on the newly created table within the `data1` database:

```
MariaDB [data1]> GRANT ALL PRIVILEGES ON data1.need TO 'swalko'@'localhost' IDENTIFIED BY "Password1" with grant option;
Query OK, 0 rows affected (0.00 sec)
```

Re-load all privileges to activate them:

```
MariaDB [data1]> FLUSH PRIVILEGES;
Query OK, 0 rows affected (0.00 sec)
```

And log out from MariaDB as the `root` user, as well as from the EC2 instance as the `ubuntu` user:

```
MariaDB [data1]> EXIT
Bye
$ logout
```

Now `ssh` back to the instance as `swalko` and access MariaDB as `swalko` with the previously set password `Password1`:

```
$ ssh -i "rstudio_mariadb.pem" swalko@ec2-52-16-189-227.eu-west-1.compute.amazonaws.com
    ... #output truncated
$ mysql -p
```

Check whether you have access to the need table by executing a number of commands:

```
MariaDB [(none)]> SHOW databases;
+--------------------+
| Database           |
+--------------------+
| data1              |
| information_schema |
+--------------------+
2 rows in set (0.00 sec)
MariaDB [(none)]> USE data1
Database changed
MariaDB [data1]> SHOW tables;
+-----------------+
| Tables_in_data1 |
+-----------------+
| need            |
+-----------------+
1 row in set (0.00 sec)
MariaDB [data1]> DESCRIBE need;
...#output truncate
50 rows in set (0.00 sec)
```

All seems to be working just fine. We may now use the data stored in the /home/swalko/ directory and upload it to the need table:

```
MariaDB [data1]> LOAD DATA LOCAL INFILE '/home/swalko/need_puf_2014.csv'
    -> INTO TABLE need
    -> FIELDS TERMINATED BY ','
    -> LINES TERMINATED BY '\n'
    -> IGNORE 1 ROWS;
Query OK, 49815 rows affected (1.27 sec)
Records: 49815  Deleted: 0  Skipped: 0  Warnings: 0
```

The output confirms that all 49,815 records have been copied successfully to the need table in the data1 database.

We can now exit MariaDB and log out from the instance as the swalko user:

```
MariaDB [data1]> EXIT
Bye
$ logout
```

We have now completed all essential preparatory tasks for MariaDB and, the data that we will be using during the R part of the tutorial. The next section will guide you through some important, but also tricky, activities that users should carry out before being able to connect RStudio with MariaDB on EC2.

Working with MariaDB from RStudio

Before we can connect to MariaDB from RStudio, we need to install a number of critical R packages, including `rJava`. Firstly, `ssh` to the EC2 instance as the `ubuntu` user:

```
$ ssh -i "rstudio_mariadb.pem" ubuntu@ec2-52-16-189-227.eu-west-1.compute.amazonaws.com
```

We then need to obtain and install Oracle Java 8 libraries for Ubuntu (there will be some Terminal outputs after each line, but we will skip them in this listing):

```
$ sudo add-apt-repository ppa:webupd8team/java
$ sudo apt-get update
$ sudo apt-get install oracle-java8-installer
```

Once installed, running `java -version` in a Terminal should allow us to double-check whether the installation was successful:

```
$ java -version
java version "1.8.0_77"
Java(TM) SE Runtime Environment (build 1.8.0_77-b03)
Java HotSpot(TM) 64-Bit Server VM (build 25.77-b03, mixed mode)
```

To automatically set up the Java environment variables, you can install the following package:

```
$ sudo apt-get install oracle-java8-set-default
```

Now we are ready to install the `rJava` package using the following commands:

```
$ sudo R CMD javareconf
$ sudo apt-get install r-cran-rjava
```

In order to check whether the installation was successful, it's best to log in to RStudio Server on `http://IP:8787`, as described previously, and try loading `rJava` from the R console:

```
> library(rJava)
```

If it fails, log out of RStudio Server, go back to the Terminal/shell and copy the `libjvm.so` file physically to the `/usr/lib/` directory:

```
$ sudo updatedb
$ locate libjvm.so
$ sudo ln -s /usr/lib/jvm/java-8-oracle/jre/lib/amd64/server/libjvm.so /usr/lib/
```

The last line of the preceding code may need updating, depending on your path to the `libjvm.so` file.

We also need to install appropriate libraries for the MariaDB client on Ubuntu (note that for MySQL, these libraries will be `libmysqlclient-dev`):

```
$ sudo apt-get install libmariadbclient-dev
```

Once that been completed, we can proceed to the installation of the new version of the `DBI` package for the `RMySQL` package. Before this happens, however, we have to download and install, `OpenSSL`, `Curl`, and `LibSSH2` libraries for Ubuntu, and then the `devtools` package for R:

```
$ sudo apt-get install libssl-dev
$ sudo apt-get install libcurl4-openssl-dev
$ sudo apt-get install libssh2-1-dev
$ sudo Rscript -e 'install.packages("devtools", repos = "http://cran.r-project.org/")'
```

The installation of all required Ubuntu libraries and the `devtools` package may take several minutes. When they finish, using GitHub repositories, install the new versions of the `Rcpp` and `DBI` packages for R:

```
$ sudo Rscript -e 'devtools::install_github("RcppCore/Rcpp")'
$ sudo Rscript -e 'devtools::install_github("rstats-db/DBI")'
```

Finally, install the `RMySQL` package and other R packages, for example `dplyr` and `ggplot2`, that may be useful in connecting to the MariaDB database:

```
$ sudo Rscript -e 'install.packages(c("RMySQL", "dplyr", "ggplot2"), repos = "http://cran.r-project.org/")'
```

It is, however, very likely that the installation of the `dplyr` and `ggplot2` packages fails if you perform it on a Free Tier Amazon EC2 instance, due to the insufficient RAM available. If that happens, you will see the following message with the unpleasant *non-zero exit status* error:

```
virtual memory exhausted: Cannot allocate memory
make: *** [dplyr.o] Error 1
ERROR: compilation failed for package 'dplyr'
* removing '/usr/local/lib/R/site-library/dplyr'

The downloaded source packages are in
        '/tmp/RtmprfgMKH/downloaded_packages'
Warning message:
In install.packages("dplyr", repos = "http://cran.r-project.org/") :
  installation of package 'dplyr' had non-zero exit status
```

In such a case, make sure to increase the size of your swap file. First, confirm the swap file is enabled on your virtual machine:

```
$ swapon -s
Filename            Type    Size Used Priority
```

If the output is not populated with any details (as shown in the preceding output), you need to enable the swap file with the following commands:

```
$ sudo dd if=/dev/zero of=/swapfile bs=1024 count=512k
524288+0 records in
524288+0 records out
536870912 bytes (537 MB) copied, 8.27689 s, 64.9 MB/s
$ sudo mkswap /swapfile
Setting up swapspace version 1, size = 524284 KiB
no label, UUID=66c0cb3c-26d4-42fd-bf46-27be8f1bcd9d
$ sudo swapon /swapfile
```

From the outputs, we can assume that a swap file of the specified size has been created (the output of the first statement) and that it has been assigned the ID string (the output of the second statement). Additionally, if you type the `swapon -s` command again, you should now see a swap file enabled:

```
$ swapon -s
Filename            Type    Size Used Priority
/swapfile           file    524284 0  -1
```

Let's now try installing the `dplyr` package again:

```
$ sudo Rscript -e 'install.packages("dplyr", repos = "http://cran.r-project.org/")'
```

After a couple of minutes, the `dplyr` package should be installed without any issues. You may now repeat a similar installation procedure for the `ggplot2` package:

```
$ sudo Rscript -e 'install.packages("ggplot2", repos = "http://cran.r-project.org/")'
```

At this stage, we have all the tools and libraries ready and prepared for us to use in RStudio Server and to query MariaDB directly from the R environment.

Navigate your web browser to `http://IP:8787` to log in to RStudio Server using the known credentials (in our case the username was `swalko`, but you may have set a different one during the process). As we have already installed the `RMySQL` package, we just need to load it into the R session. This will also load the required `DBI` package:

```
> library(RMySQL)
Loading required package: DBI
```

In order to be able to query the database, we first have to create a connection to the `data1` database in MariaDB:

```
> conn <- dbConnect(RMySQL::MySQL(), user = "swalko",
+                   password = "Password1",
+                   host = "localhost",
+                   dbname = "data1")
```

The preceding code connects R with MariaDB/MySQL; however, it is often unsafe to reveal the credentials in the production process. To hide them, you may want to create a MySQL configuration file, `.my.cnf`, which will securely store the database name, user, and password variables (and also others if needed). The file can be created using the `nano` editor. While connecting through `ssh` to the instance as `swalko`, type the following command in the terminal:

```
$ nano ~/.my.cnf
```

The nano editor will launch and within the body, type a few lines of options:

```
[dt1]
database=data1
user=swalko
password=Password1
```

The top value in the square brackets [] refers to the `group` option value in the `dbConnect()` function from the RMySQL package. Exit the nano editor by pressing the *Ctrl + X* keys, then type Y for *Yes* and press the *Enter* key to save the file. From now on in R, you can make the connection to this particular `data1` database using the following, more secure method:

```
> conn <- dbConnect(RMySQL::MySQL(), group = "dt1",
+                   host = "localhost")
```

As you may notice, the `group` parameter is set to the value (`dt1`) specified in square brackets in the `.my.cnf` file.

We can obtain confirmation of the created connection with the `summary()` function:

```
> summary(conn)
<MySQLConnection:0,1>
  User:    swalko
  Host:    localhost
  Dbname:  data1
  Connection type: Localhost via UNIX socket
Results:
```

The `dbGetInfo()` function provides a little bit more detailed output about the connection with MariaDB:

```
> dbGetInfo(conn)
...#output truncated
```

We can now obtain the names of the tables present in the database, as well as all their fields (that is, variables) using the `dbListTables()` and `dbListFields()` functions, respectively:

```
> dbListTables(conn)
[1] "need"
> dbListFields(conn, "need")
 [1] "hh_id"             "region"            "imd_eng"
 [4] "imd_wales"         "gcons2005"         "gcons2005valid"
 [7] "gcons2006"         "gcons2006valid"    "gcons2007"
[10] "gcons2007valid"    "gcons2008"         "gcons2008valid"
[13] "gcons2009"         "gcons2009valid"    "gcons2010"
[16] "gcons2010valid"    "gcons2011"         "gcons2011valid"
[19] "gcons2012"         "gcons2012valid"    "econs2005"
[22] "econs2005valid"    "econs2006"         "econs2006valid"
[25] "econs2007"         "econs2007valid"    "econs2008"
[28] "econs2008valid"    "econs2009"         "econs2009valid"
[31] "econs2010"         "econs2010valid"    "econs2011"
[34] "econs2011valid"    "econs2012"         "econs2012valid"
```

```
[37] "e7flag2012"       "main_heat_fuel"   "prop_age"
[40] "prop_type"        "floor_area_band"  "ee_band"
[43] "loft_depth"       "wall_cons"        "cwi"
[46] "cwi_year"         "li"               "li_year"
[49] "boiler"           "boiler_year"
```

Let's test whether we can perform the simplest query by calculating the total number of records in our `need` table:

```
> query.1 <- dbSendQuery(conn, "SELECT COUNT(*) AS records FROM need")
```

The standard `RMySQL` functions, such as `dbGetStatement()`, `dbColumnInfo()`, and `dbGetInfo()`, are also supported for MariaDB:

```
> dbGetStatement(query.1)
[1] "SELECT COUNT(*) AS records FROM need"
> dbColumnInfo(query.1)
     name Sclass   type length
1 records double BIGINT     21
> dbGetInfo(query.1)
$statement
[1] "SELECT COUNT(*) AS records FROM need"
$isSelect
[1] 1
...#output truncated
```

We can pull the results of the query into R as a `data.frame` object using the `dbFetch()` function:

```
> query.1.res <- dbFetch(query.1, n=-1)
> query.1.res
  records
1   49815
```

As usual, after obtaining the aggregated or processed data from the database, we need to free the resources associated with a result set by running the `dbClearResult()` function on our `MySQLResult` object:

```
> dbClearResult(query.1)
[1] TRUE
```

Let's perform the second query on the `need` table in the `data1` database in MariaDB. We will calculate the average electricity consumption in 2012 (`Econs2012`) grouped by electricity efficiency band (`ee_band`), property age (`prop_age`), and property type (`prop_type`). We will sort the results in ascending order by first the electricity efficiency band and then property type:

```
> query.2 <- dbSendQuery(conn, "SELECT EE_BAND, PROP_AGE, PROP_TYPE,
+                              AVG(Econs2012) AS AVERAGE_ELEC_2012
+                              FROM need
+                              GROUP BY EE_BAND, PROP_AGE, PROP_TYPE
+                              ORDER BY EE_BAND, PROP_TYPE ASC")
```

The `dbColumnInfo()` function can again provide us with some useful information about the expected structure of the result set:

```
> dbColumnInfo(query.2)
               name   Sclass    type length
1           EE_BAND  integer INTEGER     11
2          PROP_AGE  integer INTEGER     11
3         PROP_TYPE  integer INTEGER     11
4 AVERAGE_ELEC_2012   double  DOUBLE     23
```

We may now *fetch* the results in the standard way (note the following output has been truncated; the original results set contained 208 rows):

```
> query.2.res <- dbFetch(query.2, n=-1)
> query.2.res
  EE_BAND PROP_AGE PROP_TYPE AVERAGE_ELEC_2012
1       1      102       101          12162.500
2       1      106       101           5246.774
3       1      101       101           2650.000
4       1      104       101           4200.000
5       1      105       101           3933.333
6       1      103       101           3137.500
...#output truncated
```

After querying the data, we will free the resources and close the connection to MariaDB:

```
> dbClearResult(query.2)
[1] TRUE
> dbDisconnect(conn)
[1] TRUE
```

The implementation of the `RMySQL` package with MariaDB is almost the same as with the MySQL database, and both databases can be used interchangeably, depending on your preferences. The only difference is the database server type, its libraries, and specific installation requirements dependent on the operating system of a virtual machine.

To make this part of the tutorial a little bit more exciting, we will now try to connect to MariaDB using the `dplyr` package authored and maintained by Hadley Wickham. We didn't present `dplyr` connectivity with SQLite earlier, but it generally works very well with most open source databases. It also allows users with minimal SQL knowledge to run SQL queries without writing SQL queries explicitly; however, this functionality is also allowed in `dplyr` and enables users to perform more complex querying, as you will see in the next section of this chapter.

We have already installed `dplyr`, therefore we just need to load the package into R in the standard way:

```
> library(dplyr)
...#output truncated
```

For the MariaDB connection with R through `dplyr`, we can use the credentials (`user` and `password` values only) stored in the previously created MySQL configuration file located in `~/.my.cnf`. Note that both `user` and `password` parameters must be set to `NULL` in this case:

```
> dpl.conn <- src_mysql(dbname = 'data1',
+                       host = 'localhost',
+                       user = NULL,
+                       password = NULL,
+                       group = 'dt1')
```

By calling the name of the connection (the `dpl.conn` object), we can view the type of server used, the path to the database, and the available tables:

```
> dpl.conn
src:  mysql 10.1.13-MariaDB-1~trusty [swalko@localhost:/data1]
tbls: need
```

The `tbl()` function uses the created connection and provides a snapshot of the referenced table:

```
> need.data <- tbl(dpl.conn, "need")
> need.data
Source: mysql 10.1.13-MariaDB-1~trusty [swalko@localhost:/data1]
From: need [49,815 x 50]
    hh_id    region imd_eng imd_wales gcons2005 gcons2005valid
    (int)     (chr)   (chr)     (chr)     (chr)          (chr)
1       1 E12000007       1               35000              V
2       2 E12000002       4               19000              V
3       3 E12000002       4               22500              V
4       4 E12000005       1               21000              V
5       5 E12000003       1                                  M
```

```
     6     6 E12000007        2                              O
     7     7 E12000006        3           12000              V
     8     8 E12000005        5           18500              V
     9     9 E12000007        4           35000              V
    10    10 E12000003        2           28000              V
    ..   ...       ...      ...             ...            ...
Variables not shown: gcons2006 (chr), gcons2006valid (chr),
  gcons2007 (chr), gcons2007valid (chr), gcons2008 (chr),
  gcons2008valid (chr), gcons2009 (chr), gcons2009valid
  (chr), gcons2010 (chr), gcons2010valid (chr), gcons2011
  (chr), gcons2011valid (chr), gcons2012 (chr),
  gcons2012valid (chr), econs2005 (chr), econs2005valid
  (chr), econs2006 (chr), econs2006valid (chr), econs2007
  (chr), econs2007valid (chr), econs2008 (chr),
  econs2008valid (chr), econs2009 (chr), econs2009valid
  (chr), econs2010 (chr), econs2010valid (chr), econs2011
  (int), econs2011valid (chr), econs2012 (chr),
  econs2012valid (chr), e7flag2012 (chr), main_heat_fuel
  (int), prop_age (int), prop_type (int), floor_area_band
  (int), ee_band (int), loft_depth (int), wall_cons (int),
  cwi (chr), cwi_year (chr), li (chr), li_year (chr), boiler
  (chr), boiler_year (chr)
```

You can also obtain more detailed output on the structure of the need.data tbl_mysql object by using the generic str() function:

```
> str(need.data)
...#output truncated
```

We will now run a little bit more advanced SQL query on the NEED data. We will calculate the average electricity consumption for the years 2005 through 2012, grouped by geographical region (region) and property type (prop_type). We will order the results by region and property type.

The dplyr package requires that all activities are performed in sequence. Therefore, we first need to explicitly set the grouping variables (region and prop_type) for the table:

```
> by.regiontype <- group_by(need.data, region, prop_type)
> by.regiontype
Source: mysql 10.1.13-MariaDB-1~trusty [swalko@localhost:/data1]
From: need [49,815 x 50]
Grouped by: region, prop_type
    hh_id      region imd_eng imd_wales gcons2005 gcons2005valid
    (int)       (chr)   (chr)     (chr)     (chr)          (chr)
1       1 E12000007       1                 35000              V
2       2 E12000002       4                 19000              V
3       3 E12000002       4                 22500              V
```

```
       4      4 E12000005         1             21000              V
...#output truncated
```

Note that the preceding output contains (on the third line) information about both grouping variables added to the structure of the table. This grouped table has been stored as a new `tbl_mysql` object named `by.regiontype`. This new grouped object will now be used to calculate the average electricity consumption for each year (from 2005 until 2012) aggregated by both grouping variables defined earlier:

```
> avg.elec <- summarise(by.regiontype,
+                      elec2005 = mean(econs2005),
+                      elec2006 = mean(econs2006),
+                      elec2007 = mean(econs2007),
+                      elec2008 = mean(econs2008),
+                      elec2009 = mean(econs2009),
+                      elec2010 = mean(econs2010),
+                      elec2011 = mean(econs2011),
+                      elec2012 = mean(econs2012))
```

Finally, we can order the resulting table by region and property type. By default, the `arrange()` function sorts the values in ascending order:

```
> avg.elec <- arrange(avg.elec, region, prop_type)
> avg.elec
Source: mysql 10.1.13-MariaDB-1~trusty [swalko@localhost:/data1]
From: <derived table> [?? x 10]
Arrange: region, prop_type
Grouped by: region
       region prop_type elec2005 elec2006 elec2007 elec2008
        (chr)     (int)    (dbl)    (dbl)    (dbl)    (dbl)
1  E12000001       101 5341.386 5196.255 5298.689 4862.547
2  E12000001       102 3840.788 3757.433 3733.164 3523.888
3  E12000001       103 3734.703 3816.210 3890.868 3676.256
4  E12000001       104 3709.131 3701.773 3617.465 3372.784
5  E12000001       105 3337.374 3346.970 3278.114 3144.276
6  E12000001       106 3009.375 3010.417 2954.167 2934.635
7  E12000002       101 5276.891 5531.513 5415.006 5123.770
8  E12000002       102 4384.243 4346.923 4261.663 3912.655
9  E12000002       103 3809.194 4140.323 3954.597 3741.290
10 E12000002       104 3726.642 3715.623 3693.892 3473.204
..       ...       ...      ...      ...      ...      ...
Variables not shown: elec2009 (dbl), elec2010 (dbl), elec2011
   (dbl), elec2012 (dbl)
Warning message:
In .local(conn, statement, ...) :
   Decimal MySQL column 8 imported as numeric
```

This is the final truncated output of the results set of our query. You are probably wondering right now why we use the word *query* if we didn't even write a single SQL command during our data processing. Well, in fact we did. The `dplyr` package is extremely user friendly and it doesn't require its users to understand and know Structured Query Language – although this knowledge would undoubtedly help as some of the errors are related to how the SQL queries are constructed. However, it translates its functions, such as `group_by()`, `summarise()`, `arrange()`, and many others, into their SQL equivalents *in the background*. If you are curious, how `dplyr` does it, you may use `show_query()` or, even better, the `explain()` function, which prints the applied SQL query and its plan, for example:

```
> show_query(avg.elec)
...#output truncated
> explain(avg.elec)
<SQL>
SELECT `region`, `prop_type`, `elec2005`, `elec2006`, `elec2007`,
`elec2008`, `elec2009`, `elec2010`, `elec2011`, `elec2012`
    FROM (SELECT `region`, `prop_type`, AVG(`econs2005`) AS `elec2005`,
AVG(`econs2006`) AS `elec2006`, AVG(`econs2007`) AS `elec2007`,
AVG(`econs2008`) AS `elec2008`, AVG(`econs2009`) AS `elec2009`,
AVG(`econs2010`) AS `elec2010`, AVG(`econs2011`) AS `elec2011`,
AVG(`econs2012`) AS `elec2012`
    FROM `need`
    GROUP BY `region`, `prop_type`) AS `zzz1`
    ORDER BY `region`, `region`, `prop_type`
<PLAN>
    id select_type     table type possible_keys  key key_len
1    1    PRIMARY <derived2>  ALL              <NA> <NA>    <NA>
2    2    DERIVED       need  ALL              <NA> <NA>    <NA>
    ref  rows                  Extra
1  <NA> 49386                  Using filesort
2  <NA> 49386 Using temporary; Using filesort
```

It's important to stress that all this processing has occurred within the database, without affecting the performance of R. We may now pull the results set from the database into R as a `data.frame` object. To be precise, as we grouped the table by two grouping variables, our new object is now a grouped data frame (`grouped_df`):

```
> elec.df <- collect(avg.elec)
...#output truncated
> elec.df
Source: local data frame [60 x 10]
Groups: region [10]
        region prop_type elec2005 elec2006 elec2007 elec2008
         (chr)     (int)    (dbl)    (dbl)    (dbl)    (dbl)
1    E12000001       101 5341.386 5196.255 5298.689 4862.547
```

```
2  E12000001    102 3840.788 3757.433 3733.164 3523.888
3  E12000001    103 3734.703 3816.210 3890.868 3676.256
4  E12000001    104 3709.131 3701.773 3617.465 3372.784
5  E12000001    105 3337.374 3346.970 3278.114 3144.276
6  E12000001    106 3009.375 3010.417 2954.167 2934.635
7  E12000002    101 5276.891 5531.513 5415.006 5123.770
8  E12000002    102 4384.243 4346.923 4261.663 3912.655
9  E12000002    103 3809.194 4140.323 3954.597 3741.290
10 E12000002    104 3726.642 3715.623 3693.892 3473.204
..    ...        ...   ...      ...      ...      ...
Variables not shown: elec2009 (dbl), elec2010 (dbl), elec2011
   (dbl), elec2012 (dbl)
```

In order to convert the grouped_df into an R native data.frame, we can simply use as.data.frame() from base R:

```
> elec <- as.data.frame(elec.df)
> elec
     region prop_type elec2005 elec2006 elec2007 elec2008
1  E12000001    101 5341.386 5196.255 5298.689 4862.547
2  E12000001    102 3840.788 3757.433 3733.164 3523.888
3  E12000001    103 3734.703 3816.210 3890.868 3676.256
4  E12000001    104 3709.131 3701.773 3617.465 3372.784
5  E12000001    105 3337.374 3346.970 3278.114 3144.276
6  E12000001    106 3009.375 3010.417 2954.167 2934.635
...#output truncated
```

Of course, a large amount of aggregated electricity consumption is not too meaningful in this format. If you want to present this data graphically, you should probably transform the data from wide into narrow/long format. You can achieve this through the reshape() function:

```
> elec.l <- reshape(elec,
+     varying = c("elec2005", "elec2006", "elec2007", "elec2008",
+                 "elec2009", "elec2010", "elec2011", "elec2012"),
+     v.names = "electricity",
+     timevar = "year",
+     times = c("2005", "2006", "2007", "2008",
+               "2009", "2010", "2011", "2012"),
+     direction = "long")
```

The `varying` parameter refers to variables that you want to reshape from wide into long format. Putting it simply, the variables/columns specified in the `varying` option will be converted into a single variable named in the `timevar` parameter. The labels of `varying` variables will be used as categorical values of the new variable specified in `timevar`, but you can re-label them in the `times` parameter. The `v.names` parameter sets the name for the new variable, which will take over the values of average electricity consumption for each year. The resulting `data.frame` will look as follows:

```
> head(elec.1, n=6)
         region prop_type year electricity id
1.2005 E12000001       101 2005    5341.386  1
2.2005 E12000001       102 2005    3840.788  2
3.2005 E12000001       103 2005    3734.703  3
4.2005 E12000001       104 2005    3709.131  4
5.2005 E12000001       105 2005    3337.374  5
6.2005 E12000001       106 2005    3009.375  6
```

We can also do some extra tidying up of the labels for the `region` and `prop_type` variables to make them more human readable and user-friendly. For that reason, we will re-label all values for these two variables according to the data dictionary file for the NEED dataset, available at https://www.gov.uk/government/statistics/national-energy-efficiency-data-framework-need-anonymised-data-2014 (see the *Look up tables* Excel spreadsheet file for details):

```
> elec.1 <- within(elec.1, {
+    region[region=="E12000001"] <- "North East"
+    region[region=="E12000002"] <- "North West"
+    region[region=="E12000003"] <- "Yorkshire and The Humber"
+    region[region=="E12000004"] <- "East Midlands"
+    region[region=="E12000005"] <- "West Midlands"
+    region[region=="E12000006"] <- "East of England"
+    region[region=="E12000007"] <- "London"
+    region[region=="E12000008"] <- "South East"
+    region[region=="E12000009"] <- "South West"
+    region[region=="W99999999"] <- "Wales"
+ })
> elec.1 <- within(elec.1, {
+    prop_type[prop_type==101] <- "Detached house"
+    prop_type[prop_type==102] <- "Semi-detached house"
+    prop_type[prop_type==103] <- "End terrace house"
+    prop_type[prop_type==104] <- "Mid terrace house"
+    prop_type[prop_type==105] <- "Bungalow"
+    prop_type[prop_type==106] <- "Flat (incl. maisonette)"
+ })
> head(elec.1, n=6)
```

```
         region             prop_type year electricity id
1.2005 North East      Detached house 2005    5341.386  1
2.2005 North East Semi-detached house 2005    3840.788  2
3.2005 North East   End terrace house 2005    3734.703  3
4.2005 North East   Mid terrace house 2005    3709.131  4
5.2005 North East            Bungalow 2005    3337.374  5
6.2005 North East Flat (incl. maisonette) 2005 3009.375 6
```

Finally, using the `ggplot2` package, we can visualize the obtained results in a more informative way:

```
> library(ggplot2)
> ggplot(elec.1, aes(x=year, y=electricity, group=factor(prop_type),
colour=factor(prop_type))) +
+   geom_line() + geom_point() +
+   facet_wrap(~region, nrow = 2) +
+   scale_colour_discrete(name="Property Type") +
+   theme(axis.text.x = element_text(angle = 90),
+         panel.grid.major=element_line(colour = "white"),
+         panel.grid.minor=element_blank(),
+         panel.background=element_rect(fill = "#f6f7fb"),
+         strip.background = element_rect(colour = "#f6f7fb", fill =
"#d6e8ff"))
```

The preceding code produces multiple line plots of the average electricity consumption by property type across all years of NEED data, presented for each level of the geographical region variable (`facet_wrap(~region, nrow = 2)`). The final output is displayed in the following figure:

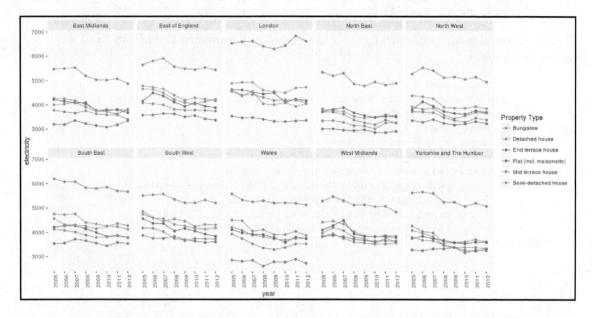

We can clearly observe some patterns in the data, for example that detached houses have been consuming on average much more electricity than other property types across several years, but this doesn't surprise us considering that detached houses usually contain more rooms, are poorly insulated, and generally accommodate more residents. It is, however, striking that in general, most households show a tendency to lower their electricity consumption from year to year. We can only speculate as to what the reason for this behavior might be. Do people become more environmentally friendly? Are we too busy to stay indoors for a long time? Are our houses better insulated? Or maybe we simply use less electricity due to increasing energy prices?

In the next section, we will use R to remotely query another interesting large dataset, but this time it will be stored in a PostgreSQL database on the Amazon **Relational Database Service (RDS)**.

PostgreSQL with R on Amazon RDS

The methods for launching various open-source SQL databases described in the preceding two sections present just one way of connecting R to data stored in these databases. An alternative approach is offered by Amazon RDS – a managed and highly-scalable solution for database management. In fact, Amazon RDS is probably the easiest and fastest method of deploying a fully-operational SQL database, as it requires very minimal input from users. It also allows support for remote connectivity from RStudio, ensuring that users can quickly connect to the database in the cloud from the comfort of their local computer.

Launching an Amazon RDS database instance

Setting up and launching the database instance on Amazon RDS is generally very user friendly. The following instructions will guide you through the process and will let you create a small, Free Usage Tier (as of April 2016) **t2.micro** RDS instance with a PostgreSQL database ready for use:

1. Go to `https://aws.amazon.com` and sign in to the console with the credentials you set up when creating an Amazon AWS account (go back to *Online Chapter, Pushing R Further* (`https://www.packtpub.com/sites/default/files/downloads/5396_6457OS_PushingRFurther.pdf`) for detailed instructions). While in the main dashboard of the console, click on the **Services** tab in the top menu bar, mouse over **Database**, and choose **RDS**, as shown in the following screenshot:

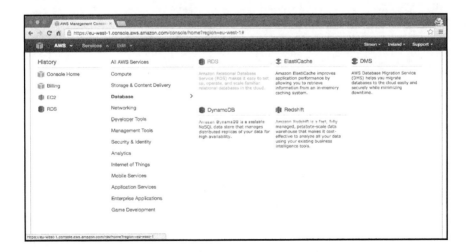

2. In the RDS Dashboard, click on the large button **Get Started Now** button located in the center of the screen:

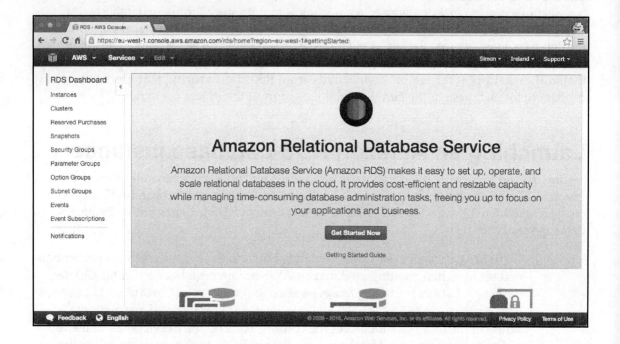

3. This will take you to the next step, in which you can select your preferred database engine. As you can see, there are a number of options available, including databases such as MySQL and MariaDB. This time, however, we will choose **PostgreSQL**– a very popular open-source and reliable relational database management system. Click the **Select** button to confirm your choice:

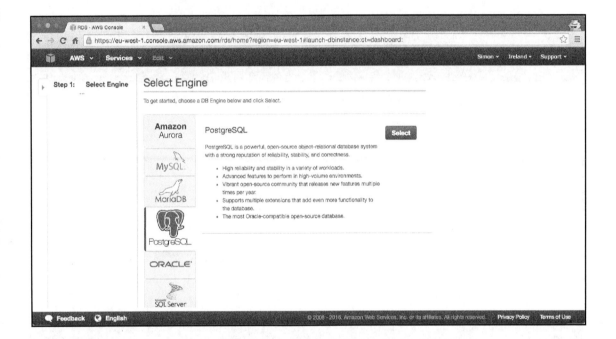

4. On the next screen, you are asked whether you plan to use this database for production purposes. If you intend to use this DB instance only for testing and development work, make sure to select the **Dev/Test** option. Selecting **Production** may end up costing you quite a lot of money, which is probably something you want to avoid when testing Amazon RDS. Click **Next Step** to proceed:

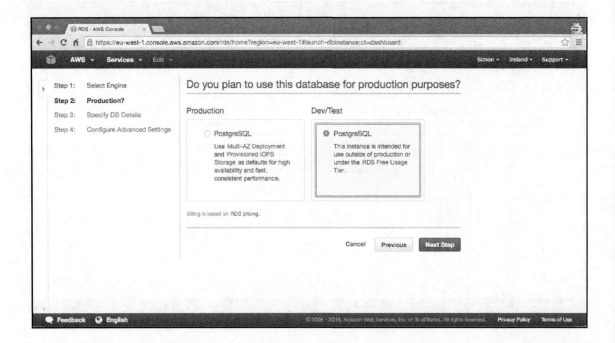

5. We can now specify our database details. If you want to keep it free, make sure you understand the rules and requirements for a Free Tier DB instance. In April 2016, the Amazon RDS Free Tier could be a single db.t2.micro instance with up to 20 GB of data storage. If you choose other specifications beyond the Free Tier limits, you will be charged extra costs – how much depends on several factors, which we explained in *Online Chapter, Pushing R Further* (https://www.packtpub.com/sites/default/files/downloads/5396_6457OS_PushingRFurther.pdf). In the **Instance Specification** form, make sure to keep **postgresql-license** as the chosen **License Model**. We will also stick to the default **9.4.7DB Engine Version**. In the **DB Instance Class**, we will go along with the Free Tier recommendation and will select **db.t2.micro** with only 1 CPU and 1 GB of RAM. It's a very small instance and most likely, for larger datasets you will have to choose a much bigger and stronger DB instance. However, for testing purposes with the sample dataset provided with this chapter, a db.t2.micro instance will work just fine. In the **Multi-AZ Deployment** select the **No** option, keep the **General Purpose (SSD)** as the **Storage Type**, but select 20 **GB** of **Allocated Storage** – it is the upper limit allowed in the Free Tier, as explained in the top section of the screen:

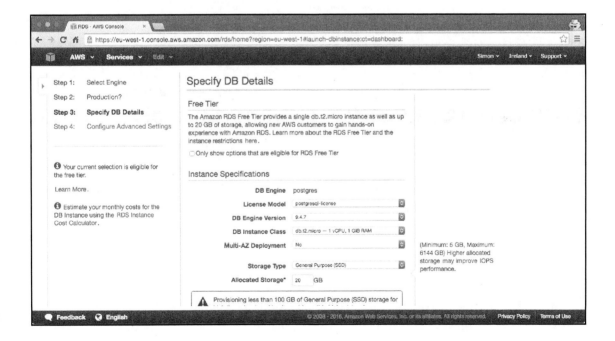

6. In the **Settings** form of the **Specify DB Details** screen, provide a unique **DB Instance Identifier***, **Master Username***, and **Master Password***. We have typed `database1` as the DB Instance Identifier and `swalko` as the Master Username. Make sure you remember the values that you provide in this form, as you will need them when connecting to the database later on. When ready, click the **Next Step** button to proceed:

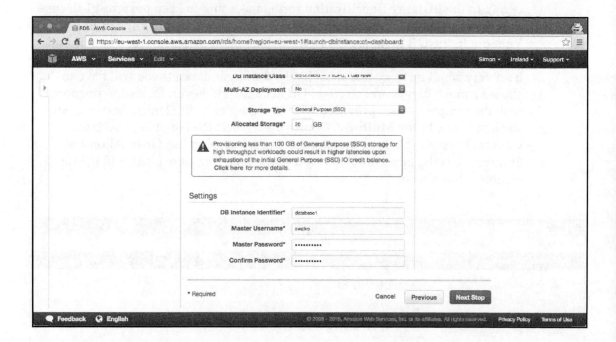

Chapter 5

7. In the **Configure Advanced Settings** screen, complete the **Network & Security** form by choosing **Create new VPC** in the **VPC*** (Virtual Private Cloud) section. Make sure to select **Yes** in the **Publicly Accessible** row. You may leave the rest without any further changes. In the **Database Options** form, provide the **Database Name** (we typed `data1` – remember the name you typed as you will need it later):

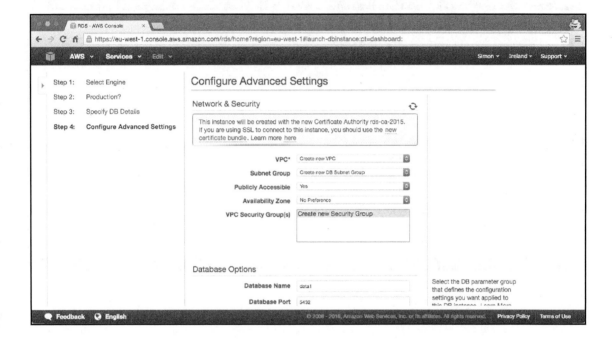

Keep the **Database Port** and all other entries in **Database Options**, **Backup**, **Monitoring**, and **Maintenance** unchanged. Click the **Launch DB Instance** button to complete:

8. At this point, your DB instance will start being created and you will see a confirmation page similar to the following one:

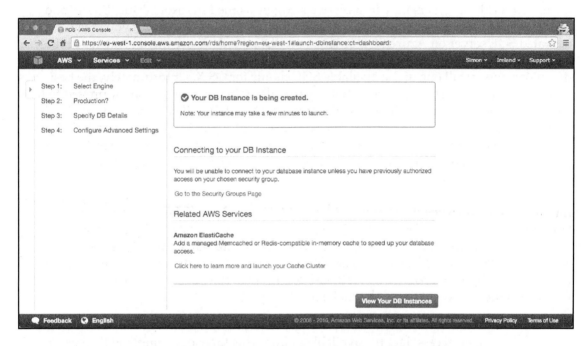

Click on **View Your DB Instances** to be transferred to the **RDS Dashboard**, where you can manage the newly created DB instance. After several minutes, the instance will be activated:

Hopefully, launching the RDS database instance was not too difficult. In the next section, we will prepare the data for use and we will import it into the database using a very useful third-party tool called RazorSQL.

Preparing and uploading data to Amazon RDS

For those users who prefer working with databases on Mac OS X, database management from their local computers may be a bit of a struggle as the Microsoft Visual Studio and SQL Server Management Studio are not supported on OS X. Especially if you create a cloud operated SQL server on Azure SQL or Amazon RDS, the database management activities will soon become quite problematic. Therefore, it is recommended that Mac OS X users download and install one of the available SQL clients for OS X. Unfortunately, the best ones on offer are not free. The SQL client that is particularly interesting and so far proved to be of a very good quality is a Java-based application for Mac OS X called **RazorSQL**, available from `http://www.razorsql.com/`. The software as of April 2016 costs around $100 per user, but you can download a 30-day free trial version to test its functionalities. RazorSQL provides very user-friendly database management capabilities and is supplied with a number of pre-installed connectors, allowing users to quickly connect, insert, and query data to/from a majority of relational (SQL) and non-relational (NoSQL) databases including (but not limited to) MySQL, MariaDB, MonetDB, Cassandra, MS SQL Server, MongoDB, SQL Azure, PostgreSQL, Teradata, SQLite, and many others. The tool is also available for Windows and Linux machines, making it a very useful alternative to SQL Server Management Studio and Microsoft Visual Studio. Having said that, we will prepare and upload the data to the PostgreSQL database on Amazon RDS using the RazorSQL app. The following guidelines will hopefully make this task easy.

In this tutorial, we will be using another sample dataset based on a large, open access data file available online. This time, our sample will come from the *Anonymised MOT tests and results* data provided by the **Driver & Vehicle Standards Agency**, and available at `https://data.gov.uk/dataset/anonymised_mot_test`. The MOT is an annual safety and exhaust emissions check of all vehicles registered to drive in Great Britain. If a vehicle fails the MOT test it may not be used on roads across the whole country. If you are interested in, specifics of the MOT test, feel free to read a bit about it on Wikipedia at `https://en.wikipedia.org/wiki/MOT_test` or just browse the Internet for details on the MOT in Great Britain – you will find a large number of links to further resources.

The complete data file for the year 2013 is 3.41 GB heavy, whereas all the data files for years 2005 through 2013 are bundled into an 8.7 GB heavy ZIP file. Of course, the original files would be far too large to use on our small RDS instance, but we are supplying a small sample of only 100,000 rows in a `mot_small_sample.csv` file (10.9 MB) available to download from the Packt Publishing website created for this book.

Chapter 5

Firstly, we will set up RazorSQL to connect to the Amazon RDS DB instance that we created earlier:

1. Download a version of RazorSQL that is appropriate for your operating system from `http://www.razorsql.com` and install it in the standard way. Upon launching the application, you will see the following screen:

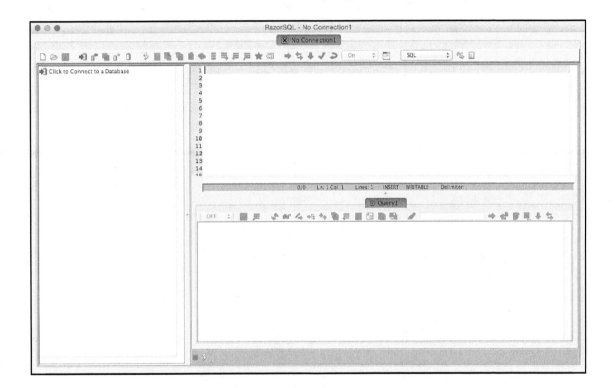

2. In order to initiate a connection between RazorSQL and our database, click on the **Click to Connect to a Database** link in the left-hand side panel. A new, empty **Connection Wizard** window will appear:

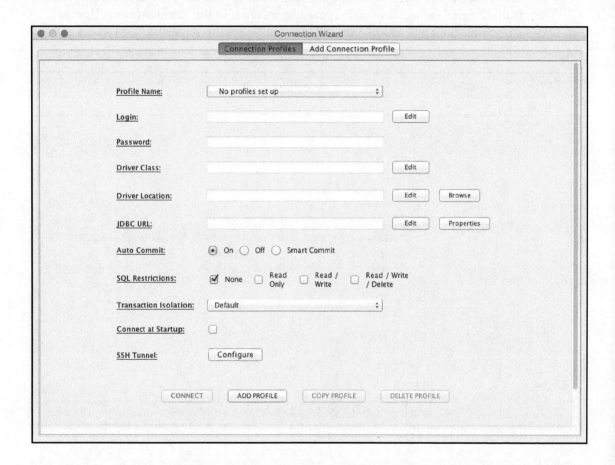

3. Click on the **Add Connection Profile** tab. A new screen with available connectors to a selection of databases will show up. From the list of database types, choose **PostgreSQL** and click the **Continue** button to proceed:

Chapter 5

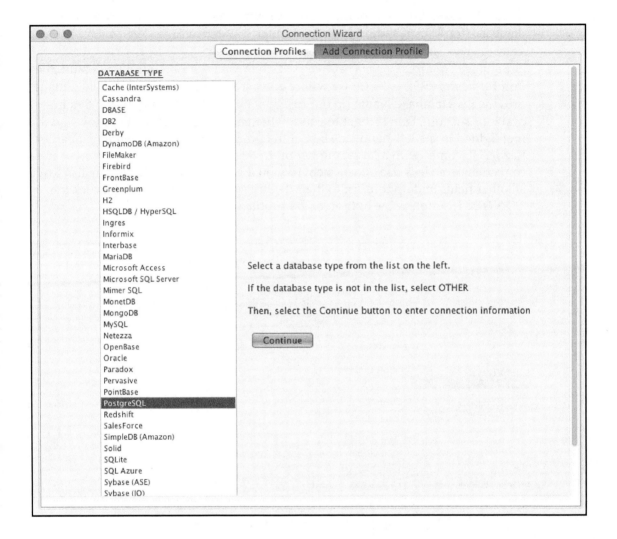

4. You will now be taken to a new **Add Connection Profile** window, which we can populate with information required to create a connection with our PostgreSQL database. First, provide the **Profile Name**, for example **postgresql_rds**. In the **Authentication** form, enter the **Login** and **Password** that you set up for the instance in step 5 in the previous section (our **Master Username** was swalko).

R with Relational Database Management Systems (RDBMSs)

In the **Database Info** part of the profile, provide the **Host or IP Address** of the database (for RDS, it should be in the format: `DBIdentifier.XYS.eu-west-1.rds.amazonaws.com`, in our case it will be `database1.cgsn1orvgmc4.eu-west-1.rds.amazonaws.com`). By default, the **Port** for PostgreSQL is `5432`; we will re-enter it in the corresponding field. Finally, provide the **Database Name** (in our case it was `data1`; as set up in step 6 in the previous section). Don't forget to check whether the **Connection Type** is highlighted in the left-hand side box of the profile window. As mentioned earlier, RazorSQL comes with a large number of drivers already pre-installed for convenience, so let's use one of them (denoted with an asterisk). We can also leave all other fields and options set to their default selections. When ready, click the **CONNECT** button at the bottom of the profile window:

5. After a second or two, RazorSQL will connect to the PostgreSQL database that we configured earlier using Amazon RDS. When it happens, you will see a slightly changed main RazorSQL window with our database data1 visible in the left-hand side directory tree and with the reference to the postgresql_rds connection profile in the top section of the screen:

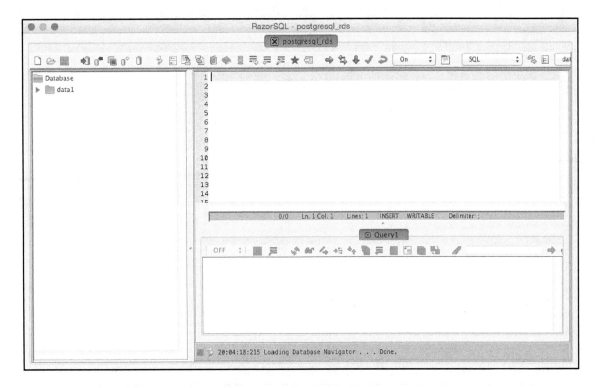

We have now been able to successfully connect to PostgreSQL from the RazorSQL app. Our next task will be to create a table with our sample dataset in CSV format.

6. In order to initialize the process of importing data, highlight the `data1` database and click on the small icon with a blue arrow turned downwards:

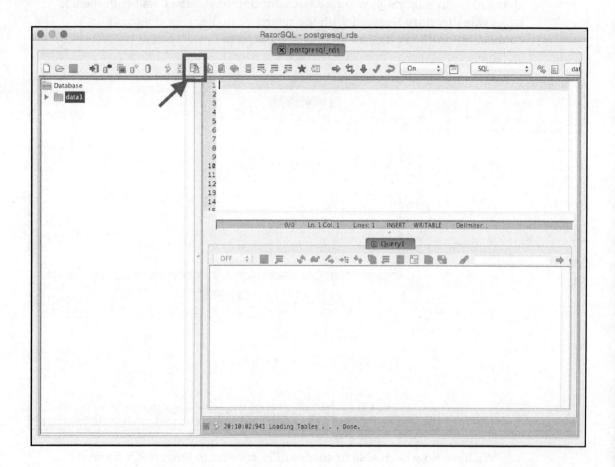

7. A new **Import Data** screen will appear. As our data is comma-delimited and comes in CSV format, select the radio button next to **Import from a delimited file, such as a CSV (comma separated value) file**. In the **Schema** field, keep it as `public`. Finally, click on the radio button next to **Create New Table**, and in the field below provide the name for the table that will hold our data – we will simply call it `mot`. Click the **Next** button to proceed:

8. In the next window, choose **<COMMA>** as the correct **Delimiter** for our data and click the **Browse** button to navigate to the data file on your local computer. As the file comes with a header on the first line that contains variable/column names, we will enter the value 2 in the field next to **Delimited File Start Row** option. We will also tick the boxes next to **Do not include column names in inserts** and **Fill with empty data when not enough values**. When ready, click the **Next** button to proceed:

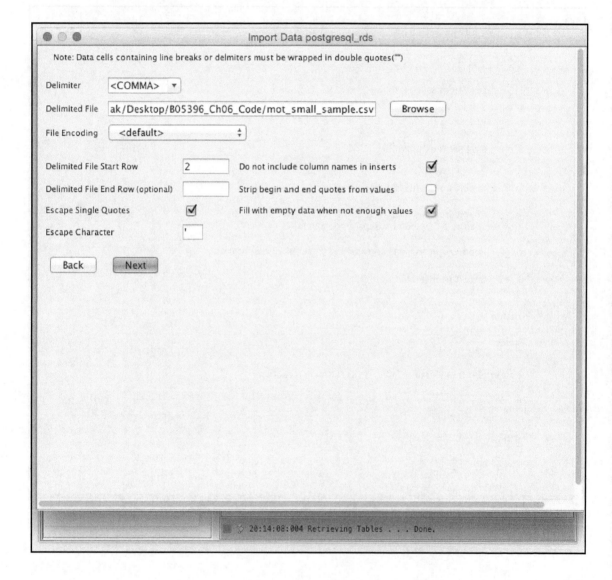

9. Before importing the data into the mot table, RazorSQL will display a **Create Table Tool** window with all the columns found in the data, their names, data types, and lengths:

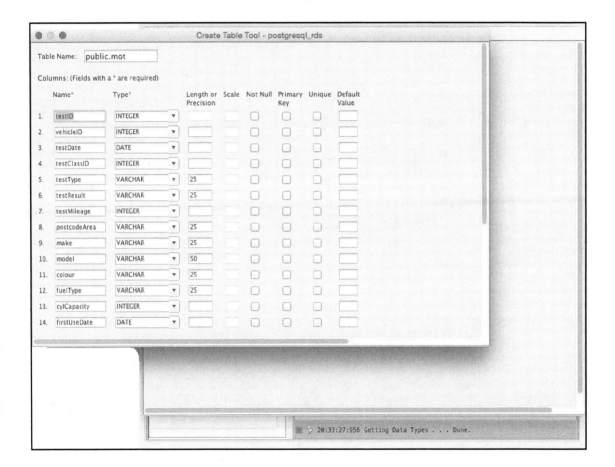

If all seems fine, scroll down, click **Generate SQL**, and then the **Execute SQL** button. The action will be confirmed by a new window. Click **OK** to proceed:

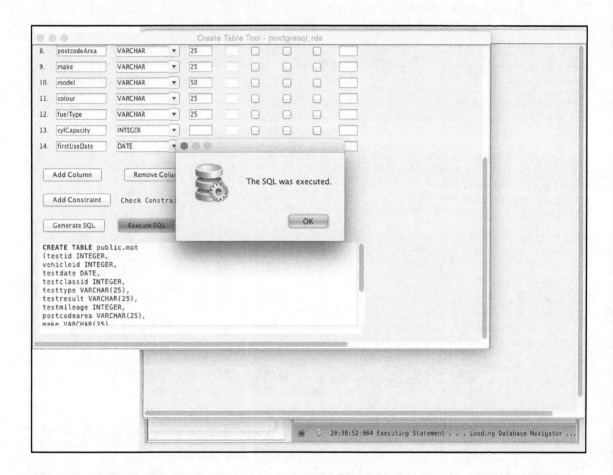

10. The **Import Data** window will appear with all variables to be imported to the table. Click **Next** to continue:

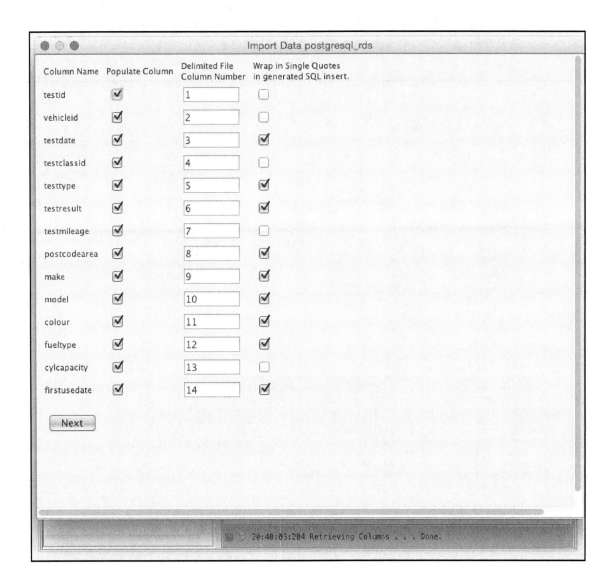

11. Finally, the SQL import code is populated. Select to **Execute as Batch** and keep the **Number of Statements per Network Call** set to **500**. Proceed by clicking on the **Execute** button. The data import job will be initialized (if you wish to save the populated SQL code to a file, make sure to select the **Save to File Only** radio button):

Chapter 5

It may take up to two or three minutes for the data to be imported into the `mot` table in the `data1` PostgreSQL database on Amazon RDS. Once finished, click the **OK** button in the confirmation window message and close the **Import Data** window.

You may verify whether the import process was successful by searching for a `mot` table in the directory tree of our database (the left-hand side box) or by running a very simple query to calculate the total number of records in the table:

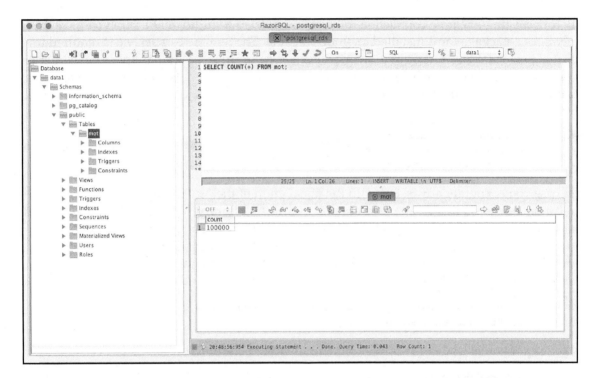

We are nearly ready to connect to the PostgreSQL database from the RStudio installed on our personal, local computer. At this stage, we should only download and install PostgreSQL on the local machine. Depending on your operating system, follow the instructions and guidelines provided at http://www.postgresql.org/download/. Remember that PostgreSQL is currently an open-source and free-of-charge platform, so you don't need to worry about license fees when using this database.

[303]

Remotely querying PostgreSQL on Amazon RDS from RStudio

Provided that you have a recent release of RStudio installed on your computer, along with all the tools and applications required to manage and run the PostgreSQL database on Amazon RDS, you are now ready to remotely crunch, process, query, and analyze the MOT data that we imported earlier.

Before creating a remote connection to PostgreSQL from RStudio, make sure you have the `devtools` package for R installed on your machine. The `devtools` package is required to install new versions of `Rcpp` and `DBI` packages. We performed a similar operation earlier on an Amazon EC2 virtual machine with Ubuntu. For the `Rcpp` and `DBI` installation, we simply need to run the following commands locally in RStudio:

```
> devtools::install_github("RcppCore/Rcpp")
...#output truncated
> devtools::install_github("rstats-db/DBI")
...#output truncated
```

In case you encounter any issues with the installation of the `DBI` package, it is recommended you detach all packages that use the `DBI` library, as well as the old version of any previously installed `DBI` package should you have such. Then re-install the `DBI` package using the preceding command. Once both packages are installed, download and install the `RPostgres` package, which will be used to facilitate the connection between R and, PostgreSQL database:

```
> devtools::install_github("rstats-db/RPostgres")
...#output truncated
```

We may now make both `DBI` and `RPostgres` ready to use in RStudio:

```
> library(DBI)
> library(RPostgres)
```

As we are ready to make a remote connection with the PostgreSQL database, make sure you remember all the credentials that you used when launching the Amazon RDS instance. You will need to provide them in the `dbConnect()` function. For our values, the connection function will look as follows (please adjust the values of all parameters according to your own credentials):

```
> con <- dbConnect(RPostgres::Postgres(), dbname = 'data1',
+                  host = 'database1.cgsn1orvgmc4.eu-west-1.rds.amazonaws.com',
+                  port = 5432,
+                  user = 'swalko',
+                  password = 'Password1')
```

The `conPqConnection` object stores information about RStudio's connection with the PostgreSQL database:

```
> con
<PqConnection> data1@database1.cgsn1orvgmc4.eu-west-1.rds.amazonaws.com:5432
```

We can list all available tables in the database using the standard `dbListTables()` command:

```
> dbListTables(con)
[1] "mot"
```

The output confirms that the data import performed in the previous section was successful and we can access the `mot` table remotely.

The `RPostgres` package is not very good at describing tables; for example, the `dbListFields()` function returns an error. This means that you need to know the structure (schema) of the table before you can query it. It is, however, likely that this function will soon become operational:

```
> dbListFields(con, "mot")
Error in (function (classes, fdef, mtable)  :
  unable to find an inherited method for function 'dbListFields' for signature '"PqConnection", "character"'
```

You may, however, run SQL queries straight from R/RStudio; for example, in the following call we will calculate the frequencies of MOT test results for all makes of vehicles in our sample:

```
> query.1 <- dbSendQuery(con, "SELECT make, testresult,
+                        COUNT(*) AS count
+                        FROM mot
+                        GROUP BY make, testresult
+                        ORDER BY make, testresult ASC")
```

The preceding statement results in the `query.1` object being created, which holds information about the performed query:

```
> query.1
<PqResult>
  SQL  SELECT make, testresult,
              COUNT(*) AS count
              FROM mot
              GROUP BY make, testresult
              ORDER BY make, testresult ASC
  ROWS Fetched: 0 [complete]
       Changed: 0
```

Similar to other R packages that support SQL databases, `RPostgres` evaluates queries lazily. The results sets can be pulled from the database using the familiar `dbFetch()` function:

```
> query.1.res <- dbFetch(query.1, n = -1)
> query.1.res
             make testresult count
1           ABARTH          F     1
2           ABARTH          P     5
3             ADLY          P     1
4            AIXAM          P     5
5              AJS          P     4
6       ALFA ROMEO        ABR     4
7       ALFA ROMEO          F    89
8       ALFA ROMEO          P   175
9       ALFA ROMEO        PRS    27
10         APRILIA        ABR     1
...#output truncated
> dbClearResult(query.1)
[1] TRUE
```

In the second example of a query, we can estimate the frequencies for failed (F) MOT tests only, and we will create a new column with the average mileage (avg_miles) for the aggregated data. This will allow us to find which vehicle makes are more likely to fail their MOT tests earlier:

```
> query.2 <- dbSendQuery(con, "SELECT make, testresult,
+                              COUNT(*) AS count,
+                              AVG(testmileage) AS avg_miles
+                              FROM mot
+                              WHERE testresult = 'F'
+                              GROUP BY make, testresult
+                              ORDER BY avg_miles DESC")
> query.2.res <- dbFetch(query.2, n = -1)
> query.2.res
              make testresult count  avg_miles
1         METROCAB          F     2  318089.50
2         CARBODIES         F     5  268627.80
3   LONDON TAXIS INT        F    36  239215.53
4      ISUZU TRUCKS         F     3  183176.67
5              IVECO        F    53  128849.68
6         IVECO-FORD        F     2  127822.00
7            LEYLAND        F     1  116182.00
8                LDV        F    63  113946.57
9              VOLVO        F   337  110105.41
10          MERCEDES        F   773  106594.80
11              AUDI        F   500  104877.09
12        LAND ROVER        F   468  102991.18
13              SAAB        F   131   99426.88
14        VOLKSWAGEN        F  1820   98700.80
15             ISUZU        F    22   97631.86
16          DAIHATSU        F    49   97255.61
17               BMW        F   783   96176.70
18        MITSUBISHI        F   261   95913.33
19            TOYOTA        F   667   93330.67
20          CHRYSLER        F    74   92662.66
...#output truncated
> dbClearResult(query.2)
[1] TRUE
> dbDisconnect(con)
[1] TRUE
```

The full output contained 128 makes of vehicle that were included in our data sample. Based on the obtained results, we can see that there is a number of low-frequency brands that, on average, fail their MOTs once they reach very high mileage. Most of these makes are purpose-built vehicles for long-range cargo transportation, such as trucks, or frequently used vehicles, such as taxi and cab services.

Among a small number of R packages suitable for querying PostgreSQL databases, the previously introduced `dplyr` deserves separate treatment. It connects to PostgreSQL using the `RPostgreSQL` dependency, so make sure to install this package before installing and loading `dplyr`:

```
> install.packages("RPostgreSQL")
...#output truncated
> library(RPostgreSQL)
> install.packages("dplyr")
...#output truncated
> library(dplyr)
...#output truncated
```

In order to make a connection with a remote database using the `dplyr` package, we need to provide all essential credentials in the `src_postgres()` function:

```
> dpl.conn <- src_postgres(dbname = 'data1',
+                          host = 'database1.cgsn1orvgmc4.eu-west-1.rds.amazonaws.com',
+                          port = 5432,
+                          user = 'swalko',
+                          password = 'Password1')
> dpl.conn
src:  postgres 9.4.7 [swalko@database1.cgsn1orvgmc4.eu-west-1.rds.amazonaws.com:5432/data1]
tbls: mot
```

The `src_postgres` connection object (`dpl.conn`) stores details about the connection and PostgreSQL version, as well as the names of the available tables. They can be retrieved as `tbl_postgres` objects using the `tbl()` function:

```
> mot.data <- tbl(dpl.conn, "mot")
> mot.data
Source: postgres 9.4.7 [swalko@database1.cgsn1orvgmc4.eu-west-1.rds.amazonaws.com:5432/data1]
From: mot [100,000 x 14]
      testid   vehicleid    testdate testclassid testtype testresult
1  295651455   12103486  2013-01-22           4        N          P
2  297385751    2987701  2013-09-11           4        N        PRS
3  302850213   18092246  2013-12-09           4        N        ABR
4  306425393   30931602  2013-11-02           4        N          P
5  211580531   34623039  2013-03-08           4        N          P
6   51149643    9343427  2013-05-08           4       PR          P
7  303246702   18902134  2013-09-20           4        N          F
8   17002174    3420544  2013-03-18           4        N          P
9   68048070   12332736  2013-08-09           4        N          P
10 188393529   31579107  2013-05-03           4        N          P
```

```
            ..       ...       ...        ...       ...       ...
      Variables not shown: testmileage (int), postcodearea (chr), make
          (chr), model (chr), colour (chr), fueltype (chr), cylcapacity
          (int), firstusedate (date)
```

Just like in the preceding examples with `dplyr`, users are only provided with a small snapshot of the full dataset. All queries are evaluated lazily and only explicit calls to pull specific results sets to the R workspace are executed. This saves a tremendous amount of resources and allows users to work on out-of-memory datasets without loading all the data into an R session. You may however look into the data and its structure using the very helpful `glimpse()` function:

```
> glimpse(mot.data)
Observations: 100000
Variables:
$ testid        (int)  295651455, 297385751, 302850213, 30642539...
$ vehicleid     (int)  12103486, 2987701, 18092246, 30931602, 34...
$ testdate      (date) 2013-01-22, 2013-09-11, 2013-12-09, 2013...
$ testclassid   (int)  4, 4, 4, 4, 4, 4, 4, 4, 4, 4, 4, 4, 4, 4,...
$ testtype      (chr)  "N", "N", "N", "N", "N", "PR", "N", "N", ...
$ testresult    (chr)  "P", "PRS", "ABR", "P", "P", "P", "F", "P"...
$ testmileage   (int)  42135, 85557, 0, 74548, 113361, 75736, 95...
$ postcodearea  (chr)  "YO", "TS", "HU", "SR", "DD", "TR", "NW",...
$ make          (chr)  "HYUNDAI", "VOLVO", "RENAULT", "FORD", "S...
$ model         (chr)  "GETZ GSI", "V50 SE", "SCENIC DYN VVT", "...
$ colour        (chr)  "BLUE", "GREEN", "BLACK", "BLUE", "SILVER...
$ fueltype      (chr)  "P", "P", "P", "P", "D", "D", "P", "P", "...
$ cylcapacity   (int)  1341, 1798, 1598, 1596, 1896, 2401, 1975,...
$ firstusedate  (date) 2005-09-09, 2005-09-30, 2007-12-31, 2001...
```

In the first example query, we will calculate the average mileage for failed MOT tests of each make, but we will only extract makes with at least 50 vehicles with failed MOTs. We will also sort the results by mileage in descending order.

Firstly, select only the records with failed MOTs:

```
> mot.failed <- filter(mot.data, testresult == "F")
> mot.failed
Source: postgres 9.4.7 [swalko@database1.cgsn1orvgmc4.eu-
west-1.rds.amazonaws.com:5432/data1]
From: mot [23,594 x 14]
Filter: testresult == "F"
      testid  vehicleid   testdate testclassid testtype testresult
1  303246702   18902134 2013-09-20           4        N          F
2   59387545   10822249 2013-04-26           4        N          F
3     320181     106195 2013-02-11           4        N          F
4  307090812   25673696 2013-11-14           4        N          F
```

```
5   185589215   31186307 2013-06-17            4           N           F
...#output truncated
```

You can see from the output that the number of filtered records decreases from the original 100,000 to 23,594. We can group the filtered `mot.failed` data by the make, as we want to calculate average mileage for each vehicle make. The estimation of mean mileage by make can be achieved using the `summarise()` function:

```
> by.make <- group_by(mot.failed, make)
> avg.mileage <- summarise(by.make,
+                          count = n(),
+                          avg = mean(testmileage))
```

Finally, we can subset only those makes that had at least 50 failed MOTs and sort them by average mileage in descending order:

```
> avg.mileage <- arrange(filter(avg.mileage, count >= 50), desc(avg))
> avg.mileage
Source: postgres 9.4.7 [swalko@database1.cgsn1orvgmc4.eu-
west-1.rds.amazonaws.com:5432/data1]
From: <derived table> [?? x 3]
Filter: count >= 50
Arrange: desc(avg)
         make count       avg
1       IVECO    53 128849.68
2         LDV    63 113946.57
3       VOLVO   337 110105.41
4    MERCEDES   773 106594.80
5        AUDI   500 104877.09
6  LAND ROVER   468 102991.18
7        SAAB   131  99426.88
8  VOLKSWAGEN  1820  98700.80
9         BMW   783  96176.70
10 MITSUBISHI   261  95913.33
..         ...   ...       ...
```

As previously mentioned, we may obtain detailed insights into how `dplyr` translates, plans, and executes the SQL queries with the `explain()` function:

```
> explain(avg.mileage)
<SQL>
SELECT "make", "count", "avg"
FROM (SELECT "make", count(*) AS "count", AVG("testmileage") AS "avg"
FROM "mot"
WHERE "testresult" = 'F'
GROUP BY "make") AS "_W1"
WHERE "count" >= 50.0
```

```
      ORDER BY "avg" DESC
      <PLAN>
      Sort  (cost=2694.05..2694.14 rows=36 width=47)
        Sort Key: (avg(mot.testmileage))
        ->  HashAggregate  (cost=2692.13..2692.76 rows=36 width=11)
              Group Key: mot.make
              Filter: ((count(*))::numeric >= 50.0)
              ->  Seq Scan on mot  (cost=0.00..2456.00 rows=23613 width=11)
                    Filter: ((testresult)::text = 'F'::text)
```

At the end of data processing, we can simply *fetch* the data into the R session in the form of a `tbl_df` object, using `collect()`:

```
> mileage.df <- collect(avg.mileage)
> mileage.df
Source: local data frame [36 x 3]
       make count       avg
1     IVECO    53 128849.68
2       LDV    63 113946.57
3     VOLVO   337 110105.41
4   MERCEDES  773 106594.80
5      AUDI   500 104877.09
...#output truncated
```

In the second example of a SQL query run on the PostgreSQL database with `dplyr`, we will calculate the average age and mileage of vehicles that either passed or failed the MOT tests for each make. As before, we will only show makes with at least 50 vehicles for each group and we will order makes alphabetically. This is a trickier exercise than the one carried out previously, as it requires us to know how to calculate the elapsed time between two dates in PostgreSQL and also how to apply inner joins.

It might be that your connection with the database drops after several minutes of inactivity. Make sure then to re-run the following commands to restart the communication with the PostgreSQL server:

```
> dpl.conn <- src_postgres(dbname = 'data1',
+                          host = 'database1.cgsn1orvgmc4.eu-
west-1.rds.amazonaws.com',
+                          port = 5432,
+                          user = 'swalko',
+                          password = 'Password1')
> mot.data <- tbl(dpl.conn, "mot")
```

Once we have the connection back, we can filter the data to select only the records with either passed or failed MOT tests:

```
> mot.pf <- filter(mot.data, testresult == "F" | testresult == "P")
> mot.pf
Source: postgres 9.4.7 [swalko@database1.cgsn1orvgmc4.eu-west-1.rds.amazonaws.com:5432/data1]
From: mot [91,912 x 14]
Filter: testresult == "F" | testresult == "P"
    testid   vehicleid   testdate   testclassid testtype testresult
1 295651455   12103486 2013-01-22             4        N          P
2 306425393   30931602 2013-11-02             4        N          P
3 211580531   34623039 2013-03-08             4        N          P
4  51149643    9343427 2013-05-08             4       PR          P
5 303246702   18902134 2013-09-20             4        N          F
...#output truncated
```

We may now focus our attention on creating a new variable (age in days), which will simply be the difference between two dates: the day of the MOT test (testdate), and the day of the first use of a particular vehicle (firstusedate). Note that in our query, we explicitly make a reference to date as a data type for both date variables of interest. The sql() function within the tbl() command allows us to run a specific SQL query using dplyr:

```
> age <- tbl(dpl.conn, sql("SELECT testid, vehicleid, testdate::date - firstusedate::date as age from mot"))
> age
Source: postgres 9.4.7 [swalko@database1.cgsn1orvgmc4.eu-west-1.rds.amazonaws.com:5432/data1]
From: <derived table> [?? x 3]
    testid   vehicleid  age
1 295651455   12103486 2692
2 297385751    2987701 2903
3 302850213   18092246 2170
4 306425393   30931602 4346
5 211580531   34623039 3812
...#output truncated
```

The resulting object age is in fact a three-column wide tbl_postgres data structure, which we can subsequently join (by common variables of testid and vehicleid) with the filtered mot.pf subset that contains both passed and failed MOT test records. This can be achieved using the inner_join() function:

```
> mot.combined <- inner_join(mot.pf, age, by = c("testid", "vehicleid"))
```

As we wish to calculate average mileage and age of passed and failed vehicles for each brand, we need to group the `mot.combined` object by `make` and `testresult` variables, using the `group_by()` function:

```
> by.maketest <- group_by(mot.combined, make, testresult)
```

We may now calculate the statistics of interest:

```
> avg.agemiles <- summarise(by.maketest,
+                           count = n(),
+                           age = mean(age/365.25),
+                           mileage = mean(testmileage))
```

Finally, we can filter the records of our results set with only 50 entries for each aggregation and order all makes alphabetically:

```
> avg.agemiles <- arrange(filter(avg.agemiles, count >= 50),
desc(make))
> avg.agemiles
Source: postgres 9.4.7 [swalko@database1.cgsn1orvgmc4.eu-
west-1.rds.amazonaws.com:5432/data1]
From: <derived table> [?? x 5]
Filter: count >= 50
Arrange: desc(make)
Grouped by: make
          make testresult count        age   mileage
1   ALFA ROMEO          F    89  10.766921  85060.93
2   ALFA ROMEO          P   175   9.756464  73652.82
3      APRILIA          P    57   9.584015  15558.89
4         AUDI          F   500   9.880728 104877.09
5         AUDI          P  2055   7.757449  83533.96
6          BMW          F   783  10.158958  96176.70
7          BMW          P  2920   8.554350  78044.34
8    CHEVROLET          F    51   6.279395  47046.96
9    CHEVROLET          P   253   5.625667  35588.53
10    CHRYSLER          P   203   9.221135  81411.67
..         ...        ...   ...        ...       ...
```

As we might expect, the average age and mileage of the vehicles that failed their MOT tests is generally much greater than for the vehicles of the same makes that passed their MOTs. However, the values for both measures differ from make to make, meaning that some vehicles, for example Chevrolets, are much more likely to fail MOT inspections earlier and with much lower mileage than some other brands, for example Alfa Romeo or BMW.

If you want to look into the complexity of such a calculation in SQL, you are more than welcome to run the `explain()` function. As the output of this call would be quite long, it won't be included in this section:

```
> explain(avg.agemiles)
...#output truncated
```

This exercise completes the chapter on using R with SQL databases. Although we had to be quite selective about which R packages and open source databases to use, hopefully the tutorials included in this chapter have allowed you to gain an array of skills that will help you in storing and processing Big Data in Relational Database Management Systems.

Summary

We began this chapter with a very gentle introduction to Relational Database Management Systems and the basics of Structured Query Language, in order to equip you with the essential skills required to manage RDBMSs on your own.

We then moved on to practical exercises that let you explore a number of techniques of connecting R with relational databases. We first presented how to query and process data locally using a SQLite database, then we thoroughly covered connectivity with MariaDB (and also MySQL, as both are very similar) installed on an Amazon Elastic Cloud Computing instance, and finally we remotely analyzed the data stored and managed in the PostgreSQL database through the Amazon Relational Database Service instance.

Throughout the sections and tutorials of this chapter, you have learned that R can be conveniently used as a tool for the processing and analysis of large, out-of-memory collection of data stored in traditional SQL-operated databases.

In the next chapter, we will continue the topic of R's connectivity with databases, but this time we will explore a diverse world of non-relational, NoSQL databases, such as MongoDB and HBase.

6
R with Non-Relational (NoSQL) Databases

In the previous chapter we showed that R connects very well with traditional, relational, SQL databases. They are still used in the majority of business scenarios, especially when dealing with standard, two-dimensional, rectangular data. However, recently a new range of non-relational or NoSQL databases have been rapidly emerged mostly in response to the growing ecosystem of applications that collect and process different types of data with more flexible, or no, schema. In these times of dynamic development of the Internet of Things, such databases are of particular interest. The growth of many NoSQL databases, especially the open source ones, are also passionately supported by extremely vibrant and dynamic communities of developers, many of whom are R users at the same time. In this chapter, we will guide you through a number of tutorials to help you achieve the following objectives:

- Understanding data models, basic NoSQL commands, and the aggregation pipeline framework of MongoDB
- Installing and running MongoDB on a Linux Ubuntu virtual machine
- Connecting to MongoDB on EC2 and processing data (including complex NoSQL queries) using `rmongodb`, `RMongo` and `mongolite` packages for R
- Using the HBase database on HDInsight Microsoft Azure as a source for processing and analysis of large datasets using R

We will begin with a gentle introduction to NoSQL databases, what they are, and examples of their use cases, and then we will proceed to practical tutorials on how to launch and operate a very popular non-relational database system called MongoDB. We will end the chapter with a quick guide on how to install and connect to HBase.

Introduction to NoSQL databases

We already know basic features and characteristics of the traditional **Relational Database Management Systems (RDBMSs)**, which we presented in Chapter 5, *R with Relational Database Management Systems (RDBMSs)*. We are also well aware of their limitations and specific requirements, for example, that they contain predefined schema, are vertically scalable which results in constant hardware upgrades to catch up with data growth, and they generally do not support unstructured or hierarchical data.

Non-relational or NoSQL databases attempt to fill these gaps and some of them are specialized in certain aspects more than others. In the next section, we will briefly present several NoSQL databases and their particular use cases.

Review of leading non-relational databases

To say that NoSQL databases are what SQL databases are not, may be a bit of an over-simplification. However, this statement is true to some extent and the following characteristics can shed some light on differences between non-relational and relational databases for those data scientists who have just begun their adventure with databases:

1. In general, NoSQL databases are *non-relational* and *distributed* collections of *documents*, *key-value pairs*, *XML*, *graphs* or other data structures with no standard, predefined schema. In simple words, they may not follow a typical two-dimensional, rectangular structure of a standard SQL table. This feature gives NoSQL databases a real advantage over relational stores on the basis of much greater flexibility and generally improved performance.
2. They are designed to handle massive amounts of data as they tend to be *horizontally scalable*. The number of servers they use can be easily added (or reduced) depending on the actual traffic and processing load. This flexibility is extremely valued in the industry, where the needs of specific applications can vary daily and sometimes even more frequently.
3. These massive amounts of information that NoSQL databases are trying to organize and process can be of differing types, formats, speed, length, and many more. Many non-relational databases specialize in *unstructured data*, time series, textual data, and others, all coming from a variety of data sources and a large pool of available formats. We also can't forget to mention that some NoSQL databases have already become industry-standard solutions for *(near) real-time analytics* and *streaming data processing*—especially in finance, banking, retail, and a wide selection of modern social media applications, where the speed of processing of large quantities of data is of utmost importance.

4. A possible trade-off of the above advantages may be the *expressiveness of NoSQL queries*. Firstly, the commands and methods used for managing and querying data may vary quite considerably from one NoSQL database to another making it difficult for data scientists or even database architects to move quickly between available database systems. Secondly, in general, the NoSQL flavors are less expressive than standard SQL queries, which simply means that complex queries are not handled very well in NoSQL. However, this aspect is changing and some non-relational databases (for example, MongoDB) already offer quite dynamic, fast, and expressive NoSQL implementations.

As we mentioned before, some of the NoSQL databases specialize in different aspects of data management and processing. Below we present selected non-relational databases that are considered the leaders in the Big Data industry, including two databases that we will be exploring further in this chapter during the practical tutorials, **MongoDB** and **HBase**:

- **MongoDB** (https://www.mongodb.org/): A NoSQL document-based database. Its name comes from *humongous* – a reference to the ease of Big Data processing through horizontal scalability. MongoDB's documents are stored in BSON format (a binary representation of popular JSON) in *collections*, which are groups of related documents that contain shared indices. It features a dynamic schema, which allows flexibility without altering the data. Its query language is probably not as simple as SQL, but it's quite expressive. Additionally, multiple drivers allow the use of numerous programming languages including R. Speed of processing makes it a good choice for (near) real-time analytics – many MongoDB users come from retail, marketing, and financial services. MongoDB is an open source project, with great community support, and is extremely popular among application and web developers as well as in business.
- **HBase** (http://hbase.apache.org/): A non-relational, key-value database offering a highly-optimized storage with no default schema. As an open source Apache project, HBase runs on top of the Hadoop Distributed File System and often works as the input or output for Hadoop MapReduce jobs. Although HBase cannot be queried using SQL commands, it provides connectivity through the Java API and also other APIs, for example, REST, Avro, and Thrift, allowing developers to connect and manage data using other languages, for example R. The HBase database is well suited for fast, high throughput read/write operations on large pools of unstructured or multi-dimensional data, for example, in global messaging services or large-scale time series analysis.

- **CouchDB** (http://couchdb.apache.org/): Yet another open source Apache project, this time a document-based database, quite similar in its functionalities to MongoDB. However, it differs from MongoDB in how it stores the data (eventual consistency in CouchDB versus strict consistency in MongoDB) and applications in which it can be used. CouchDB is more preferred for mobile applications or offline tools which need to sync to update the changes, whereas MongoDB is suitable for processing on the server. Also, many users find the querying of data much easier and more user-friendly in MongoDB due to its resemblance to SQL. On the other hand, CouchDB requires creating separate MapReduce jobs for each view.
- **Cassandra** (http://cassandra.apache.org/): Again an Apache project, Cassandra is a Java-based open source, fault-tolerant, non-relational database that just like HBase, belongs to the family of key-value stores. It is very well-known and popular for its proven high performance and is used by leading technology giants including Apple, Netflix, Instagram, and eBay. It uses its own **Cassandra Query Language** (**CQL**), which consists of SQL-like commands, to query the data. Through a number of drivers, it can be easily connected to other applications or programming languages, including R.
- **Neo4j** (http://neo4j.com/): Probably the most popular example of a NoSQL database that stores data in the form of graphs – displaying relationships between labeled or unlabeled nodes and edges and their attributes. Due to a lack of indexed adjacencies, it's fast and highly scalable. It's used in social networks, graph-based search engines, graph visualizations, and fraud detection applications.

The presented five databases are very popular choices amongst database engineers and architects, but of course there are many other NoSQL databases that may be even more suitable for your individual business and research purposes. Many of them are being developed by very vibrant communities of contributors and the most commonly used NoSQL databases generally connect very well with R and other languages. As an example of such connectivity, in the next section we will guide you through a practical tutorial on how to query data stored in MongoDB using R.

MongoDB with R

After the short introduction provided earlier, you should now be able to define the basic characteristics of a variety of NoSQL databases. In this part of the book, we will explore the features and practical applications of MongoDB.

Introduction to MongoDB

MongoDB is one of the examples of non-relational data storage systems, and it also supports a number of data processing and analytics frameworks such as complex aggregations and even MapReduce jobs. All these operations are carried out by the means of MongoDB NoSQL queries – an alternative to the standard SQL language for querying relational databases. As you will soon find out, MongoDB NoSQL commands are very expressive and quite simple to learn. The only problem that most users encounter is the quite convoluted syntax (**BSON** format) for complex aggregations and queries, but we will explore this issue in the following sections.

MongoDB data models

One of the reasons for this difficulty in writing very complex aggregations using MongoDB's flavor of NoSQL is the fact that it stores the data and its records, or we should say **collections** and **documents**, to follow MongoDB's naming conventions, in a BSON format. This, however, allows users to flexibly design data structures depending on the actual relationships between documents in the application. MongoDB offers two **data models** that represent these relationships:

- References, also known as normalized data models
- Embedded documents

References (or **normalized data models**) may contain several separate documents that can be linked together using shared indexes. In the example depicted in the following figure, the collection designed as a normalized data model consists of four separate documents that hold topically unrelated information about a dummy medical patient. These documents can be extracted and linked to one another through a shared index of `ID1`, which is the main identification number for the document storing the first name and the surname of a patient:

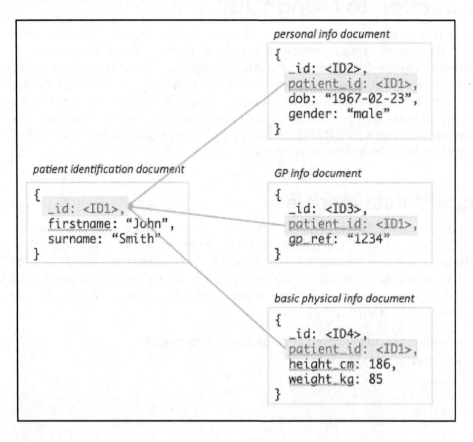

The normalized data model allows users to easily attach additional documents to a particular patient as long as they can be linked using a common ID. This flexibility in designing the structure of documents within a collection also means that, for large datasets, the querying may be restricted to specific, requested documents, and it doesn't require processing the whole dataset. This, on the other hand, may considerably improve the speed and performance of operations.

Another advantage of references in MongoDB is that the data can be continually and dynamically updated with new incoming information without rebuilding the schema of the collection whenever new data is to be added. Such an approach is particularly useful in modern, near real-time applications, which are susceptible to frequent updates and alterations in their structures. The obvious drawbacks of normalized data models are that by creating several linked documents, certain fields (for example `patient_id` in our preceding example) are duplicated, which results in greater resources being used to store the data and may also cause a potential latency in retrieving the linked documents due to the time spent on the search of shared indices.

If the two disadvantages of normalized data models are too much of a compromise, users can turn to more traditional embedded documents. An example of an embedded data model based on the previously used dummy medical patient records can be seen in the following figure:

```
{
    _id: <ID1>,
    firstname: "John",
    surname: "Smith",

    personal: {
                dob: "1967-02-23",
                gender: "male"
              },

    gp: {
          gp_ref: "1234"
        },

    physical: {
                height_cm: 186,
                weight_kg: 85
              }
}
```

As the name suggests, in the embedded model all individual documents are embedded into one document and cannot be separated without physically changing the schema of a collection. As long as the structure of the documents is consistent throughout and at different points of data collection, this design should not cause any issues. However, if the data requirements change, the whole collection and its data structure has to be re-built to reflect the modifications.

Knowing the basics of two models representing data relationships in MongoDB, you may now take a step further to learn how to launch MongoDB on the server as well as how to process and query the data using MongoDB's implementation of SQL-like commands.

Installing MongoDB with R on Amazon EC2

In this section we will install MongoDB database on an Amazon EC2 instance with Linux Ubuntu as an operating system. In order to successfully achieve this goal, you should first launch a new (for example, a Free Tier) EC2 instance with Ubuntu. There are two ways of doing it. You can either go back to *Online Chapter, Pushing R Further* (https://www.packtpub.com/sites/default/files/downloads/5396_6457OS_PushingRFurther.pdf), and follow its instructions to start a fresh instance, or alternatively you may just re-use the virtual machine that we created in the preceding chapter. If you go for the first solution you must also install **Java Development Kit (JDK)** and the `rJava` package for R in the same way we did in `Chapter 5`, *R with Relational Database Management Systems (RDBMSs)*. If you followed the contents of this book, you should already know how to do this. However, it may be easier for you to re-use the instance from the previous chapter as it is partially pre-configured to support communication between MongoDB and R – it already contains most of the essential Ubuntu libraries and several useful R packages. Let's then make a shortcut to save time and let's assume that we will continue working on the Amazon EC2 virtual machine from the previous chapter.

If you stopped the instance at the end of the previous chapter, make sure to restart it before continuing. Upon restart of the virtual machine, it will be assigned a new public DNS address which you should use to `ssh` into the instance as the `ubuntu` user. Open a new Terminal/shell window and `cd` to a directory where you are storing the key pair `pem` file for the instance (in our case it was the `Downloads` folder and the filename was `rstudio_mariadb.pem`):

```
$ cd Downloads/
$ ssh -i "rstudio_mariadb.pem" ubuntu@ec2-52-19-205-207.eu-west-1.compute.amazonaws.com
```

Type `yes` and confirm by pressing the Enter key when prompted. You should now see the instance welcome message with basic information about the resources being used by the virtual machine.

At this stage we need to import the public key for MongoDB. In our data processing activities within this chapter we will be using the 3.2 version of MongoDB Community Edition, which is an open source distribution of the database:

```
$ sudo apt-key adv --keyserver hkp://keyserver.ubuntu.com:80 --recv EA312927
```

After issuing the preceding command you should see a similar output to the following one:

```
Executing: gpg --ignore-time-conflict --no-options --no-default-keyring --homedir /tmp/tmp.KBWdrMQVH6 --no-auto-check-trustdb --trust-model always --keyring /etc/apt/trusted.gpg --primary-keyring /etc/apt/trusted.gpg --keyring /etc/apt/trusted.gpg.d/webupd8team-java.gpg --keyserver hkp://keyserver.ubuntu.com:80 --recv EA312927
gpg: requesting key EA312927 from hkp server keyserver.ubuntu.com
gpg: key EA312927: public key "MongoDB 3.2 Release Signing Key <packaging@mongodb.com>" imported
gpg: Total number processed: 1
gpg:               imported: 1  (RSA: 1)
```

We may now create a list file for MongoDB for our specific distribution of Ubuntu – we are using the 14.04 version:

```
$ echo "deb http://repo.mongodb.org/apt/ubuntu trusty/mongodb-org/3.2 multiverse" | sudo tee /etc/apt/sources.list.d/mongodb-org-3.2.list
```

Once complete, we should update the local Ubuntu libraries:

```
$ sudo apt-get update
...#output truncated
```

After several seconds we are ready to install the recent stable version of MongoDB:

```
$ sudo apt-get install -y mongodb-org
```

It may take another several seconds for the MongoDB installation files to download and install on the Ubuntu Amazon EC2 instance. Once the installation process is finished, you may issue the following command to start MongoDB:

```
$ sudo service mongod start
```

R with Non-Relational (NoSQL) Databases

A similar output to the following line should follow:

```
mongod start/running, process 1211
```

Don't worry if your output contains a different process number. The digits at the end of the output are simply dependent on the number and the sequence of processes in operation on your instance.

You can check what version of MongoDB is currently installed on your virtual machine and on which port it is running by opening the `mongod.log` file in the Nano editor:

```
$ nano /var/log/mongodb/mongod.log
```

While in the Nano editor, you should see the following entries:

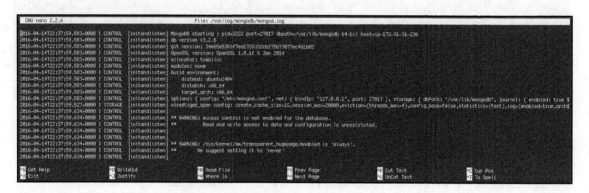

Among several useful details provided in the file, from its contents we can clearly see that our MongoDB is listening on port `27017`, which is the default port for MongoDB, and its database version is 3.2.5. To leave the Nano editor press *Ctrl* + *X* and, as we didn't edit any existing entries in the file, press *N* to close the editor without saving any changes.

To enter the MongoDB shell, which you can use to run all queries or data management processes, just issue the `mongo` command:

```
$ mongo
```

Upon login you will see a number of welcome and warning notifications, as shown in the following output:

```
MongoDB shell version: 3.2.5
connecting to: test
Server has startup warnings:
2016-04-14T22:37:59.634+0000 I CONTROL  [initandlisten]
2016-04-14T22:37:59.634+0000 I CONTROL  [initandlisten] ** WARNING: Access control is not enabled
r the database.
2016-04-14T22:37:59.634+0000 I CONTROL  [initandlisten] **          Read and write access to data
d configuration is unrestricted.
2016-04-14T22:37:59.634+0000 I CONTROL  [initandlisten]
2016-04-14T22:37:59.634+0000 I CONTROL  [initandlisten]
2016-04-14T22:37:59.634+0000 I CONTROL  [initandlisten] ** WARNING: /sys/kernel/mm/transparent_hugep
age/enabled is 'always'.
2016-04-14T22:37:59.634+0000 I CONTROL  [initandlisten] **          We suggest setting it to 'never'
2016-04-14T22:37:59.634+0000 I CONTROL  [initandlisten]
2016-04-14T22:37:59.634+0000 I CONTROL  [initandlisten] ** WARNING: /sys/kernel/mm/transparent_hugep
age/defrag is 'always'.
2016-04-14T22:37:59.634+0000 I CONTROL  [initandlisten] **          We suggest setting it to 'never'
2016-04-14T22:37:59.634+0000 I CONTROL  [initandlisten]
>
```

To exit the MongoDB shell simply use the `quit()`, command:

```
> quit()
```

Note that the command prompt in MongoDB looks the same as in the R console. Also, it is interesting that MongoDB functions, for example `quit()` very much resemble the functions from R.

Processing Big Data using MongoDB with R

At this stage we may start to think about connecting MongoDB with R. If you are running the EC2 instance from the previous chapter it should already have the RStudio Server environment installed. If you have just created a new Amazon virtual machine, make sure to follow the instructions from the previous chapter and prepare your instance, R, and RStudio Server to include the JDK and `rJava` package for R. Also, ensure that both `libssl` and `libsasl2` libraries for Ubuntu are installed. In case both libraries are not present, you can install them by issuing the following commands:

```
$ sudo apt-get install libssl-dev libsasl2-dev
...#output truncated
```

Once the preceding installations complete we may now configure our R environment by installing several R packages that will be used for carrying out data processing activities on a data stored and managed in MongoDB directly from the R console. These packages will include `mongolite`, `rmongodb`, and `RMongo`:

```
$ sudo Rscript -e 'install.packages(c("mongolite", "RMongo", "rmongodb"),
repos="https://cran.r-project.org/")'
...#output truncated
```

Also, we can install other R packages which may prove useful later (note that some of these packages have probably already been installed by you if you are using the same EC2 instance as earlier in Chapter 5, *R with Relational Database Management Systems (RDBMSs)*):

```
$ sudo Rscript -e 'install.packages(c("Rcpp", "RJSONIO", "bitops",
"digest", "functional", "stringr", "psych", "plyr", "reshape2", "caTools",
"R.methodsS3", "Hmisc", "memoise", "lazyeval", "rjson", "ggplot2",
"jsonlite", "data.table", "lubridate"), repos =
"http://cran.r-project.org/")'
...#output truncated
```

The installation process should take a few minutes. After it completes, you should have MongoDB and RStudio Server fully configured and ready to use.

The next section will guide you through essential steps of importing data into MongoDB and will also be a very good opportunity to practice querying and processing data using MongoDB shell commands.

Importing data into MongoDB and basic MongoDB commands

While connecting through `ssh` to the instance as the `ubuntu` user, log in to MongoDB and view current databases and collections:

```
> show dbs
local    0.000GB
```

As we have just installed the MongoDB database, there is currently no data in either the local database or any other database. Exit MongoDB and import the data:

```
> quit()
```

The data that we will be using in this tutorial is the **Land Registry Price Paid Data**, which is regularly published by the Land Registry in the United Kingdom and is available for public use under the Open Government Licence. According to the description of the original data downloadable from `https://data.gov.uk/dataset/land-registry-monthly-price-paid-data`, the **Price Paid Data** (**PPD**) contains prices of residential properties sold in England and Wales and other essential information about these properties such as their tenure (freehold/leasehold), transaction date, specific address of the property (for example street name, postcode, town, country, and other details), the type of the property for example whether a detached, terraced house or a flat, and several other variables. The original full data includes nearly 25 million records (as of April 2016) and covers the period from January 1995 until the current month. The data files are being updated monthly. As the overall size of the full dataset is larger than the Free Tier allowances on a `t2.micro` Amazon EC2 instance, we will be using the PPD Data for the year 2015 only. It will provide us with a complex enough variety of variables and types of data. You can download the 2015 Price Paid Data from `https://data.gov.uk/dataset/land-registry-monthly-price-paid-data`. Scroll down to the section with files covering 2015 and choose **2015 Price Paid Data – YTD** in `CSV` format. Click on the arrow to the right and then on the **Download** button from the drop-down menu. The 2015 PPD file is 169 MB in size and contains just under one million rows.

Once you download the data to a memorable directory, in the new Terminal window `cd` to the location where you store the key pair `pem` file to the virtual machine (for example, `Downloads` folder):

```
$ cd Downloads/
```

Copy the downloaded data file from the known location (for example the `~/Desktop/B05396_Ch07_Code` directory) to the local area of your preferred Amazon EC2 instance, which you might have already set up if you followed the instructions from `Chapter 5`, *R with Relational Database Management Systems (RDBMSs)*. In our case, the username is `swalko` (please alter the paths and names of files/user accordingly for the following code to work):

```
$ scp -r -i "rstudio_mariadb.pem" ~/Desktop/B05396_Ch07_Code/pp-2015.csv
swalko@ec2-52-19-205-207.eu-west-1.compute.amazonaws.com:~/
```

It should take up to a few minutes (depending on the quality of your Internet connection) to upload the data from the local machine to the EC2 instance.

Secondly, because the data file lacks any variable names, we have to provide them in a new file called `pp-2015-variables.csv`, which we have prepared for your convenience and stored in a folder with R scripts and data files that you can download from the Packt Publishing website created for this book.

Preparing a file with variable names

If you want to prepare a similar file for your own data in the future, make sure to put each variable name on a single line only and save it (for example using Microsoft Excel, but you can also do it efficiently in R) to a comma-separated values format (`CSV`). If you are using a non-Windows machine, ensure that you save the file in the Windows comma-separated format – it is a standard Windows-friendly `CSV` file. It is important as there seems to be an issue that tends to skip all but the last variable name in files created in Excel on non-Windows operating systems. This happens even if they are completely fine and operational standard format files such as CSV or TXT.

Once ready, copy the variable names file to `swalko` (or another previously created and preferred user) area on the Amazon EC2 virtual machine:

```
$ scp -r -i "rstudio_mariadb.pem" ~/Desktop/B05396_Ch07_Code/pp-2015-
variables.csv swalko@ec2-52-19-205-207.eu-west-1.compute.amazonaws.com:~/
```

When the upload completes, in the old terminal window (with the `ubuntu` user already logged in) issue the following command to import the data to MongoDB:

```
$ mongoimport --db houses --collection prices --type csv --fieldFile
/home/swalko/pp-2015-variables.csv --file /home/swalko/pp-2015.csv
```

A few words of explanation are needed to understand the structure of the preceding statement. The `mongoimport` command is a self-explanatory command initializing the import of the data to MongoDB, and to be precise, the data will be uploaded to a new database called `houses` and it will also constitute the contents of the collection called `prices`. By using option `--type` we indicate the format of our variable names and data files (both as CSV). The `--fieldFile` simply points to the location of the file which contains the variable labels, whereas the `--file` option indicates the full path to the data file. Once run, the command should produce the following output:

```
2016-04-15T18:47:07.843+0000    connected to: localhost
2016-04-15T18:47:10.838+0000    [#.......................] houses.prices    11.6 MB/161.2 MB (7.2%)
2016-04-15T18:47:13.838+0000    [###.....................] houses.prices    23.5 MB/161.2 MB (14.6%)
2016-04-15T18:47:16.844+0000    [#####...................] houses.prices    35.8 MB/161.2 MB (22.2%)
2016-04-15T18:47:19.845+0000    [######..................] houses.prices    48.1 MB/161.2 MB (29.9%)
2016-04-15T18:47:22.839+0000    [########................] houses.prices    59.8 MB/161.2 MB (37.1%)
2016-04-15T18:47:25.842+0000    [##########..............] houses.prices    71.5 MB/161.2 MB (44.3%)
2016-04-15T18:47:28.842+0000    [############............] houses.prices    83.4 MB/161.2 MB (51.7%)
2016-04-15T18:47:31.843+0000    [##############..........] houses.prices    94.9 MB/161.2 MB (58.9%)
2016-04-15T18:47:34.841+0000    [###############.........] houses.prices   106.4 MB/161.2 MB (66.0%)
2016-04-15T18:47:37.846+0000    [#################.......] houses.prices   118.2 MB/161.2 MB (73.4%)
2016-04-15T18:47:40.842+0000    [###################.....] houses.prices   129.5 MB/161.2 MB (80.4%)
2016-04-15T18:47:43.848+0000    [#####################...] houses.prices   141.3 MB/161.2 MB (87.7%)
2016-04-15T18:47:46.839+0000    [#######################..] houses.prices  153.0 MB/161.2 MB (95.0%)
2016-04-15T18:47:49.030+0000    [########################] houses.prices   161.2 MB/161.2 MB (100.0%)
2016-04-15T18:47:49.030+0000    imported 971038 documents
```

The last line of the output confirms the final number of imported documents (971,038 for our data).

Now we may log back in to MongoDB shell and check whether a new database named `houses` and its collection named `prices` have been successfully created:

```
$ mongo
...#output truncated
> show dbs
houses   0.134GB
local    0.000GB
```

The result of the `show dbs` command confirms that a new database called `houses` has been created with a total size of 0.134 GB. We can indicate that we want to use this database in further operations:

```
> use houses
switched to db houses
```

We may also view the stored collections of documents within the houses databases by invoking the `show collections` statement:

```
> show collections
prices
```

You can easily check the number of documents (records) in the collection using the following command:

```
> db.prices.find().count()
971038
```

The preceding query is a standard MongoDB shell statement that contains the `find()` function that works just like the `SELECT` method in SQL, and it also includes the `count()` function – similar to the `length()` function from R, which simply calculates the number of rows in the data. The output confirms the total number of documents imported to MongoDB.

Now, it's also a good time to practice a few more complex MongoDB queries. For example, let's see how the single entry of the 100th document looks like. In this case, we skip the first 99 records of the matching documents and will limit the results to only one record using the `limit()` function, which will in fact be our requested 100th document:

```
> db.prices.find().skip(99).limit(1)
{ "_id" : ObjectId("571146f1a533ea616ef91883"), "uniqueID" :
"{49B60FF0-2827-4E8A-9087-6B3BE97D810F}", "price" : 500000, "transferDate"
: "2015-07-06 00:00", "postcode" : "N13 5TD", "propType" : "T", "oldNew" :
"N", "tenure" : "F", "PAON" : 18, "SAON" : "", "street" : "CRAWFORD
GARDENS", "locality" : "", "town" : "LONDON", "district" : "ENFIELD",
"county" : "GREATER LONDON", "ppdCat" : "A", "recordStatus" : "A" }
```

As you can see from the preceding listing, the output includes an index variable called `_id` – it's a default MongoDB indexing field that is assigned to each document in a collection when the data is imported.

Let's now calculate how many residential properties have been registered in the Land Registry database in 2015 in Manchester:

```
> db.prices.find({town: "Manchester"}).count()
```

Is it really true that there are no entries for Manchester? Note the spelling of the values in the `town` variable – MongoDB is case sensitive, so let's try `MANCHESTER` instead:

```
> db.prices.find({town: "MANCHESTER"}).count()
15973
```

We may want to aggregate average property prices for each county in England and Wales, and sort them in decreasing order from the most expensive counties to the most affordable ones. In this case we will use an aggregation pipeline framework for MongoDB:

```
> db.prices.aggregate([ { $group : { _id: "$county", avgPrice: { $avg:
"$price" } } }, { $sort: { avgPrice: -1 } }])
{ "_id" : "GREATER LONDON", "avgPrice" : 635409.3777145812 }
{ "_id" : "WINDSOR AND MAIDENHEAD", "avgPrice" : 551625.1565737051 }
{ "_id" : "SURREY", "avgPrice" : 504670.81007234025 }
{ "_id" : "WOKINGHAM", "avgPrice" : 470926.5928970733 }
{ "_id" : "BUCKINGHAMSHIRE", "avgPrice" : 451121.9522551808 }
{ "_id" : "HERTFORDSHIRE", "avgPrice" : 412466.64173755137 }
{ "_id" : "OXFORDSHIRE", "avgPrice" : 393092.78162926744 }
{ "_id" : "BRACKNELL FOREST", "avgPrice" : 387307.5234061525 }
{ "_id" : "WEST BERKSHIRE", "avgPrice" : 379225.775168979 }
{ "_id" : "READING", "avgPrice" : 373193.85240310075 }
{ "_id" : "BRIGHTON AND HOVE", "avgPrice" : 365359.7676169984 }
{ "_id" : "ISLES OF SCILLY", "avgPrice" : 356749.77777777775 }
{ "_id" : "POOLE", "avgPrice" : 343563.04866850324 }
{ "_id" : "BATH AND NORTH EAST SOMERSET", "avgPrice" : 338789.1119791667 }
{ "_id" : "HAMPSHIRE", "avgPrice" : 335520.61001046794 }
{ "_id" : "WEST SUSSEX", "avgPrice" : 333314.89596505347 }
{ "_id" : "CAMBRIDGESHIRE", "avgPrice" : 306048.801517067 }
{ "_id" : "MONMOUTHSHIRE", "avgPrice" : 301507.2670191672 }
{ "_id" : "ESSEX", "avgPrice" : 300157.60139321594 }
{ "_id" : "DORSET", "avgPrice" : 298921.8835030079 }
Type "it" for more
```

We could iterate the results further to include more affordable regions in England and Wales by simply typing the `it` command after the MongoDB prompt sign. After several iterations you will be presented with the most affordable counties as measured by the average residential property prices paid in 2015:

```
> it
{ "_id" : "TORFAEN", "avgPrice" : 141739.59360730593 }
{ "_id" : "COUNTY DURHAM", "avgPrice" : 134352.2187266849 }
{ "_id" : "MERTHYR TYDFIL", "avgPrice" : 132071.93098958334 }
{ "_id" : "CAERPHILLY", "avgPrice" : 131986.50344827585 }
{ "_id" : "STOKE-ON-TRENT", "avgPrice" : 130703.83509234828 }
{ "_id" : "NORTH EAST LINCOLNSHIRE", "avgPrice" : 128346.43109831345 }
{ "_id" : "HARTLEPOOL", "avgPrice" : 127779.55261371352 }
{ "_id" : "CITY OF KINGSTON UPON HULL", "avgPrice" : 125241.97961238358 }
{ "_id" : "NEATH PORT TALBOT", "avgPrice" : 123241.74938574938 }
{ "_id" : "RHONDDA CYNON TAFF", "avgPrice" : 120097.68402684564 }
{ "_id" : "BLACKPOOL", "avgPrice" : 109637.72557077625 }
{ "_id" : "BLAENAU GWENT", "avgPrice" : 89731.54028436019 }
```

Finally, we will aggregate the property prices for each type of property and town name in Essex, and as before we will sort the results in descending order based on the average price paid. We will also limit the results to the top 10 highest averages only:

```
> db.prices.aggregate([ { $match: { county: "ESSEX" } }, { $group : { _id:
{ town: "$town", propType: "$propType" }, avgPrice: { $avg: "$price" } } },
{ $sort: { avgPrice: -1 } }, { $limit: 10 }])
{ "_id" : { "town" : "LOUGHTON", "propType" : "O" }, "avgPrice" :
2094992.375 }
{ "_id" : { "town" : "CHELMSFORD", "propType" : "O" }, "avgPrice" :
2010153.9259259258 }
{ "_id" : { "town" : "EPPING", "propType" : "O" }, "avgPrice" :
1860545.7142857143 }
{ "_id" : { "town" : "BRENTWOOD", "propType" : "O" }, "avgPrice" :
1768124.8421052631 }
{ "_id" : { "town" : "CHIGWELL", "propType" : "O" }, "avgPrice" : 1700000 }
{ "_id" : { "town" : "HOCKLEY", "propType" : "O" }, "avgPrice" : 1504721 }
{ "_id" : { "town" : "COLCHESTER", "propType" : "O" }, "avgPrice" :
1416718.9090909092 }
{ "_id" : { "town" : "WICKFORD", "propType" : "O" }, "avgPrice" :
1216833.3333333333 }
{ "_id" : { "town" : "FRINTON-ON-SEA", "propType" : "O" }, "avgPrice" :
1052118 }
{ "_id" : { "town" : "BUCKHURST HILL", "propType" : "O" }, "avgPrice" :
1034000 }
```

The most expensive residential properties in Essex are the ones with the property type set to `"O"`, which denotes `"Other"`. This category includes properties that are not officially recognized as detached, semi-detached, or terraced houses, flats, and maisonettes. They may then include some other unclassified properties (for example, hotels or farmhouses, and others). The inclusion of such properties can skew the results, so we will make sure that these properties are not taken into account by providing an exclusion clause in the `redact` method and adding it to the previously run query:

```
> db.prices.aggregate([ { $match: { county: "ESSEX" } }, { $redact: {
$cond: { if: { $eq: [ "$propType", "O" ] }, then: "$$PRUNE", else:
"$$DESCEND" } } } ,{ $group : { _id: { town: "$town", propType: "$propType"
}, avgPrice: { $avg: "$price" } } }, { $sort: { avgPrice: -1 } }, { $limit:
10 }])
{ "_id" : { "town" : "CHIGWELL", "propType" : "D" }, "avgPrice" :
1016799.9230769231 }
{ "_id" : { "town" : "LONDON", "propType" : "D" }, "avgPrice" :
948331.6666666666 }
{ "_id" : { "town" : "LOUGHTON", "propType" : "D" }, "avgPrice" :
884204.6702127659 }
{ "_id" : { "town" : "BUCKHURST HILL", "propType" : "D" }, "avgPrice" :
845851.0645161291 }
```

```
{ "_id" : { "town" : "INGATESTONE", "propType" : "D" }, "avgPrice" :
812831.4032258064 }
{ "_id" : { "town" : "EPPING", "propType" : "D" }, "avgPrice" :
753521.4864864865 }
{ "_id" : { "town" : "ROYSTON", "propType" : "D" }, "avgPrice" :
705833.3333333334 }
{ "_id" : { "town" : "ONGAR", "propType" : "D" }, "avgPrice" :
679749.0566037736 }
{ "_id" : { "town" : "BRENTWOOD", "propType" : "D" }, "avgPrice" :
664917.34375 }
{ "_id" : { "town" : "ROMFORD", "propType" : "D" }, "avgPrice" :
652149.9666666667 }
```

As you can see from the preceding examples, the queries can get extremely convoluted and sometimes very difficult to read and understand. However, the MongoDB aggregation pipeline is extremely powerful and fast – it can flexibly and quickly produce aggregations and calculations depending on the user's needs of Big Data within seconds. In the next section, we will attempt to perform similar crunching operations on the same data stored in MongoDB, but this time directly from the RStudio Server's console.

MongoDB with R using the rmongodb package

We will begin our review of R packages supporting connectivity with MongoDB with the `rmongodb` package which provides the interface through the MongoDB C-driver. The package has been developed by MongoDB, Inc. and Markus Schmidberger from MongoSoup (https://www.mongosoup.de/) with Dmitriy Selivanov acting as the package maintainer. The CRAN link to `rmongodb` is located at https://cran.r-project.org/web/packages/rmongodb/index.html and the GitHub project repository is available at https://github.com/mongosoup/rmongodb.

We have already installed the package on the virtual machine earlier in this chapter. Log in to RStudio Server by pointing to the public address of your Amazon EC2 instance (in our case it would be http://52.19.205.207:8787 and provide your correct credentials). Upon successfully signing in, load the `rJava` package and then `rmongodb`:

```
> library(rJava)
> library(rmongodb)
...#output truncated
```

Create a connection to the MongoDB server running on `localhost` using the `mongo.create()` function:

```
> m <- mongo.create()
> m
```

```
[1] 0
attr(,"mongo")
<pointer: 0x3351280>
attr(,"class")
[1] "mongo"
attr(,"host")
[1] "127.0.0.1"
attr(,"name")
[1] ""
attr(,"username")
[1] ""
attr(,"password")
[1] ""
attr(,"db")
[1] "admin"
attr(,"timeout")
[1] 0
```

You can easily check whether the connection is working by issuing the `mongo.is.connected()` function:

```
> mongo.is.connected(m)
[1] TRUE
```

We may now extract the names of all databases from the created connection with MongoDB:

```
> mongo.get.databases(m)
[1] "houses"
```

The database called `houses` is present, so we may try to find out whether the collection that stores all PPD is also available:

```
> mongo.get.database.collections(m, "houses")
character(0)
```

The `mongo.get.database.collections()` function returns an empty character `vector`, but don't worry, it does not mean the data has not been imported correctly to MongoDB. In fact, the preceding output is a result of a known bug in the `rmongodb` package that prevents the correct extraction of the names of collections within a specified database. There is a workaround available that will let you get the name of the collection by applying a custom-made MongoDB command using the following statement:

```
> mongo.command(mongo = m, db = "houses", command =
list(listCollections=1))
    cursor : 3
        id : 18    0
```

```
      ns : 2     houses.$cmd.listCollections
      firstBatch : 4
        0 : 3
          name : 2     prices
          options : 3
...#output truncated
```

From the output, we can see that the collection is named `prices`.

At this point, you can start querying the data. We will firstly run a very simple operation using the `mongo.count()` function to calculate how many documents are within the specified collection. Note the format of the database and collection names used in the `rmongodb` functions (`db.collection`):

```
> mongo.count(m, "houses.prices")
[1] 971038
```

The `mongo.count()` function is extremely useful for Big Data processing with R as it can also be used to calculate the length of the output of a specific query without importing the results of the query as a data object into the R environment. For example, we may estimate the number of returned documents for the query that finds all detached houses in Surrey:

```
> mongo.count(m, "houses.prices", query = '{"county":"SURREY",
"propType":"D"}')
[1] 6040
```

For categorical variables, it is often useful to list all their values. In the `rmongodb` package, we can achieve it through either the `mongo.distinct()` or `mongo.get.values()` functions. For example, if we want to list all names of counties from the `county` variable, we may use the following two statements:

```
> mongo.distinct(m, "houses.prices", "county")
   [1] "WARWICKSHIRE"            "NORFOLK"
   [3] "STAFFORDSHIRE"           "GREATER LONDON"
   [5] "DEVON"                   "WINDSOR AND MAIDENHEAD"
   [7] "DERBYSHIRE"              "BLACKPOOL"
   [9] "KENT"                    "SOUTHAMPTON"
...#output truncated
> mongo.get.values(m, "houses.prices", "county")
   [1] "WARWICKSHIRE"            "NORFOLK"
   [3] "STAFFORDSHIRE"           "GREATER LONDON"
   [5] "DEVON"                   "WINDSOR AND MAIDENHEAD"
   [7] "DERBYSHIRE"              "BLACKPOOL"
   [9] "KENT"                    "SOUTHAMPTON"
...#output truncated
```

In total, there are 112 levels of the nominal `county` variable.

R with Non-Relational (NoSQL) Databases

We may now extract the first document that contains data on a single property in Surrey. From the output of the `mongo.find.one()` function, we can easily find, for example, the town of this record:

```
> surrey <- mongo.find.one(m, "houses.prices", '{"county":"SURREY"}')
> surrey
    _id : 7       571146f1a533ea616ef91854
    uniqueID : 2      {C9C0A867-C3AD-4285-A661-131809006279}
    price : 16        350000
    transferDate : 2      2015-09-17 00:00
    postcode : 2      GU2 8BL
    propType : 2      S
    oldNew : 2        N
    tenure : 2        F
    PAON : 16         172
    SAON : 2
    street : 2        ALDERSHOT ROAD
    locality : 2
    town : 2          GUILDFORD
    district : 2      GUILDFORD
    county : 2        SURREY
    ppdCat : 2        A
    recordStatus : 2      A
```

The property found by the query is located in Guildford. Also, you are probably intrigued by the numbers located next to the names of the fields. They simply denote the types of data, for example, `2` indicates a string and `16` is an integer.

The returned object `surrey` is a `BSON` data structure, and it is not recommended to use this format in R. We may then convert it to a much more preferable `list`:

```
> mongo.bson.to.list(surrey)
$`_id`
{ $oid : "571146f1a533ea616ef91854" }
$uniqueID
[1] "{C9C0A867-C3AD-4285-A661-131809006279}"
$price
[1] 350000
$transferDate
[1] "2015-09-17 00:00"
$postcode
[1] "GU2 8BL"
...#output truncated
```

From a list you can easily extract the sought-after values in R using standard notation.

As already mentioned, BSON is not very user-friendly, so you can also use lists to create specific queries, for example:

```
> query1 <- mongo.bson.from.list(list("county"="SURREY", "propType"="D"))
> query1
    county : 2    SURREY
    propType : 2    D
```

Such a prepared query can now be easily used in the find() method:

```
> surrey <- mongo.find.one(m, "houses.prices", query = query2)
> surrey
    _id : 7    571146f1a533ea616ef91973
    uniqueID : 2    {00CFB7C3-0AED-4B17-8BA6-0F7CF8B510B6}
    price : 16    600000
    transferDate : 2    2015-02-27 00:00
    postcode : 2    RH19 2LY
    propType : 2    D
    oldNew : 2    N
    tenure : 2    F
    PAON : 2    HIGH BANK
    SAON : 2
    street : 2    FURZEFIELD CHASE
    locality : 2    DORMANS PARK
    town : 2    EAST GRINSTEAD
    district : 2    TANDRIDGE
    county : 2    SURREY
    ppdCat : 2    A
    recordStatus : 2    A
```

As you have probably just noticed, we can pass several conditions in the find() query. The preceding example returns the first matching document with a property being a detached house ("propType"="D") and located in Surrey ("county"="SURREY").

In a similar way, we can create a BSON object using buffer functions. Firstly, we need to create an empty buffer:

```
> mbuf1 <- mongo.bson.buffer.create()
> mbuf1
[1] 0
attr(,"mongo.bson.buffer")
<pointer: 0x51b9c30>
attr(,"class")
[1] "mongo.bson.buffer"
```

We will then append components of a query to the `buffer`:

```
> mongo.bson.buffer.append(mbuf1, "county", "SURREY")
[1] TRUE
```

Finally, we can create a query from the `buffer` function and apply it in the `find()` function:

```
> query3 <- mongo.bson.from.buffer(mbuf1)
> query3
    county : 2    SURREY
> surrey <- mongo.find.one(m, "houses.prices", query = query3)
> surrey
    _id : 7        571146f1a533ea616ef91854
    uniqueID : 2    {C9C0A867-C3AD-4285-A661-131809006279}
    price : 16     350000
    transferDate : 2    2015-09-17 00:00
    postcode : 2    GU2 8BL
    propType : 2    S
    oldNew : 2      N
    tenure : 2      F
    PAON : 16      172
    SAON : 2
    street : 2      ALDERSHOT ROAD
    locality : 2
    town : 2        GUILDFORD
    district : 2    GUILDFORD
    county : 2      SURREY
    ppdCat : 2      A
    recordStatus : 2    A
```

To find all documents matching the query use the `mongo.find.all()` function, but be careful as this may return a large object if you are working with Big Data. Before running this statement it is recommended to first estimate the actual number of documents that will be returned using the already known `mongo.count()` function:

```
> mongo.count(m, "houses.prices", query = query3)
[1] 21703
```

If you want to create a BSON object with 21,703 records you may run the following:

```
> surrey <- mongo.find.all(m, "houses.prices", query = query3)
> surrey
...#output truncated
```

As a result, a new BSON object called surrey (62.4 MB) has been created in the R workspace. It is always important to check the size of such outputs before running the mongo.find.all() function, especially if you are working with large datasets on relatively low-resource machines.

In the rmongodb package you may also use the skip() and limit() methods:

```
> surrey <- mongo.find.all(m, "houses.prices", query = query3, skip = 100, limit=100)
> length(surrey)
[1] 100
```

The package allows users to specify the fields of interests to be returned (just like in the project method in MongoDB shell). In the following example we would like to extract the price and oldNew fields only and we also want to suppress the _id index variable. In order to achieve it we need to define our requested fields first:

```
> fields1 <- mongo.bson.from.list(list("price"=1, "oldNew"=1, "_id"=0))
```

We will then use the fields1 object as a value of the field parameter in the mongo.find.all() function. We will also sort the resulting documents in decreasing order based on the value of the price field (hence sort = '{"price": -1}'):

```
> surrey <- mongo.find.all(m, "houses.prices", query = query3,
+                          skip = 100, limit=100,
+                          fields = fields1,
+                          sort = '{"price": -1}')
> surrey
[[1]]
[[1]]$price
[1] 3110000
[[1]]$oldNew
[1] "N"
[[2]]
[[2]]$price
[1] 3100000
[[2]]$oldNew
[1] "N"
[[3]]
[[3]]$price
[1] 3100000
```

```
[[3]]$oldNew
[1] "N"
...#output truncated
```

As the output is a `list`, you can use it in R directly or *unlist* it to a `data.frame` in a standard way:

```
> df <- data.frame(matrix(unlist(surrey), nrow=100,
byrow=T),stringsAsFactors=FALSE)
```

The resulting `data.frame`, called `df`, has 100 rows over two variables. We can also add the variable names in the normal way:

```
> names(df) <- c("price", "oldNew")
> head(df, n=10)
     price oldNew
1  3110000      N
2  3100000      N
3  3100000      N
4  3100000      N
5  3100000      Y
6  3100000      N
7  3093750      N
8  3050000      N
9  3050000      N
10 3033000      N
```

We can now manually create a more complex query using `buffer` functions. In the following example we will extract all documents that match the conditions where the price paid (`price`) was lower than £300,000, the sold property was a detached house (`propType`), and it was located (`county`) in Greater London (that is a county in England inside the M25 motorway):

```
> mbuf2 <- mongo.bson.buffer.create()
> mongo.bson.buffer.start.object(mbuf2, 'price')
[1] TRUE
> mongo.bson.buffer.append(mbuf2, '$lt', 300000)
[1] TRUE
> mongo.bson.buffer.finish.object(mbuf2)
[1] TRUE
> mongo.bson.buffer.start.object(mbuf2, 'propType')
[1] TRUE
> mongo.bson.buffer.append(mbuf2, '$eq', "D")
[1] TRUE
> mongo.bson.buffer.finish.object(mbuf2)
[1] TRUE
> mongo.bson.buffer.start.object(mbuf2, 'county')
```

```
[1] TRUE
> mongo.bson.buffer.append(mbuf2, '$eq', "GREATER LONDON")
[1] TRUE
> mongo.bson.buffer.finish.object(mbuf2)
[1] TRUE
```

The resulting query will look as follows:

```
> query4 <- mongo.bson.from.buffer(mbuf2)
> query4
  price : 3
    $lt : 1      300000.000000
  propType : 3
    $eq : 2    D

  county : 3
    $eq : 2    GREATER LONDON
```

As we don't need to return all fields for all matched documents, we will make sure that only the `price`, `propType`, `county`, and `district` variables are shown. We will also suppress the `_id` field:

```
> fields2 <- mongo.bson.from.list(list("price"=1, "propType"=1, "county"=1,
"district"=1, "_id"=0))
> system.time(mfind <- mongo.find(m, 'houses.prices',
+                                  query = query4,
+                                  fields = fields2,
+                                  limit = 1000))
   user  system elapsed
  0.000   0.000   0.601
```

The created `mfind` object doesn't contain any data – it merely functions as a pointer (or a MongoDB cursor). To retrieve the matching documents, we need to use the cursor object to extract values for each requested field and therefore we have to create several empty vectors that will hold the data:

```
> Price <- Prop_Type <- County <- District <- NULL
> while (mongo.cursor.next(mfind)) {
+   value <- mongo.cursor.value(mfind)
+   Price <- rbind(Price, mongo.bson.value(value, 'price'))
+   Prop_Type <- rbind(Prop_Type, mongo.bson.value(value, 'propType'))
+   County <- rbind(County, mongo.bson.value(value, 'county'))
+   District <- rbind(District, mongo.bson.value(value, 'district'))
+ }
```

The created vectors can now be converged into a `data.frame` called `housesLondon`:

```
> housesLondon <- data.frame(Price, Prop_Type, County, District)
> summary(housesLondon)
     Price        Prop_Type           County           District
 Min.   :  7450   D:157      GREATER LONDON:157   HAVERING  :45
 1st Qu.:235000                                   BEXLEY    :26
 Median :260000                                   BROMLEY   :16
 Mean   :242287                                   CROYDON   : 9
 3rd Qu.:280000                                   REDBRIDGE : 9
 Max.   :298000                                   HILLINGDON: 6
                                                  (Other)   :46
```

From the output, we can see that the cheapest detached properties in London are located in the Havering, Bexley, and Bromley districts.

The `rmongodb` package also supports a creation of complex aggregations that follow the MongoDB aggregation pipeline framework syntax. In the following example, we will aggregate the data by creating `BSON` queries from `JSON` statements using the `mongo.bson.from.JSON()` function. Our goal is to calculate the average price paid aggregated for each town in Surrey. We want to return only the top five highest average property prices and we would like the results to be displayed in descending order (from the highest to the lowest average of the top five results). We can achieve this goal by creating single queries for individual methods. Firstly, we will filter all documents and extract only those properties that are located in Surrey:

```
> agg1 <- mongo.bson.from.JSON('{"$match":
+                              {"county":"SURREY"}}')
> agg1
    $match : 3
        county : 2      SURREY
```

Secondly, we will calculate the average property price for each town among the already filtered documents:

```
> agg2 <- mongo.bson.from.JSON('{"$group":
+                              {"_id":"$town",
+                              "avgPrice": {"$avg":"$price"}}}')
> agg2
    $group : 3
        _id : 2     $town
        avgPrice : 3
            $avg : 2     $price
```

Thirdly, we will sort the results in descending order based on the newly created measure of the average paid price (`avgPrice`):

```
> agg3 <- mongo.bson.from.JSON('{"$sort":
+                               {"avgPrice": -1}}')
> agg3
  $sort : 3
    avgPrice : 16    -1
```

Lastly, we can display only five of the returned documents – they are already sorted in the appropriate order, so we will be shown five records with the highest average paid prices only:

```
> agg4 <- mongo.bson.from.JSON('{"$limit": 5}')
> agg4
  $limit : 16    5
```

In the last stage of the aggregation pipeline creation, we need to amalgamate all aggregation components into a `list` object:

```
> listagg <- list(agg1, agg2, agg3, agg4)
```

The `listagg` object contains the final format of the aggregation we want to perform:

```
> listagg
[[1]]
  $match : 3
    county : 2      SURREY
[[2]]
  $group : 3
    _id : 2         $town
    avgPrice : 3
      $avg : 2      $price
[[3]]
  $sort : 3
    avgPrice : 16    -1
[[4]]
  $limit : 16    5
```

When ready, we can finally use it as a parameter in the `mongo.aggregation()` function:

```
> output <- mongo.aggregation(m, 'houses.prices', listagg)
> output
  waitedMS : 18    0
  result : 4
    0 : 3
      _id : 2       VIRGINIA WATER
      avgPrice : 1    1443789.184211
```

```
         1 : 3
           _id : 2       COBHAM
           avgPrice : 1      1080538.623077
         2 : 3
           _id : 2       WEYBRIDGE
           avgPrice : 1      893551.667969
         3 : 3
           _id : 2       ESHER
           avgPrice : 1      867048.614362
         4 : 3
           _id : 2       OXTED
           avgPrice : 1      804640.173770
    ok : 1      1.000000
```

It is clear (and somewhat unsurprising) that the most expensive residential properties in Surrey are located in Virginia Water with the average price in 2015 reaching a whopping £1,443,789.

Of course, just like earlier, you can convert the unfriendly BSON output to a list:

```
> mongo.bson.to.list(output)
$waitedMS
[1] 0
$result
$result[[1]]
$result[[1]]$`_id`
[1] "VIRGINIA WATER"
$result[[1]]$avgPrice
[1] 1443789
$result[[2]]
$result[[2]]$`_id`
[1] "COBHAM"

$result[[2]]$avgPrice
[1] 1080539
...#output truncated
```

After all the processing and data crunching, we can close the connection to the MongoDB database server:

```
> mongo.disconnect(m)
[1] 0
attr(,"mongo")
<pointer: 0x6416030>
attr(,"class")
[1] "mongo"
attr(,"host")
[1] "127.0.0.1"
```

```
attr(,"name")
[1] ""
attr(,"username")
[1] ""
attr(,"password")
[1] ""
attr(,"db")
[1] "admin"
attr(,"timeout")
[1] 0
```

If you need to reconnect to the disconnected MongoDB connection, you can do it very quickly using the `mongo.reconnect()` function:

```
> mongo.reconnect(m)
...#output truncated
```

However, you can't reconnect if you *kill* the connection through the `mongo.destroy()` function:

```
> mongo.destroy(m)
NULL
> mongo.reconnect(m)
Error in mongo.reconnect(m) :
  mongo connection object appears to have been destroyed.
```

Note that the created R objects are available for further analysis or visualizations even when the connection to MongoDB is closed (it is because they simply reside in RAM):

```
> summary(housesLondon)
     Price          Prop_Type           County              District
 Min.   :  7450     D:157       GREATER LONDON:157     HAVERING   :45
 1st Qu.:235000                                        BEXLEY     :26
 Median :260000                                        BROMLEY    :16
 Mean   :242287                                        CROYDON    : 9
 3rd Qu.:280000                                        REDBRIDGE  : 9
 Max.   :298000                                        HILLINGDON : 6
                                                       (Other)    :46
```

In the next section, we will explore the functionalities of the RMongo package.

MongoDB with R using the RMongo package

The second most frequently used R package for processing data residing in MongoDB is RMongo. The package authored by Tommy Chheng provides the interface through the MongoDB Java driver. The CRAN link to the package is located at https://cran.r-project.org/web/packages/RMongo/index.html and the GitHub project URL is http://github.com/tc/RMongo.

It is advisable that before starting this part of the tutorial you remove all non-essential objects from your R workspace. In case you forgot, the following command removes all objects in the environment:

```
> rm(list=ls())
```

Load the RMongo package and create a connection to the MongoDB database called houses on the localhost:

```
> library(RMongo)
> m <- mongoDbConnect("houses", port=27017)
> m
An object of class "RMongo"
Slot "javaMongo":
[1] "Java-Object{rmongo.RMongo@51c8530f}"
```

We can straightaway create a basic query that will use the connection m and extract 1,000 documents from the prices collection in which the price paid (the price field) was lower than £500,000. We will also skip the first 1,000 matching records and will only retrieve the next 1,000 appropriate documents. In RMongo, we run queries using the dbGetQuery() function:

```
> system.time(subset1 <- dbGetQuery(m, "prices", "{'price':{$lt:500000}}",
skip=1000, limit=1000))
   user  system elapsed
  0.205   0.000   0.209
```

The resulting object subset1 is a data.frame (480 KB in size) with 1,000 rows and 17 variables:

```
> str(subset1)
'data.frame':    1000 obs. of  17 variables:
 $ town       : chr  "STOKE-ON-TRENT" "STOKE-ON-TRENT" "STOKE-ON-TRENT"
"WALSALL" ...
 $ ppdCat     : chr  "A" "B" "A" "A" ...
 $ postcode   : chr  "ST2 7HE" "ST8 6SH" "ST1 5JE" "WS1 3EJ" ...
 $ locality   : chr  "" "KNYPERSLEY" "" "" ...
 $ county     : chr  "STOKE-ON-TRENT" "STAFFORDSHIRE" "STOKE-ON-TRENT"
```

```
"WEST MIDLANDS" ...
 $ SAON         : chr  "" "" "" "" ...
 $ transferDate: chr  "2015-09-11 00:00" "2015-10-09 00:00" "2015-10-12
00:00" "2015-09-04 00:00" ...
...#output truncated
```

As `RMongo` returns outputs in the format of a native R `data.frame`, it is very simple to use the result sets in further calculations or data crunching, for example:

```
> summary(subset1$price)
   Min.  1st Qu.  Median    Mean  3rd Qu.    Max.
  28500   131000  191200  212700   280600  500000
```

The following is a slightly more complex example of a query that will extract all documents with the price paid value below £500,000, but only for properties located in Greater London. We will limit the output to the first 10,000 matching records:

```
> system.time(subset2 <- dbGetQuery(m, "prices",
+                    "{'price':{$lt:500000},
+                     'county':{$eq:'GREATER LONDON'}}",
+                           skip=0, limit = 10000))
   user  system elapsed
  0.555   0.000   0.729
```

This time, it took just a little bit longer to produce the output, which resulted in a `data.frame` object of 3.8 MB in size. As you can probably notice, the preceding two queries returned documents with all fields present. We can restrict the output to include desired fields only by applying a query using the `dbGetQueryForKeys()` function:

```
> system.time(subset3 <- dbGetQueryForKeys(m, "prices",
+                    "{'price':{$lt:500000},
+                     'county':{$eq:'GREATER LONDON'}}",
+                    "{'district':1, 'price':1, 'propType':1}",
+                           skip=0, limit = 50000))
   user  system elapsed
  0.391   0.012   0.828
```

The results set contains only the specified fields in addition to the `X_id` indexing variable:

```
> str(subset3)
'data.frame':   50000 obs. of  4 variables:
 $ district: chr  "MERTON" "BRENT" "BARNET" "EALING" ...
 $ price   : int  250000 459950 315000 490000 378000 243000 347000 350000
303500 450000 ...
 $ propType: chr  "F" "F" "F" "S" ...
 $ X_id    : chr  "571146f1a533ea616ef9182e" "571146f1a533ea616ef9183a"
"571146f1a533ea616ef91847" "571146f1a533ea616ef9184b" ...
```

```
> head(subset3, n=10)
     district  price propType                       X_id
1      MERTON 250000        F 571146f1a533ea616ef9182e
2       BRENT 459950        F 571146f1a533ea616ef9183a
3      BARNET 315000        F 571146f1a533ea616ef91847
4      EALING 490000        S 571146f1a533ea616ef9184b
5      BARNET 378000        S 571146f1a533ea616ef91861
6      NEWHAM 243000        F 571146f1a533ea616ef91867
7   REDBRIDGE 347000        T 571146f1a533ea616ef9188b
8    ISLINGTON 350000       F 571146f1a533ea616ef9188d
9       BEXLEY 303500       T 571146f1a533ea616ef91897
10      NEWHAM 450000       T 571146f1a533ea616ef918a5
```

Similarly to the `rmongodb` package, `RMongo` allows users to carry out complex, custom-made aggregations in accordance with the MongoDB's aggregation pipeline framework. The `dbAggregate()` function takes a `vector` of pipeline commands in `JSON` format as a query and applies on a specified collection using the R and MongoDB connection. In the following example, we will aggregate the average price paid for properties located in Surrey by the individual towns in this county. We will also order the resulting average prices decreasingly and return only the top five matches based on the price paid:

```
> houses.agr <- dbAggregate(m, "prices",
+                    c('{"$match": {"county": "SURREY"}}',
+                      '{"$group": {"_id": "$town",
+                         "avgPrice": {"$avg": "$price"}}}',
+                      '{"$sort": {"avgPrice": -1}}',
+                      '{"$limit": 5}'))
```

The resulting object is a character `vector` that contains `JSON` strings for each matched record:

```
> houses.agr
[1] "{ "_id" : "VIRGINIA WATER" , "avgPrice" : 1443789.1842105263}"
[2] "{ "_id" : "COBHAM" , "avgPrice" : 1080538.6230769232}"
[3] "{ "_id" : "WEYBRIDGE" , "avgPrice" : 893551.66796875}"
[4] "{ "_id" : "ESHER" , "avgPrice" : 867048.6143617021}"
[5] "{ "_id" : "OXTED" , "avgPrice" : 804640.1737704918}"
```

Unfortunately, `JSON` output is not very user friendly, so we will now convert it into a `data.frame` using the `RJSONIO` package, which we installed earlier along with other useful R packages while configuring the virtual machine and RStudio Server:

```
> require(RJSONIO)
Loading required package: RJSONIO
```

In order to achieve our goal, for each record in the `vector` we will use the `lapply()` command to iterate the `fromJSON()` function for every single JSON string:

```
> datalist <- lapply(houses.agr, FUN=fromJSON)
> datalist
[[1]]
[[1]]$`_id`
[1] "VIRGINIA WATER"
[[1]]$avgPrice
[1] 1443789
[[2]]
[[2]]$`_id`
[1] "COBHAM"
[[2]]$avgPrice
[1] 1080539
...#output truncated
```

Finally, we will convert the resulting transformation into a `data.frame`, but if it's easier and more convenient for you, feel free to use the above `list` object instead:

```
> data.df <- data.frame(matrix(unlist(datalist), nrow=5, byrow=T),
+                       stringsAsFactors=FALSE)
> names(data.df) <- c("town", "price")
> data.df
            town           price
1 VIRGINIA WATER 1443789.18421053
2         COBHAM 1080538.62307692
3      WEYBRIDGE  893551.66796875
4          ESHER  867048.614361702
5          OXTED  804640.173770492
```

To finalize the processing of data using `RMongo`, make sure to disconnect from MongoDB:

```
> dbDisconnect(m)
```

To summarize, `RMongo` offers similar functionalities to the previously reviewed `rmongodb` package. Some of the functions are actually a little bit easier to implement using `RMongo`. However, both packages are not regularly maintained, which may become problematic once the MongoDB engine develops even further. An alternative approach to MongoDB querying from R is offered by a popular and well maintained `mongolite` package, which we present in the next section.

MongoDB with R using the mongolite package

For the reasons stated at the end of the preceding section, there has been a strong need for a convenient, lightweight, and flexible R package that would offer a user-friendly interface for management and processing of data stored in MongoDB. The `mongolite` package, authored by Jeroen Ooms and MongoDB, Inc., fulfills this role very well. You can access its vignettes and all its help files through CRAN at https://cran.r-project.org/web/packages/mongolite/index.html, and the development versions are available at GitHub: http://github.com/jeroenooms/mongolite.

As was the case with other MongoDB packages with R, we have already installed `mongolite` on our virtual machine. We simply need to load it to prepare for the first use during the R session:

```
> library(mongolite)
```

It's always recommended to obtain some basic information about available functions before beginning to work with any specific R packages. It may be surprising to notice that, in fact, the `mongolite` package only contains one function called `mongo()`. This however, allows users to apply a number of specific methods responsible for performing operations and queries on the data.

In the first step, we need to create the usual connection to a specified database and collection on the local MongoDB:

```
> m <- mongo(collection = "prices", db = "houses", url = "mongodb://localhost")
```

The created connection object m displays all possible methods (and their arguments/parameters) of data processing using `mongolite`:

```
> m
<Mongo collection> 'prices'
 $aggregate(pipeline = "{}", handler = NULL, pagesize = 1000)
 $count(query = "{}")
 $distinct(key, query = "{}")
 $drop()
 $export(con = stdout(), bson = FALSE)
 $find(query = "{}", fields = "{"_id":0}", sort = "{}", skip = 0, limit = 0, handler = NULL, pagesize = 1000)
 $import(con, bson = FALSE)
 $index(add = NULL, remove = NULL)
 $info()
 $insert(data, pagesize = 1000)
 $iterate(query = "{}", fields = "{"_id":0}", sort = "{}", skip = 0, limit = 0)
```

```
$mapreduce(map, reduce, query = "{}", sort = "{}", limit = 0, out = NULL,
scope = NULL)
 $remove(query, multiple = FALSE)
 $rename(name, db = NULL)
 $update(query, update = "{"$set":{}}", upsert = FALSE, multiple = FALSE)
```

We may, for instance, begin from a simple calculation of the total number of documents in the collection, just like we did with other MongoDB-related R packages. We will achieve it by issuing the following command:

```
> m$count()
[1] 971038
```

This first example shows you how easy it is to use `mongolite` – it retrieves all essential information about the database and its collection from the connection object, thus simplifying the syntax and increasing the performance of the code.

In order to query the data, we can use the `find()` method. From the previously shown output of the connection object, we can see that the `find()` method comes with some default parameters; for example, the indexing variable `_id` is suppressed in all result sets, the data is not sorted, and there is no limit on the amount of returned documents. Of course, we can simply override these defaults, which we are going to do later. However, for the time being, let's print all records for which the property paid price was lower than £100,000 and all residential properties of interest were detached houses:

```
> subset1 <- m$find('{"price":{"$lt":100000},
+                    "propType":{"$eq":"D"}}')
 Imported 1739 records. Simplifying into dataframe...
```

While creating the output, the `mongolite` package provides the user with very handy information on the total number of returned documents. It also informs us that the output will be automatically simplified into a `data.frame` object. This is a very useful functionality that definitely saves a bit of processing time. We can now inspect the structure of the resulting object using the standard `str()` command:

```
> str(subset1)
'data.frame':   1739 obs. of  16 variables:
 $ uniqueID     : chr  "{561101F1-A127-487C-A945-04D23222E4AE}"
"{1D559F3C-1770-432C-B304-6B4B628E0117}" "{B7E3429D-6C1C-4A19-
BB59-5D247D5606CE}" "{25EA59F9-86CB-4D50-E050-A8C0630562D0}" ...
 $ price        : int  54000 75000 80000 16500 80000 95000 90000 90000 45500
90000 ...
 $ transferDate: chr  "2015-02-12 00:00" "2015-09-09 00:00" "2015-03-27
00:00" "2015-06-29 00:00" ...
 $ postcode     : chr  "CF39 9SE" "PE25 1SD" "PE12 9DD" "CA26 3SB" ...
 $ propType     : chr  "D" "D" "D" "D" ...
```

```
 $ oldNew      : chr  "N" "N" "N" "N" ...
...#output truncated
```

Of course, when dealing with large datasets it's unlikely that you would want to retrieve all the variables of the data. But just like in the `rmongodb` and `RMongo` packages, in `mongolite` you can easily specify which fields you would like to include in the results set. Apart from projections, you may also use other commands known from the MongoDB shell, for example, `sort`, `skip`, or `limit`. The good thing about how these other methods are implemented in the `find` query in `mongolite` is that you don't have to set them using the JSON format, they simply work in the same way as parameters of any R function. The only exception to that rule is the `sort` parameter, which takes a short JSON entry to define the variable for which data is going to be sorted and the direction of sorting.

In the following example we will return all documents with detached properties and prices lower than £100,000. We will, however, only include the `price` and `town` fields in the results set and we will order the prices from the most expensive to the cheapest. Just in case, we will limit the output to the first 10,000 matched documents:

If you are working with extremely large datasets, it is a good practice to limit the output of result sets to a small amount of documents, especially during the early stages of code testing. Once you are sure that your code produces the desired output you may increase the size of the returned objects depending on your available resources.

```
> subset2 <- m$find('{"price":{"$lt":100000}, "propType":{"$eq":"D"}}',
+                   fields = '{"_id":0, "price":1, "town":1}',
+                   sort = '{"price":-1}', skip = 0, limit = 10000)
 Imported 1739 records. Simplifying into dataframe...
> str(subset2)
'data.frame':  1739 obs. of  2 variables:
 $ price: int  99995 99995 99995 99995 99995 99995 99995 99955 99950 99950
...
 $ town : chr  "HARTLEPOOL" "SPALDING" "CLACTON-ON-SEA" "DURHAM" ...
> head(subset2, n=5)
  price           town
1 99995      HARTLEPOOL
2 99995        SPALDING
3 99995  CLACTON-ON-SEA
4 99995          DURHAM
5 99995  CLACTON-ON-SEA
```

By restricting the structure of the final output to just two variables of interest, we have essentially decreased the size of the returned object from 650.4 KB in `subset1` to as little as 43.6 KB in `subset2`.

In `mongolite`, we can also perform typical MongoDB-style aggregations using the `aggregate()` method. Here, we need to pass the full aggregation pipeline in JSON format. The following example calculates two basic statistics: the number of records (a new `count` variable will be created) and the average price for all properties in each of the 112 counties in England and Wales (a new `avgPrice` variable will be created). The sorted results set in the form of a `data.frame` can be obtained as follows:

```
> houses.agr <- m$aggregate('[{"$group": {"_id":"$county",
+                           "count":{"$sum":1},
+                           "avgPrice":{"$avg":"$price"} }},
+                           {"$sort":{"avgPrice": -1} }]')
  Imported 112 records. Simplifying into dataframe...
> head(houses.agr, n=10)
                      _id  count  avgPrice
1          GREATER LONDON 123776  635409.4
2   WINDSOR AND MAIDENHEAD  2510  551625.2
3                  SURREY  21703  504670.8
4               WOKINGHAM   3041  470926.6
5         BUCKINGHAMSHIRE   9844  451122.0
6           HERTFORDSHIRE  20926  412466.6
7              OXFORDSHIRE  11453  393092.8
8         BRACKNELL FOREST   2243  387307.5
9           WEST BERKSHIRE   2811  379225.8
10                 READING   3225  373193.9
```

As expected, the aggregation returned 112 records in total. From the results, we clearly see, that the most expensive properties are located in Greater London, followed by the counties of Windsor and Maidenhead, and Surrey – all stereotypically associated with upper-class residents and (less stereotypically) high living costs.

Very often, when dealing with categorical variables, it is useful to view all possible values that are contained within such variables. This can be achieved in `mongolite` by the `distinct()` method. For example, if you wish to list all the unique values for two fields of `county` and `propType`, you can do it as follows:

```
> m$distinct("county")
  [1] "WARWICKSHIRE"          "NORFOLK"
  [3] "STAFFORDSHIRE"         "GREATER LONDON"
  [5] "DEVON"                 "WINDSOR AND MAIDENHEAD"
  [7] "DERBYSHIRE"            "BLACKPOOL"
  [9] "KENT"                  "SOUTHAMPTON"
...#output truncated
> m$distinct("propType")
[1] "T" "S" "D" "F" "O"
```

The `mongolite` package also offers one extremely powerful functionality that is missing in both `rmongodb` and `RMongo`. Strictly speaking, it can perform MapReduce operations just like the MapReduce jobs we introduced you to earlier when discussing Big Data analytics with Hadoop and R in Chapter 4, *Hadoop and MapReduce Framework for R*. However, the tricky thing about it is that the `mapreduce()` method, which is responsible for MapReduce implementation through the `mongolite` package, requires the mapper and reduce functions to be written in JavaScript – a skill that may sometimes be beyond the comfort zone for a large number of data scientists. The following is a very simple example of a MapReduce job using the `mongolite` package that calculates frequencies for two crossed factors of `county` and property type (`propType`):

```
> houses.xtab <- m$mapreduce(
+    map = "function(){emit({county:this.county, propType:this.propType},
1)}",
+    reduce = "function(id, counts){return Array.sum(counts)}"
+ )
> houses.xtab
                 _id.county _id.propType value
1   BATH AND NORTH EAST SOMERSET              D    739
2   BATH AND NORTH EAST SOMERSET              F    808
3   BATH AND NORTH EAST SOMERSET              O     39
4   BATH AND NORTH EAST SOMERSET              S    771
5   BATH AND NORTH EAST SOMERSET              T   1099
6                        BEDFORD              D    950
7                        BEDFORD              F    441
8                        BEDFORD              O     48
9                        BEDFORD              S    907
10                       BEDFORD              T    871
...#output truncated
```

Of course, you are more than welcome to give it a try with more complex MapReduce calculations.

In the previous sections we have provided you with quite thorough reviews of three essential R packages that support communication with MongoDB and allow users to perform complex data processing, queries, aggregations, and even MapReduce operations directly from the R environment on data stored within the MongoDB database.

In the next part of this chapter, we will briefly guide you through some basic functionalities of the `rhbase` package, which belongs to a group of R packages collectively known as `rhadoop`. We have already discussed two packages of that group when dealing with the Hadoop and HDFS frameworks in Chapter 4, *Hadoop and MapReduce Framework for R*: `rmr2` and `rhdfs`. This time, however, we will show you how to connect to data stored in a non-relational Hadoop database called HBase using R.

HBase with R

The Apache HBase database allows users to store and process non-relational data on top of HDFS. Inspired by Google's BigTable, HBase is an open source, distributed, consistent, and scalable database that facilitates real-time read and write access to massive amounts of data. It is in fact a columnar or key-column-value data store that lacks any default schema and can be defined by users at any time.

The following tutorial will present a sequence of essential activities that will allow you to import our previously used Land Registry Price Paid Data into the HBase store on the Microsoft Azure HDInsight cluster and then retrieve specific slices of data using RStudio Server.

Azure HDInsight with HBase and RStudio Server

The process of launching the fully operational Microsoft Azure HDInsight cluster with HBase database is very similar to the one described in Chapter 4, *Hadoop and MapReduce Framework for R*, where we guided you through the creation of a multi-node HDInsight Hadoop cluster. There are, however, some minor differences, which we will highlight in this section.

As before, begin your work with setting up a new **Resource Group**. Simply follow the instructions provided in the *Creating a new Resource Group* section in Chapter 4, *Hadoop and MapReduce framework for R*. You may just want to give it a new name, for example, `hbasecluster`. Remember to correctly select your subscription and the resource group location. Also note that, as with the Hadoop HDInsight cluster, it is very likely that you will be charged some fees (they might be high!) for using the service. Again, how much exactly depends on your personal circumstances, the geographical location of the servers used, and many other factors.

R with Non-Relational (NoSQL) Databases

Once the new Resource Group has been created, proceed to the deployment of a new Virtual Network. Follow the instructions given to you in Chapter 4, *Hadoop and MapReduce Framework for R*. Give it a name that can be easily assigned to the created HBase HDInsight service for example `hbaseVN`. Make sure to choose the appropriate **Resource group** which you created in the preceding step (`hbasecluster` in our case). Your completed **Create virtual network** form should resemble the information in the following figure:

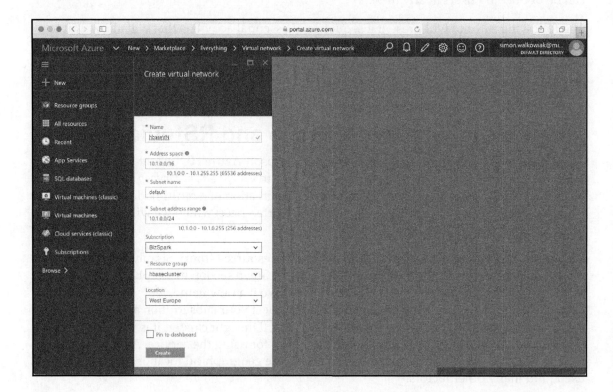

Continue following the instructions in Chapter 4, *Hadoop and MapReduce Framework for R*, to set up a new Network Security Group. Again, you should ideally give it a new name for example `hbaseNSG` and provide other details, as in the following form, making sure to alter the entries depending on your individual settings:

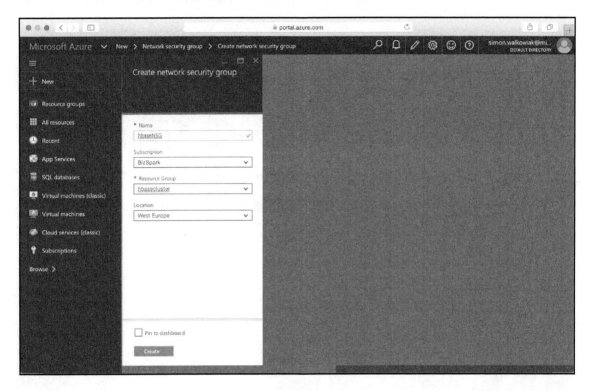

At this point, we have already launched a new Resource Group named `hbasecluster`, a Virtual Network named `hbaseVN`, and a Network Security Group – `hbaseNSG`. Having all three in place, we may now move on the actual deployment and configuration of the HBase HDInsight cluster.

R with Non-Relational (NoSQL) Databases

Start from following the steps provided in the *Setting up and configuring an HDInsight cluster* section in `Chapter 4`, *Hadoop and MapReduce Framework for R*. Our cluster will be named `hbasecluster.azurehdinsight.net`. Choose the correct **Subscription** package. In the **Select Cluster Type** option, choose **HBase** as a **Cluster Type**, **HBase 1.1.2 (HDI 3.4)** as its **Version**, and **Linux** as an **Operating System**. Select the **Standard Cluster Tier** as shown in the following screenshot:

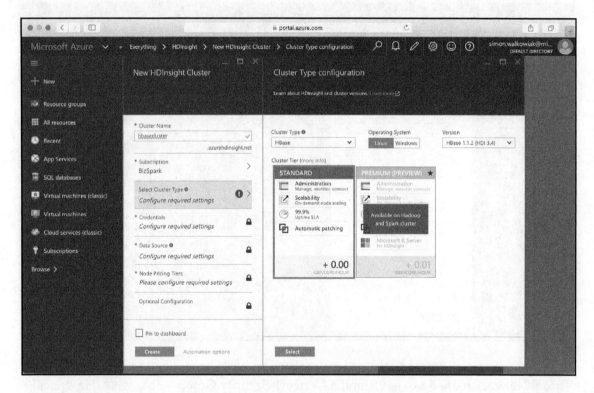

As before, provide memorable credentials. Just like in Hadoop HDInsight, we will keep `admin` as a **Cluster Login Username** and `swalko` as **SSH Username**.

In the **Data Source** tab, **Create a new storage account** and name it, for example, `hbasestorage1` with the **Default Container** set to `store1`:

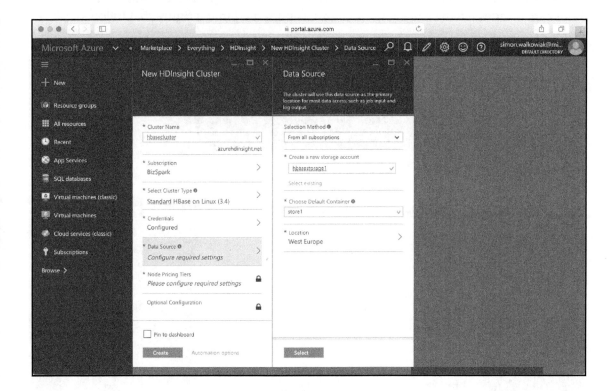

R with Non-Relational (NoSQL) Databases

For the HBase cluster, we will go for a slightly cheaper configuration of nodes (**Node Pricing Tiers**) than in Hadoop HDInsight. We will keep 4 as a default value of **Region nodes**. They will be pooled from `D3 Pricing Tier`, whereas 2 **Head nodes** and 3 **Zookeeper nodes** will be `A3 Pricing Tier` as presented in the figure that follows:

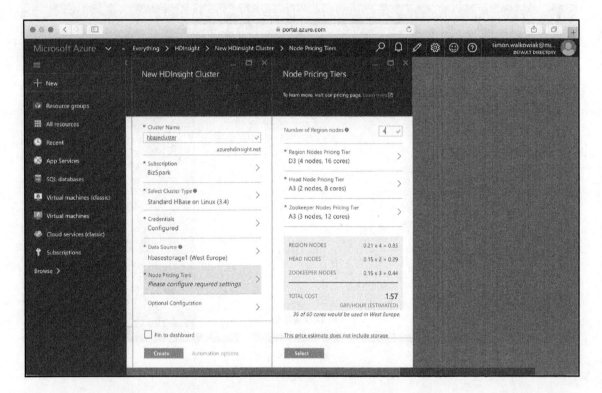

Finally, in the **Optional Configuration** tab, repeat the activities from `Chapter 4`, *Hadoop and MapReduce Framework for R*. Make sure to select the correct name of the preferred **Virtual Network** for this HDInsight service – in our case it will be `hbaseVN` with the `default` **Subnet**. Also, in the **Script Actions** tab use the same link to install R files and tick all three types of node. As before, leave **PARAMETERS** blank. As the last action, select the appropriate **Resource Group** (for example, `hbasecluster`) in the **New HDInsight Cluster** view and click the **Create** button to start deploying the cluster.

Again, it may take up to an hour to create the cluster. Once the operation completes, we may install RStudio Server in the same way as described in the *Connecting to the HDInsight cluster and installing RStudio Server* section in Chapter 4, *Hadoop and MapReduce Framework for R*. Note that as we've changed the name of the cluster, this has to be reflected in the `ssh` command. In our case we will `ssh` to the cluster using the following statement:

```
$ ssh swalko@hbasecluster-ssh.azurehdinsight.net
```

Continue installing RStudio Server, add a new Inbound Security Rule for port 8787 and edit the Virtual Network's public IP address for the head node exactly as instructed in Chapter 4, *Hadoop and MapReduce Framework for R*. After several minutes you should have an HBase HDInsight cluster with RStudio Server ready to work. Unfortunately, this is not enough for us to connect to HBase directly from RStudio Server. As you probably remember, in the Hadoop HDInsight cluster all required R packages and Java environment were already available. Although R packages such as `rJava`, `rhdfs`, and `rmr2` are provided in the HBase cluster as well, we need some additional libraries to be able to communicate with HBase through the **Thrift** server. Moreover, we need a correct version of Thrift (0.9.0) to install too, as the `rhbase` package for R that we are going to use hasn't been optimized to work with the newer versions yet.

Let's then firstly download and install the Linux Ubuntu libraries that are required for the Thrift server:

```
$ sudo apt-get install libboost-dev libboost-test-dev libboost-program-options-dev libevent-dev automake libtool flex bison pkg-config g++ libssl-dev
...#output truncated
```

It may take several minutes until all libraries are ready and the command prompt becomes available. Once it happens, we may download the Thrift 0.9.0 version tarball:

```
$ sudo wget http://archive.apache.org/dist/thrift/0.9.0/thrift-0.9.0.tar.gz
...#output truncated
```

Then extract it to a preferred location:

```
$ sudo tar -xvf thrift-0.9.0.tar.gz
```

cd to a directory with the unpacked Thrift 0.9.0:

```
$ cd thrift-0.9.0
```

Configure the system libraries and prepare to build Thrift:

```
$ sudo ./configure
...#output truncated
```

In the same directory, build and install Thrift by issuing the following commands:

```
$ sudo make
...#output truncated
$ sudo make install
...#output truncated
```

Once installed, you can quickly double-check the version of Thrift:

```
$ thrift -version
Thrift version 0.9.0
```

Another recommended way to verify the Thrift installation is to test whether the `pkg-config` path is correct. It should return `-I/usr/local/include/thrift` rather than `-I/usr/local/include`:

```
$ pkg-config --cflags thrift
-I/usr/local/include/thrift
```

Finally, cd to the `/usr/local/lib/` directory and check its contents – we are particularly interested in a Thrift library named `libthrift-0.9.0.so`. Copy it to `/usr/lib/`:

```
$ cd /usr/local/lib/
$ ls
libthrift-0.9.0.so    libthriftnb.a     libthriftz-0.9.0.so   pkgconfig
libthrift.a           libthriftnb.la    libthriftz.a          python2.7
libthrift.la          libthriftnb.so    libthriftz.la         python3.4
libthriftnb-0.9.0.so  libthrift.so      libthriftz.so         R
$ sudo cp /usr/local/lib/libthrift-0.9.0.so /usr/lib/
```

It's done! The Thrift server should now be configured and ready to work. We will start it later after we install the `rhbase` package.

Instead of using the `rhbase` package version from the GitHub of **Revolution Analytics** (`https://github.com/RevolutionAnalytics/rhbase`), the authors and maintainers of `rhadoop` libraries for R, we will use the forked version of the same package available from Aaron Benz's GitHub at `https://github.com/aaronbenz/rhbase`. Benz's version of `rhbase` is well maintained and generally quite user friendly. Also, it's easy to download and install on Linux Ubuntu. His GitHub also includes a very informative tutorial on how to use `rhbase`, which we recommend you go through when you finish reading this chapter.

In order to install `rhbase` from GitHub, we firstly need to download and install the required Ubuntu dependencies:

```
$ sudo apt-get install libssl-dev libcurl4-openssl-dev libssh2-1-dev
...#output truncated
```

Then install the `devtools` package for R in order to be able to download and install a forked `rhbase` package from `https://github.com/aaronbenz/rhbase`:

```
$ sudo Rscript -e 'install.packages("devtools",
repos="https://cran.r-project.org/")'
...#output truncated
```

The `devtools` installation process will take 2-3 minutes. After that, install `rhbase` from a specified GitHub repository:

```
$ sudo Rscript -e 'devtools::install_github("aaronbenz/rhbase")'
...#output truncated
```

The preceding command will also install ten other R packages, including: `data.table`, `dplyr`, `tidyr`, `DBI`, `Rcpp`, and others. It may take 10-15 minutes to complete.

We should have now all the essential components of our system prepared for data processing. Now we just need to read the data into HDFS and HBase.

Importing the data to HDFS and HBase

For this tutorial we will continue using the Land Registry Price Paid Data, which was introduced to you earlier in this chapter. You already know how to copy a file from your computer to a HDInsight cluster – if not, adjust the paths, filenames, and usernames of the following command according to your settings and run it in a new terminal window:

```
$ scp -r ~/Desktop/B05396_Ch07_Code/pp-2015.csv swalko@hbasecluster-ssh.azurehdinsight.net:~/
...#output truncated
```

The `pp-2015.csv` file should now reside in the home directory of the `swalko` user on the head node to which you are connected through `ssh`:

```
$ ls
pp-2015.csv   R   rstudio-server-0.99.896-amd64.deb   thrift-0.9.0
thrift-0.9.0.tar.gz
```

From this directory (with the data file), move the data to HDFS and to the `/user/swalko/` directory:

```
$ hadoop fs -copyFromLocal pp-2015.csv /user/swalko/pp-2015.csv
```

Check whether the data is in HDFS:

```
$ hadoop fs -ls /user/swalko
Found 2 items
drwx------   - swalko supergroup          0 2016-04-20 20:40
/user/swalko/.Trash
-rw-r--r--   1 swalko supergroup  169004671 2016-04-20 20:53
/user/swalko/pp-2015.csv
```

All seems fine at the moment. As we have the data in HDFS, we may now import its contents from HDFS to the HBase database. For that purpose, start the HBase shell:

```
$ hbase shell
```

After several seconds you will see the confirmation (and some warning) messages as well as the HBase welcome output, as follows:

```
2016-04-20 20:57:46,420 INFO  [main] impl.MetricsSinkAdapter: Sink azurefs2 started
2016-04-20 20:57:46,589 INFO  [main] impl.MetricsSystemImpl: Scheduled snapshot period at 60 second(s).
2016-04-20 20:57:46,590 INFO  [main] impl.MetricsSystemImpl: azure-file-system metrics system started
2016-04-20 20:57:46,944 WARN  [main] conf.Configuration: hbase-site.xml:an attempt to override final param
eter: dfs.support.append;  Ignoring.
HBase Shell; enter 'help<RETURN>' for list of supported commands.
Type "exit<RETURN>" to leave the HBase Shell
Version 1.1.2.2.4.1.1-3, r31b5c303cc0515995a2da7b3297abad1c50fb2d9, Tue Apr  5 19:05:54 UTC 2016

hbase(main):001:0>
```

We will now create a table in HBase which will hold our data. Unfortunately, as we haven't collected this particular data ourselves, we had no influence on the shape and format of the final dataset. The data doesn't contain a friendly ID variable for each row, so extracting specific rows in R will be a very difficult task. However, our main goal in this tutorial is to guide you through the most essential processes of installation and configuration of all components required by R, enable you to connect it to HBase through the Thrift server, and retrieve some data using the R environment. Having said this, while in HBase shell, let's create a table named `bigTable` that will include three distinct families of columns: `transaction`, `geography` and `property`. In our example, it makes sense to create three such column families as they are topically related to the variables present in the data. However, very often only one family suffices, and this approach is generally recommended in most cases:

```
hbase(main):001:0> create 'bigTable','transaction','geography','property'
0 row(s) in 4.8500 seconds
=> Hbase::Table - bigTable
```

The created table is currently empty, but it has been deployed. To double-check, you can issue the `list` command, which will print all available tables in the database:

```
hbase(main):002:0> list
TABLE
bigTable
1 row(s) in 0.0530 seconds
=> ["bigTable"]
```

Once `bigTable` is created, we can exit the HBase shell:

```
hbase(main):003:0> exit
...#output truncated
```

We should now be back to the `swalko` area on the head node of our cluster. We will then populate the `bigTable` with the data that currently resides in HDFS. In this case we will use the `importtsv` method, which is nothing more than a pre-defined MapReduce job that has been created for our convenience. Before we can run the `importtsv` job, we need to make sure we specify a correct variable separator (for CSV files it should generally be a comma sign) and identify all columns that have to be read in. As you can see, we have provided a reference to previously defined column families for each column in our data. Also, the first column (which is `uniqueID`) will actually function as the HBase `row key`. Finally, we determine the name of the table in HBase, which will be populated with this data and the location of the data file in HDFS:

```
$ hbase org.apache.hadoop.hbase.mapreduce.ImportTsv -Dimporttsv.separator=,
-
```

```
Dimporttsv.columns="HBASE_ROW_KEY,transaction:price,transaction:transferDat
e,geography:postcode,property:propType,property:oldNew,property:tenure,geog
raphy:PAON,geography:SAON,geography:street,geography:locality,geography:tow
n,geography:district,geography:county,transaction:ppdCat,transaction:record
Status" bigTable /user/swalko/pp-2015.csv
...#output truncated
```

Note that, as this is in fact a MapReduce job, you will see a standard MapReduce output with the progress of Mapper and Reducer functions (although in the `importtsv` job, only the Mapper is used). It may take several minutes for the job to run.

When the job completes successfully, we can go back to the HBase shell and check whether the dataset has been imported:

```
$ hbase shell
...#output truncated
hbase(main):005:0> scan 'bigTable'
...#output truncated
"{0009132D-E1E9-4BAB-AD94-   column=geography:PAON, timestamp=1461187474121,
value="10"
     DB4F356E351A}"
 "{0009132D-E1E9-4BAB-AD94-   column=geography:SAON, timestamp=1461187474121,
value=""
     DB4F356E351A}"
 "{0009132D-E1E9-4BAB-AD94-   column=geography:county,
timestamp=1461187474121, value="TYNE AND WEAR"
     DB4F356E351A}"
 "{0009132D-E1E9-4BAB-AD94-   column=geography:district,
timestamp=1461187474121, value="SOUTH TYNESIDE"
     DB4F356E351A}"
 "{0009132D-E1E9-4BAB-AD94-   column=geography:locality,
timestamp=1461187474121, value=""
     DB4F356E351A}"
...#output truncated
```

You can exit the listing and the HBase shell by pressing *Ctrl + C*.

There is only one last task we need to complete before proceeding to RStudio. We have to start the Thrift server:

```
$ hbase thrift start
...#output truncated
```

It may take several seconds to start it. When ready you should see the following output:

```
2016-04-20 22:25:56,008 INFO  [main] http.HttpServer: Jetty bound to port 9095
2016-04-20 22:25:56,014 INFO  [main] mortbay.log: jetty-6.1.26.hwx
2016-04-20 22:25:57,132 INFO  [main] mortbay.log: Started SelectChannelConnector@0.0.0.0:9095
2016-04-20 22:25:57,134 DEBUG [main] thrift.ThriftServerRunner: Using binary protocol
2016-04-20 22:25:57,430 INFO  [main] thrift.ThriftServerRunner: starting TBoundedThreadPoolServer on /0.0.0.0:9090; min worker threads=16, max worker threads=1000, max queued requests=1000
```

Note there is no command prompt available as the server is in use, but remember to press *Ctrl* + *C* to exit at any time when needed. Once Thrift is operational, we may go to RStudio Server and try to connect with HBase.

Reading and querying HBase using the rhbase package

Point your preferred web browser to the public IP address of the head note with port `8787`, which we set up earlier in this chapter, and log in to RStudio Server as usual. Upon login, just like in the standard Hadoop cluster, we need to define environment variables:

```
> cmd <- system("which hadoop", intern=TRUE)
> cmd
[1] "/usr/bin/hadoop"
> Sys.setenv(HADOOP_CMD=cmd)
> stream <- system("find /usr -name hadoop-streaming*jar", intern=TRUE)
...#output truncated
> stream
[1] "/usr/hdp/2.4.1.1-3/hadoop-mapreduce/hadoop-streaming-2.7.1.2.4.1.1-3.jar"
[2] "/usr/hdp/2.4.1.1-3/hadoop-mapreduce/hadoop-streaming.jar"
[3] "/usr/hdp/2.4.1.1-3/oozie/share/lib/mapreduce-streaming/hadoop-streaming-2.7.1.2.4.1.1-3.jar"
attr(,"status")
[1] 1
> Sys.setenv(HADOOP_STREAMING=stream[1])
> Sys.getenv("HADOOP_CMD")
[1] "/usr/bin/hadoop"
> Sys.getenv("HADOOP_STREAMING")
[1] "/usr/hdp/2.4.1.1-3/hadoop-mapreduce/hadoop-streaming-2.7.1.2.4.1.1-3.jar"
```

R with Non-Relational (NoSQL) Databases

Load the `rmr2` and `rhdfs` packages to be able to check HDFS:

```
> library(rmr2)
> library(rhdfs)
```

Start HDFS using the `hdfs.init()` function:

```
> hdfs.init()
```

Check whether the data set is in HDFS:

```
> hdfs.ls("/user/swalko")
  permission  owner      group       size
1 drwx------  swalko supergroup          0
2 -rw-r--r--  swalko supergroup  169004671
           modtime                         file
1 2016-04-20 20:40         /user/swalko/.Trash
2 2016-04-20 20:53 /user/swalko/pp-2015.csv
```

Based on the output, everything works well and we can see the data in HDFS. Let's now connect to the HBase database. Firstly, load the `rhbase` package and set up the host address and port for the Thrift server (the default is `9090`):

```
> library(rhbase)
> hostLoc = '127.0.0.1'
> port = 9090
```

Start the connection with HBase through the Thrift server:

```
> hb.init(hostLoc, port, serialize = "character")
<pointer: 0x167450d0>
attr(,"class")
[1] "hb.client.connection"
```

We may print all available tables in HBase using the `hb.list.tables()` function:

```
> hb.list.tables()
$bigTable
              maxversions compression inmemory
geography:              1        NONE    FALSE
property:               1        NONE    FALSE
transaction:            1        NONE    FALSE
              bloomfiltertype bloomfiltervecsize
geography:                ROW                  0
property:                 ROW                  0
transaction:              ROW                  0
              bloomfilternbhashes blockcache
geography:                      0       TRUE
```

[368]

```
property:                        0       TRUE
transaction:                     0       TRUE
                 timetolive
geography:       2147483647
property:        2147483647
transaction:     2147483647
```

As we expected, there is only the `bigTable` table, which we created earlier. The output shows several basic default settings and the three column families which we defined when creating the table in the HBase shell.

In case you have several tables available, you can print the preceding output for a named table using the `hb.describe.table()` command:

```
> hb.describe.table("bigTable")
              maxversions compression inmemory
geography:         1          NONE     FALSE
property:          1          NONE     FALSE
transaction:       1          NONE     FALSE
...#output truncated
```

We may inspect the regions of the table using the `hb.regions.table()` function:

```
> hb.regions.table("bigTable")
[[1]]
[[1]]$start
[1] NA
[[1]]$end
[1] ""{288DCE29-5855"
[[1]]$id
[1] 1.461188e+12

[[1]]$name
[1] "bigTable,,1461187589128.d422aab7c3bd51a502e7a39d9271ce3d."
[[1]]$version
[1] 01
...#output truncated
```

Depending on how you defined row keys and column families, you may extract requested values for specific transactions in the dataset. As explained before, the row keys for the Price Paid Data are particularly unfriendly and they are not very informative. The following snippet retrieves the data related to the `property` column family including values for `oldNew`, `propType`, and `tenure` columns for one specific `row key` (which is in fact a single, individual transaction):

```
> hb.pull("bigTable",
+         column_family = "property",
```

```
+          start = ""{23B6165E-FED6-FCF4-E050-A8C0620577FA}"",
+          end = ""{23B6165E-FED6-FCF4-E050-A8C0620577FA}"",
+          batchsize = 100)
                                                    rowkey
1: "{23B6165E-FED6-FCF4-E050-A8C0620577FA}"
2: "{23B6165E-FED6-FCF4-E050-A8C0620577FA}"
3: "{23B6165E-FED6-FCF4-E050-A8C0620577FA}"
       column_family  column  values
1:     property:oldNew    NA   "N"
2:     property:propType  NA   "S"
3:     property:tenure    NA   "F"
```

Based on how we created our table earlier, we can also extract just one particular value for the row key; in the following example we will retrieve the paid price for the specific property:

```
> hb.pull("bigTable",
+          column_family = "transaction:price",
+          start = ""{23B6165E-FED6-FCF4-E050-A8C0620577FA}"",
+          end = ""{23B6165E-FED6-FCF4-E050-A8C0620577FA}"",
+          batchsize = 100)
                                                    rowkey
1: "{23B6165E-FED6-FCF4-E050-A8C0620577FA}"
       column_family   column   values
1:  transaction:price    NA    "355000"
```

Using the `hb.scan()` function we may retrieve values for specified columns across multiple transactions from the range defined by the `startrow` and `end` parameters. The `while()` loop allows us to iterate the scan for all transactions within this range:

```
> iter <- hb.scan("bigTable",
+          startrow = ""{23B6165E-FED6-FCF4-E050-A8C0620577FA}"",
+          end = ""{23B6165F-0452-FCF4-E050-A8C0620577FA}"",
+          colspec = "transaction:price")
> while( length(row <- iter$get(1))>0 ){
+     print(row)
+ }
...#output truncated
[[1]]
[[1]][[1]]
[1] ""{23B6165F-0450-FCF4-E050-A8C0620577FA}""
[[1]][[2]]
[1] "transaction:price"
[[1]][[3]]
[[1]][[3]][[1]]
[1] ""211000""

> iter$close()
```

The `hb.get()` command, on the other hand, allows users to extract values for all columns for a list of defined row keys. As the output is returned as a `list`, it's easy to retrieve values of interest in further calculations:

```
> hb.get("bigTable",
+       list(""{23B6165E-FED6-FCF4-E050-A8C0620577FA}"",
+            ""{23B6165F-0452-FCF4-E050-A8C0620577FA}""))
[[1]]
[[1]][[1]]
[1] ""{23B6165E-FED6-FCF4-E050-A8C0620577FA}""
 [[1]][[2]]
 [1] "geography:PAON"
 [2] "geography:SAON"
 [3] "geography:county"
 [4] "geography:district"
 [5] "geography:locality"
 [6] "geography:postcode"
 [7] "geography:street"
 [8] "geography:town"
 [9] "property:oldNew"
[10] "property:propType"
[11] "property:tenure"
[12] "transaction:ppdCat"
[13] "transaction:price"
[14] "transaction:recordStatus"
[15] "transaction:transferDate"
[[1]][[3]]
[[1]][[3]][[1]]
[1] ""23""
[[1]][[3]][[2]]
[1] """"
[[1]][[3]][[3]]
[1] ""CARDIFF""
[[1]][[3]][[4]]
[1] ""CARDIFF""
...#output truncated
```

If the data is no longer needed you can delete the entire table from HBase:

```
> hb.delete.table("bigTable")
[1] TRUE
```

Apart from the above functionalities, the `rhbase` package supports other data management and processing operations that go slightly beyond the limitations of this chapter. Depending on your data and how you define its schema, other queries and aggregations are possible. The `rhbase` tutorial created by Aaron Benz, the author of the package, available at https://github.com/aaronbenz/rhbase/blob/master/examples/rhbase_tutorial.Rmd, will guide you through the remaining methods of data management and values extraction on a different sample of data.

Just a final word of caution. When working with large datasets, you should probably create tables through the HBase shell, as was shown during the tutorial in this chapter. Building a table that will hold Big Data from R is counter-productive, as in that case the data needs to be read into RAM beforehand anyway.

Summary

This chapter was entirely dedicated to non-relational databases. At the very beginning, we introduced you to the general concept of highly-scalable, NoSQL databases with flexible schemas. We have discussed their major features and presented practical applications in which they excel when compared with standard relational SQL databases.

We then began a series of tutorials that were aimed at explaining how to read, manage, process, and query data stored in a very popular, open source NoSQL database called MongoDB. We reviewed three leading R packages that allow users to implement a variety of techniques and methods in MongoDB directly from the R environment.

Finally, we introduced you to an open source, non-relational and distributed HBase database that operates on top of the Hadoop Distributed File System. We guided you through some time-consuming installation procedures to enable you to connect to HBase using R. We then showed you several examples of how to apply functions in the `rhbase` package for R to manage, process, and query data stored in HBase deployed on a multi-node Azure HDInsight cluster.

In the next chapter, we will continue working on a Distributed File System through the Apache Spark engine connected to R.

7
Faster than Hadoop - Spark with R

In `Chapter 4`, *Hadoop and MapReduce Framework for R*, you learned about Hadoop and MapReduce frameworks that enable users to process and analyze massive datasets stored in the **Hadoop Distributed File System** (**HDFS**). We launched a multi-node Hadoop cluster to run some heavy data crunching jobs using R language which would not be otherwise achievable on an average personal computer with any of the R distributions installed. We also said that although Hadoop is extremely powerful, it is generally recommended for data that greatly exceeds the memory limitations due to its rather slow processing. In this chapter we would like to present Apache Spark engine—a faster way to process and analyze Big Data. After reading this chapter, you should be able to:

- Understand and appreciate Spark characteristics and functionalities
- Deploy a fully-operational, multi-node Microsoft Azure HDInsight cluster with Hadoop, Spark, and Hive fully-configured and ready to use
- Import data from HDFS into Hive tables and use them as a data source for processing in Spark engine
- Connect RStudio Server to HDFS resources including Hadoop and Spark and run simple and more complex Spark data transformations, aggregations, and analyses using Bay Area Bike Share open data

Spark for Big Data analytics

Spark is often considered as a new, faster, and more advanced engine for Big Data analytics that could soon overthrow Hadoop as the most widely used Big Data tool. In fact, there is already a visible trend for many businesses to opt for Spark rather than Hadoop in their daily data processing activities. Undoubtedly, Spark has several selling points that make it a more attractive alternative to the slightly complicated, and sometimes clunky Hadoop:

- It's pretty fast and can reduce the processing time by up to 100 times when run in memory as compared to standard Hadoop MapReduce jobs or up to 10 times if run on disk.
- It's a very flexible tool that can run as a standalone application, but also can be deployed on top of Hadoop and HDFS, and other distributed file systems.
- It can use a variety of data sources from standard relational databases, through HBase, Hive, to Amazon S3 containers. It may also be launched in the cloud. In fact, in the tutorial in this chapter, we will show you how to run Spark in the cloud on top of HDFS using Hive as a data course and RStudio Server as a front-end application.
- Spark is easy to use and may appeal to a large number of developers and data scientists as it supports multiple popular programming languages such as Java, Python, Scala, and of course our favorite R language. These can also further connect through numerous drivers to other languages, for example SQL, allowing users to run queries directly from the R console or Python shell.

Since 2012, Spark has developed from a university run project initiated by Matei Zaharia at the University of California Berkeley to the largest open-sourced Big Data solution, with hundreds of contributors and a vibrant community of active users. Over this period, Spark has also transformed into a general application that supports not only simple data processing of massive datasets, but also allow users to carry out complex analytics using a large selection of in-built machine learning algorithms through its **MLlib** library, graph data analysis through the GraphX component, SQL querying using **Spark SQL** library, and ability to build streaming data applications through **Spark Streaming**. These modules can be integrated together to create even more powerful Big Data analytics stacks powered by the Spark engine as described at its project website: `http://spark.apache.org/`.

In the next section, we will launch a multi-node Hadoop and Spark cluster and we will run several Big Data transformations and analyses using the `SparkR` package for R.

Spark with R on a multi-node HDInsight cluster

Although Spark can be deployed in single-node, standalone mode, its powerful capabilities are best fit for multi-node applications. With this in mind, we will dedicate most of this chapter to practical Big Data crunching with Spark and R on a Microsoft Azure HDInsight cluster. As you should already be familiar with the deployment process of HDInsight clusters, our Spark workflows will contain one additional twistâ©©the Spark framework will process the data straight from the **Hive** database, which will be populated with tables from HDFS. The introduction of Hive is a useful extension of the concepts covered in `Chapter 5`, *R with Relational Database Management Systems (RDBMSs)* and `Chapter 6`, *R with Non-Relational (NoSQL) Databases*, where we discussed the connectivity of R with relational and non-relational databases. But before we can use it, we should firstly launch a new HDInsight cluster with Spark and RStudio.

Launching HDInsight with Spark and R/RStudio

After reading `Chapter 4`, *Hadoop and MapReduce Framework for R* and `Chapter 6`, *R with Non-Relational (NoSQL) Databases*, you are already familiar with the deployment process of HDInsight clusters. In this section, we will only very briefly explain several alterations to the names of created Azure resources and specific options to select for the purpose of Spark installation. Despite these little modifications, you should follow the previously explained instructions from the HDInsight – A multi-node Hadoop cluster on Azure section in `Chapter 4`, *Hadoop and MapReduce Framework for R*.

> It is also very important for you to understand that, by following this tutorial you may be charged additional fees depending on your individual Microsoft Azure subscription. As usual, the exact amount of these fees depends on many factors such as the location of servers you choose, the amount of worker and head nodes and their configuration, the size of storage space, and many others. Please check the Microsoft Azure web pages for further details on fees, charges, and subscription options.

Faster than Hadoop - Spark with R

Just as we did earlier in Chapter 4, *Hadoop and MapReduce Framework for R*, and Chapter 6, *R with Non-Relational (NoSQL) Databases*, we will begin by creating a new **Resource Group**. This time we will give it the name `sparkcluster` and we will select an appropriate location, **West Europe** in our case. The resulting **Resource group** may look like the following one:

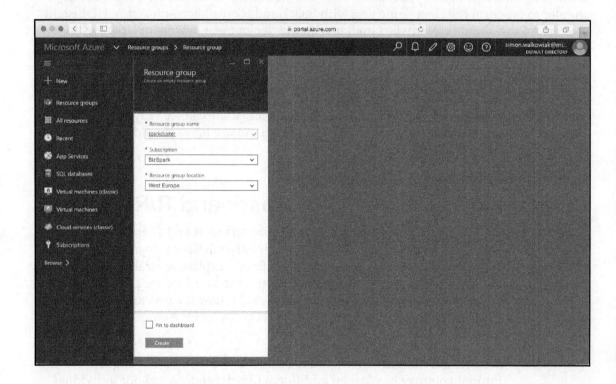

Chapter 7

Secondly, we will create a new Virtual Network. Its name will be `sparkVN`. Make sure to select your correct location and choose the **Resource group** that you set up in the preceding step (for example `sparkcluster`). Once complete, click the **Create** button:

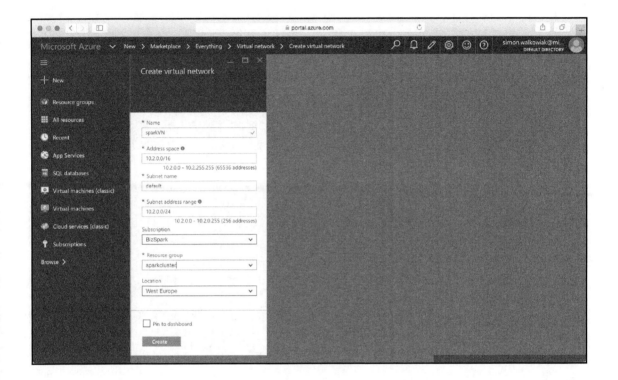

Thirdly, we may set up a new Network Security Group. Again, we will give it a name, for example `sparkNSG`, and identify its **Resource Group** (for example `sparkcluster`) and the **Location** (for example `West Europe`):

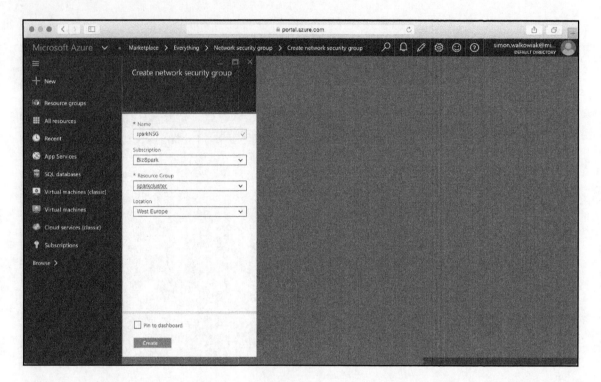

At this stage, we already have three basic Microsoft Azure resources set up and ready for linking with the HDInsight cluster. These resources include the **Resource Group** called `sparkcluster`, the **Virtual Network** named `sparkVN`, and finally the **Network Security Group** called `sparkNSG`. We may now move on to creating a new HDInsight cluster.

Chapter 7

In the **New HDInsight Cluster** window, provide the **Cluster Name**, for example `swalkospark`, and select your **Subscription** type. Click on the **Select Cluster Type** tab and choose **Spark (Preview)** from the **Cluster Type** option menu. As you may notice, the options contain Microsoft R Server, but as it is a new R distribution and we hadn't introduced it in the earlier chapters we will continue using RStudio Server, which has to be installed separately after the cluster's deployment. Make sure that `Linux` is chosen as the **Operating System** and choose a recent **Version** of Spark release, in our case it will be `Spark 1.6.0 (HDI 3.4)`. Finally, select the `Standard`, **Cluster Tier** without the Microsoft R Server for HDInsight. Click on the **Select** button when ready:

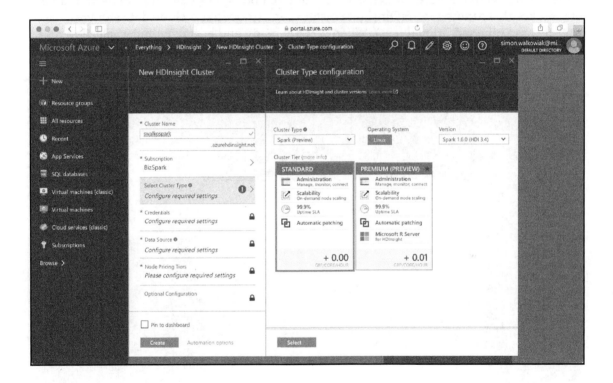

Faster than Hadoop - Spark with R

Provide Ambari and `ssh` credentials by completing the **Credentials** form just as you did last time (go back to `Chapter 4`, *Hadoop and MapReduce Framework for R* if you need help with it). Then create a new data storage, for example `swalkostorage1` in the appropriate geographical region (for example `West Europe`). Again, the **Data Source** configuration is quite straightforward and we have covered it already in `Chapter 4`, *Hadoop and MapReduce Framework for R*. When both the **Credentials** and **Data Source** forms are completed, proceed to the **Node Pricing Tiers** tab. Here, we will keep the recommended 4 Worker Nodes, but as we don't deal with enormously large datasets in this tutorial we will scale them down to `D12` nodes, each with 4 cores and 28 GB of RAM. We will also select the same type of virtual machines for both our Head Nodes. The selection of node types will result in re-estimation of your hourly spending for HDInsight cluster usage (it doesn't include data storage, transfer, and other additional fees). When you are ready and happy with the selection, proceed by clicking the **Select** button:

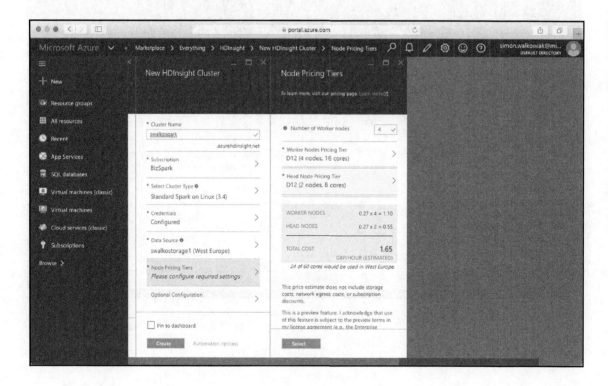

Chapter 7

As we want to link the HDInsight with previously created resources and install R, we need to complete the **Optional Configuration** form in the same manner we did last time. Make sure to associate the cluster with the **Virtual Network** created before, for example `sparkVN`, and choose `default` in the **Subnet** option. In the **Script Actions** provide a **NAME** for the script and a correct **SCRIPT URI**, which will install core R. The link is the same as in the previous chapters:
`https://hdiconfigactions.blob.core.windows.net/linuxrconfigactionv01/r-installer-v01.sh`. Make sure to tick all the fields for **HEAD**, **WORKER** and **ZOOKEEPER**, and leave **PARAMETERS** empty. Confirm your configurations by pressing the **Select** button until you reach the main **New HDInsight Cluster** window. Then, finally, assign the cluster to the appropriate **Resource Group**, which you created earlier, for example `sparkcluster`. When everything is configured, press **Create** to start the deployment process of this HDInsight cluster:

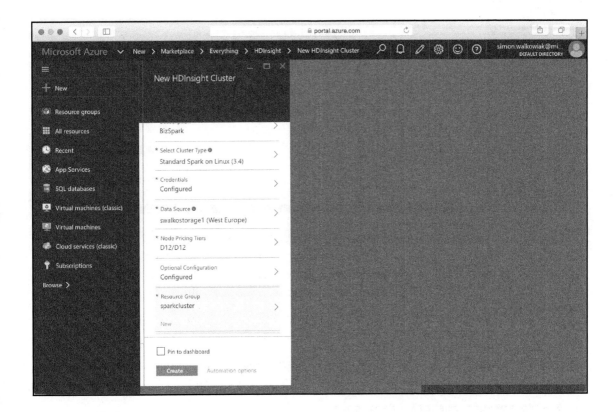

[381]

Faster than Hadoop - Spark with R

It may take half an hour or longer to launch. Once you receive the notification that the deployment succeeded, you may create a new public IP address. Go back to the Resource Group (for example `sparkcluster`), find the Head Node 0 and enable the public IP address for this node. We did this previously in Chapter 4, *Hadoop and MapReduce Framework for R*♦♦please re-visit it if you need a revision of detailed instructions on how to do it. A properly set up public IP address window in Azure should look as follows:

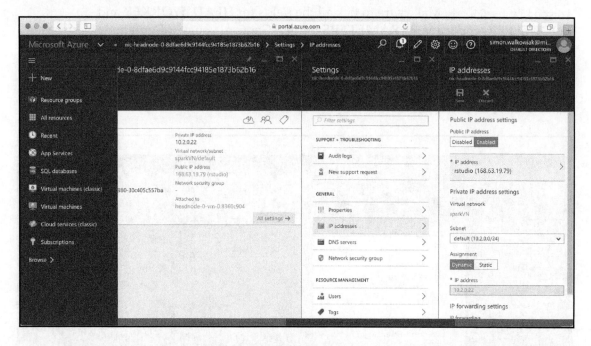

As you probably remember, assigning a public IP address allows us to access the installed RStudio Server from a browser. But before we can do that, we firstly need to `ssh` to the Head Node of the cluster and install RStudio Server. In the new Terminal window, `ssh` to your cluster (alter the user and cluster names according to the values you set up in the preceding step):

```
$ ssh swalko@swalkospark-ssh.azurehdinsight.net
```

Then install RStudio Server on the Spark HDInsight cluster as explained on the RStudio website at https://www.rstudio.com/products/rstudio/download-server-2/:

```
$ sudo apt-get install gdebi-core
...#output truncated
$ wget https://download2.rstudio.org/rstudio-server-0.99.896-amd64.deb
...#output truncated
```

```
$ sudo gdebi rstudio-server-0.99.896-amd64.deb
...#output truncated
```

We can now access RStudio Server by directing our web browser to the assigned public IP address for the Head Node and port `8787`. If you followed all the instruction guidelines from Chapter 4, *Hadoop and MapReduce Framework for R* very cautiously, you might have already added port `8787` to the Inbound Security Rules in the Network Security Group for this cluster (we named it `sparkNSG`). If you didn't do it, make sure to complete this step before pointing your web browser to the public IP address with port `8787`. Once this is done, you can log in to the RStudio Server by providing your `ssh` credentials. Then you may proceed to the next section of this chapter.

Reading the data into HDFS and Hive

Hopefully, the preceding section wasn't too difficult. At this point you should have the HDInsight cluster configured and prepared for the import of the data. As explained earlier, you can also install other R packages, which you may find useful during the data processing in the later stages. We had already explained how to do this in the previous chapters of this book and so we will skip these instructions in this part.

In this tutorial we will be using the **Bay AreaBike Share** open data, available for download (as a ZIP file) from `http://www.bayareabikeshare.com/open-data`. To be precise, we will only use the Year 2 data, which covers bike trips and related events between September 1, 2014 and August 31, 2015. Bike share schemes are increasingly popular, cheap (and sometimes even completely free!), healthy, and environmentally-friendly public transportation systems in major cities around the world. In order for these schemes to operate and provide a good quality service, their operators need to collect and analyze quite considerable amounts of data on a daily basis.
Each year a ZIP file of the Bay Area Bike Share open data contains four CSV files with information about the bike and dock availability (status data files), station metadata (station data files), details of each individual bike trip (trip data files), and weather data for each day and city (weather data files). The ZIP bundle also includes a standard `README.txt` file that describes the contents of the downloaded files. The largest of all four data files is the `201508_status_data.csv` file which contains nearly 37 million rows of bike and dock availability by minute for each station (~1.1 GB in total).

Each docking station comes with its own `station_id` value, which can be used to merge the status data file with more detailed station metadata provided in the `201508_station_data.csv` (only 5 KB in size). Apart from the `station_id` variable, the station data file includes the actual full name of each station and its latitudinal and longitudinal co-ordinates—a useful piece of information for geo-spatial visualizations, but also a total number of docks available at each station, the city which the station belongs to (for example San Francisco, Redwood City, Palo Alto, Mountain View, or San Jose) and the original date of the station's installation. Knowing the names of the stations and the cities, the status and station data files can be easily combined with information from the two remaining files that contain individual bike trip details and weather data. The trip data (`201508_trip_data.csv`) provides fine-grained information on approximately 354,000 individual bike trips (43 MB in total) such as the duration of each trip in seconds, both start and end dates and docking stations for the trip, the type of bike scheme member (whether a subscriber, who has either an annual or 30-day pass, or a short-term customer with either a 24-hour or 3-day pass), the Zip Code of a subscriber, and even the identification number of the bike that was used during the trip. On the other hand, the weather data file (`201508_weather_data.csv`, 1,825 records, ~159 KB in size) includes daily weather information for each city such as the temperature, dew point, precipitation level, cloud cover, gust, and wind speed, humidity, pressure, visibility, and even the wind direction. The Zip Code variable makes it possible to combine the data geographically with other Bay Area Bike Share datasets.

Of course, all data files can be relatively easily explored and analyzed on personal computers as there is only one file over the 1GB size threshold. However, note that as you create more variables or merge datasets with one another, the actual memory used during all data processing activities or calculations may far exceed your available RAM. The combined sizes of all Bay Area Bike Share files make them very suitable for processing in HDFS using either Hadoop or Spark frameworks. If you want to follow the examples in this chapter, make sure to download the Year 2 data from `http://www.bayareabikeshare.com/open-data`.

Getting the data into HDFS

Assuming that you have downloaded the zipped Year 2 data files from the Bay Area Bike Share website, you may now either unpack them on your personal machine or copy the files across as a ZIP file to your user area on the HDInsight cluster and unzip it there. We have unpacked the bundle locally and renamed the folder, data.

Provided that you did similarly, and your user and cluster names are the same as ours, you may now open a new Terminal window and copy all files across to the HDInsight cluster:

```
$ scp -r ~/Desktop/data/ swalko@swalkospark-ssh.azurehdinsight.net:~/data/
```

As always, you will be asked to provide the password for the specified user. The process of copying of all files may take up to several minutes depending on your Internet connection speed. Once complete, you may check whether the directory has been copied successfully. In the old Terminal window type the following:

```
$ ls
data   R   rstudio-server-0.99.896-amd64.deb
```

We can see that the data folder has indeed been created. Use the already introduced cd and ls commands to inspect the folder to check whether it has been populated with all files.

We may also double-check the current path to the user storage area on HDInsight. This information will be needed later when we copy the files from the user area to the Hadoop Distributed File System (HDFS):

```
$ pwd
/home/swalko
```

Once we have all our data files on the virtual disk, we may copy them to HDFS. But before we are able to do that, we have to create a new directory on HDFS for the swalko user. We can achieve this with the hadoop fs commands and mkdir option, which you have already learned in Chapter 4, *Hadoop and MapReduce Framework for R*:

```
$ hadoop fs -mkdir /user/swalko
$ hadoop fs -mkdir /user/swalko/data
```

In the preceding code, we have also created a folder named data, which will store all our data files. Let's now check whether this operation was successful:

```
$ hadoop fs -ls
Found 1 items
drwxr-xr-x   - swalko supergroup          0 2016-04-30 10:27 data
```

The empty `data` folder is now visible in HDFS for the `swalko` user. We may now copy the files from the local user area to this newly created folder on HDFS:

```
$ hadoop fs -copyFromLocal data/ /user/swalko/
$ hadoop fs -ls /user/swalko/data
Found 4 items
-rw-r--r--   1 swalko supergroup       5272 2016-04-30 10:42 /user/swalko/data/201508_station_data.csv
-rw-r--r--   1 swalko supergroup 1087241932 2016-04-30 10:42 user/swalko/data/201508_status_data.csv
-rw-r--r--   1 swalko supergroup   43012650 2016-04-30 10:42 /user/swalko/data/201508_trip_data.csv
-rw-r--r--   1 swalko supergroup     158638 2016-04-30 10:42 /user/swalko/data/201508_weather_data.csv
```

The preceding output confirms that we have completed copying all data files from the local storage area on HDInsight to the `swalko` user area within HDFS. The output also provides information on the size of files, their creation timestamps, and most importantly, their individual paths on HDFS.

Importing data from HDFS to Hive

As mentioned in the introduction to this chapter, we will be processing the data that will be read from tables created in the Hive database. Although initially originating from Facebook, Hive is one of the many Apache projects that is usually built on top of HDFS and other compatible Big Data file systems like Amazon S3, and others. The Hive engine can be thought of as a relational database management system that supports HDFS storage. In fact, its operations and queries can be performed using an SQL-like language called **HiveQL**. HiveQL is very similar to standard SQL, however it only offers limited support for indexes and sub-queries. A comparison between SQL and HiveQL commands can be found on Hortonworks's website at `http://hortonworks.com/blog/hive-cheat-sheet-for-sql-users/`. Note that the cheat sheet was created in the second half of 2013, hence it doesn't include commands for newer releases of Hive, but in essence all presented functions should still work fine. The full HiveQL language manual is available at the Confluence page created by Apache at `https://cwiki.apache.org/confluence/display/Hive/LanguageManual`. Despite some already identified limitations, Hive can provide users with fast data querying and analysis of massive datasets stored within HDFS. It does it by converting the queries into MapReduce or Spark, which are then processed by Hadoop.

Chapter 7

You can start the Hive shell straight from the command line using the `hive` function:

```
$ hive
WARNING: Use "yarn jar" to launch YARN applications.
Logging initialized using configuration in file:/etc/hive/2.4.1.1-3/0/hive-log4j.properties
```

After several seconds you will be able to see the Hive shell command prompt:

```
hive>
```

As the HiveQL basic management functions are very similar to SQL commands, we may easily start creating tables for our data in Hive. Firstly, let's see if there are any databases already created in Hive:

```
hive> show databases;
OK
default
Time taken: 0.309 seconds, Fetched: 1 row(s)
```

There is only a `default` Hive database included. We will then create a new one called `bikes`, which will hold all four Bay Area Bike Share data files:

```
hive> create database bikes;
OK
Time taken: 0.693 seconds
```

We may test whether it has been created:

```
hive> show databases;
OK
bikes
default
Time taken: 0.241 seconds, Fetched: 2 row(s)
```

As with other database engines, we should indicate which specific database we want to use to create new tables within:

```
hive> use bikes;
OK
Time taken: 0.193 seconds
```

Although the bikes database should now be empty, we can always check its contents using the `show tables;` command:

```
hive> show tables;
OK
Time taken: 0.177 seconds
```

Having prepared the database, we are ready to import our data files from HDFS. Before we can do that, we must specify a schema for each table.

> The README.txt file that comes with the ZIP bundle of the Bay Area Bike Share open data contains information on variable names for most of the files. From the contents of this file we can also deduce the data type of each variable, however, sometimes it is not that straightforward. It is slightly disappointing that most of the public use files available online do not include any details on variables, their types, or even the size of the downloadable files. In the times of Big Data and the Internet of Things, users should be provided with essential information about the composition of data files and their contents before downloading the files or applying for API access. Fortunately, in the United Kingdom there are several initiatives to regulate and standardize the distribution and access to open dataâ⊚⊚this also includes the access to large datasets or governmental and disclosure data. One of the organizations standing at the very forefront of this activity is the **Open Data Institute** http://theodi.org/ founded by Sir Tim Berners-Lee and Sir Nigel Shadbolt. It is, however, still a very new area of interest for many organizations, academics, and businesses, but we hope this aspect of data science can soon become more standardized across most of the publically available data sources such as online repositories, digital archives, governmental, and business open data services.

We have explored all of the files for you, so you don't have to do it yourself. The following code will create a table with a schema containing four variables. The table (named status_data) will hold values extracted from the 201508_status_data.csv file, however we will load them once the table is created:

```
hive> create external table status_data(station_id int, bikes_available
int, docks_available int, time string) row format delimited FIELDS
terminated by "," LINES terminated by "\n" stored as textfile
tblproperties("skip.header.line.count"="1");
OK
Time taken: 1.138 seconds
```

From the preceding code, you can see that we have set the field separator to a comma as our data comes in CSV format. Also, as each row is on a new line we had to indicate that the lines are terminated by "\n". Finally, our data contained variable labels in the first row, so we had to skip the header line when creating the schema for our table. This however will not work for other tools, for example with Spark run through RStudio Serverâ€”the header information will continue to be included as the actual values of the data, but we will show you how to remove it later.

Once the schema for the first file is defined, we may now populate it with our data stored in HDFS. Remember that the path to HDFS files starts from /user/swalko/, and not /home/swalko/-the latter is simply the path to the storage area for swalko on the HDInsight disk:

```
hive> load data inpath '/user/swalko/data/201508_status_data.csv' into
table status_data;
Loading data to table bikes.status_data
Table bikes.status_data stats: [numFiles=1, totalSize=1087241932]
OK
Time taken: 3.432 seconds
```

The output of the import confirms the name of the table and some basic statistics about the number of imported files and their total size. We may repeat the same procedure for the remaining three Bay Area Bike Share data files. The following commands will create three separate tables with their individual schemas and will import the data from 201508_station_data.csv, 201508_trip_data.csv, and 201508_weather_data.csv:

```
hive> create external table station_data(station_id int, name string, lat
double, long double, dockcount int, landmark string, installation string)
row format delimited FIELDS terminated by "," LINES terminated by "\n"
stored as textfile tblproperties("skip.header.line.count"="1");
OK
Time taken: 0.975 seconds
hive> load data inpath '/user/swalko/data/201508_station_data.csv' into
table station_data;
Loading data to table bikes.station_data
Table bikes.station_data stats: [numFiles=1, totalSize=5272]
OK
Time taken: 3.233 seconds
hive> create external table trip_data(trip_id int, duration int, start_date
string, start_st string, start_term int, end_date string, end_st string,
end_term int, bike_no int, cust_type string, zip int) row format delimited
FIELDS terminated by "," LINES terminated by "\n" stored as textfile
tblproperties("skip.header.line.count"="1");
OK
Time taken: 0.997 seconds
```

```
hive> load data inpath '/user/swalko/data/201508_trip_data.csv' into table
trip_data;
Loading data to table bikes.trip_data
Table bikes.trip_data stats: [numFiles=1, totalSize=43012650]
OK
Time taken: 3.147 seconds
hive> create external table weather_data(daydate string, max_temp int,
mean_temp int, min_temp int, max_dewpoint int, mean_dewpoint int,
min_dewpoint int, max_humid int, mean_humid int, min_humid int,
max_sealevpress double, mean_sealevpress double, min_sealevpress double,
max_visib int, mean_visib int, min_visib int, max_windspeed int,
mean_windspeed int, max_gustspeed int, precipitation double, cloudcover
int, events string, winddirection int, zip int) row format delimited FIELDS
terminated by "," LINES terminated by "\n" stored as textfile
tblproperties("skip.header.line.count"="1");
OK
Time taken: 1.227 seconds
hive> load data inpath '/user/swalko/data/201508_weather_data.csv' into
table weather_data;
Loading data to table bikes.weather_data
Table bikes.weather_data stats: [numFiles=1, totalSize=158638]
OK
Time taken: 3.655 seconds
```

After creating all four tables we can find them in the `bikes` database:

```
hive> show tables;
OK
station_data
status_data
trip_data
weather_data
Time taken: 0.16 seconds, Fetched: 4 row(s)
```

The `describe` command displays variable information for a chosen table, for example:

```
hive> describe status_data;
OK
station_id              int
bikes_available         int
docks_available         int
time                    string
Time taken: 1.106 seconds, Fetched: 4 row(s)
```

Note that in our tables we set the `time` field as a string variable. Hive contains date and timestamp data types, but their default formats didn't work with our data (however they can be defined separately). Setting atypical formats of dates or timestamps to strings will allow us to convert them to the actual date or timestamp values later using R language and functionalities of the `SparkR` package.

Using HiveQL, we can query the data straight from the Hive shell. In the following example, we will extract the first five rows of records from the `status_data` table:

```
hive> select * from status_data limit 5;
OK
2    15    12    "2014-09-01 00:00:03"
2    15    12    "2014-09-01 00:01:02"
2    15    12    "2014-09-01 00:02:02"
2    15    12    "2014-09-01 00:03:03"
2    15    12    "2014-09-01 00:04:02"
Time taken: 0.537 seconds, Fetched: 5 row(s)
```

As previously mentioned, users can calculate statistics with HiveSQL just as in standard SQL queries. In the following statement we will calculate the overall number of records in the `status_data` table:

```
hive> select count(*) from status_data;
```

Note that Hive runs this query as a MapReduce job:

Based on the output, the `status_data` table contains exactly 36,647,622 records. You may also notice that the whole operation took 50 seconds to complete.

We can now repeat the `describe` command and a simple query to print the first five rows of data for the `weather_data` table:

```
hive> describe weather_data;
OK
daydate                 string
max_temp                int
mean_temp               int
min_temp                int
max_dewpoint            int
mean_dewpoint           int
min_dewpoint            int
max_humid               int
mean_humid              int
min_humid               int
max_sealevpress         double
mean_sealevpress        double
min_sealevpress         double
max_visib               int
mean_visib              int
min_visib               int
max_windspeed           int
mean_windspeed          int
max_gustspeed           int
precipitation           double
cloudcover              int
events                  string
winddirection           int
zip                     int
Time taken: 0.83 seconds, Fetched: 24 row(s)
```

The schema for the `weather_data` table was the most complex one out of the four Bay Area Bike Share datasets, as it contained 24 variables of differing data types, but it seems we were successful in creating the table correctly. As promised, the following query will print the first five rows of the data:

```
hive> select * from weather_data limit 5;
OK
9/1/2014   83   70   57   58   56   52   86   64   42   29.86   29.82   29.76   10   10
8    16   7    20   0.0   0        290   94107
9/2/2014   72   66   60   58   57   55   84   73   61   29.87   29.82   29.79   10   10
7    21   8    NULL 0.0   5        290   94107
9/3/2014   76   69   61   57   56   55   84   69   53   29.81   29.76   29.72   10   10
10   21   8    24   0.0   4        276   94107
```

```
9/4/2014   74  68  61   57   57  56  84   71  57  29.81  29.76  29.72  10  10
8   22   8   25  0.0   5       301   94107
9/5/2014   72  66  60   57   56  54  84   71  57  29.92  29.87  29.81  10   9   7
18   8   32  0.0   4       309   94107
```

As we have completed the import of data from HDFS to Hive, we may now exit the Hive shell and return to the Terminal command line for our HDInsight cluster using the `quit` function:

```
hive> quit;
```

In the next section we will connect the RStudio Server session with Spark and Hive. We will use the former as a data processing engine, whereas the latter will serve as a data source.

Bay Area Bike Share analysis using SparkR

You will have probably noticed that we didn't install any additional R packages before this point. Although you may want to install some other packages, this is not needed as the Spark installation by default includes an R package called `SparkR`, which allows connectivity between Spark and R. In fact, you don't even have to download and install RStudio. `SparkR` can be easily initialized from the command line; in that case you will be able to use core R and all downloaded and installed R packages from a Terminal window. However, this is not the most user-friendly approach, hence we decided to make it easier for you and to launch RStudio Server, through a web browser. The only drawback of this solution is that RStudio doesn't know where to look for the `SparkR` package. When you log in to RStudio Server try to extract the value of the `SPARK_HOME` environment variable:

```
> Sys.getenv("SPARK_HOME")
[1] ""
```

The empty character value means that it has not been defined. Also, loading the `SparkR` package through the generic `library()` function will not work. An easy and fast way to resolve this issue is to start core R and the `SparkR` package from the command line. When connecting through `ssh` to your HDInsight cluster, start `SparkR` with the `sparkR` command:

```
$ sparkR
...#output truncated
```

A long output will follow with standard core R welcome messages and notifications about Hadoop and Spark frameworks being initialized. At the end of the output, you will eventually see a familiar R console command prompt:

```
>
```

We may now extract the already configured environmental variable SPARK_HOME:

```
> Sys.getenv("SPARK_HOME")
[1] "/usr/hdp/2.4.1.1-3/spark"
```

Make sure to note down or copy the resulting string containing the path to Spark. You may now exit the R shell in the Terminal window and go back to your RStudio Server session:

```
> quit()
...#output truncated
```

In RStudio Server, define the SPARK_HOME variable using the copied path to Spark:

```
> Sys.setenv(SPARK_HOME = "/usr/hdp/2.4.1.1-3/spark")
```

Let's check whether this has been recorded by R:

```
> Sys.getenv("SPARK_HOME")
[1] "/usr/hdp/2.4.1.1-3/spark"
```

At this point we should load rJava and SparkR packages. Note that when loading SparkR we need to define the location of library files using the SPARK_HOME variable determined previously. Make sure to always double-check the exact value of SPARK_HOME for your release of Hadoop and Spark before loading the SparkR package in the way we did in the preceding step:

```
> library(rJava)
> library(SparkR, lib.loc = c(file.path(Sys.getenv("SPARK_HOME"), "R", "lib")))
...#output truncated
```

Finally, we may initialize a new `SparkContext` using the `sparkR.init()` function. The `SparkContext` is a connection between R (or we should say any R distribution, for example core R, RStudio, and others) and a Spark cluster. As we are connecting to Spark from RStudio using the Hadoop/Spark cluster and reading the data from HDFS and Hive, we have to set the Spark master URL (`master` argument) to `"yarn-client"` and we also need to specify a character vector of `jar` files to pass to the worker nodes (in the `sparkJars` argument):

```
> sc <- sparkR.init(master="yarn-client",
+        appName="SparkRStudio",
+        sparkJars = c("/usr/hdp/2.4.1.1-3/hadoop/hadoop-nfs.jar,
+                      /usr/hdp/2.4.1.1-3/hadoop/hadoop-azure.jar,
+                      /usr/hdp/2.4.1.1-3/hadoop/lib/azure-
storage-2.2.0.jar"))
...#output truncated
```

In the preceding statement, we have also given a value to the `appName` parameter. You can give it any arbitrary name. In terms of `sparkJars` values, they may differ depending on your Hadoop and Spark version. The easiest way to check their paths is to use the path of the `SPARK_HOME` variable as a reference, `cd` in the terminal window, to the appropriate Hadoop directory and double-check the names of `hadoop-nfs`, `hadoop-azure`, and `azure-sturage` `jar` files. After executing the preceding code, quite a long output will follow that may contain numerous warning and information messages. As long as the output does not consist of any error notifications it means that we have defined all parameters of the `SparkContext` correctly.

From the newly created `SparkContext` (stored as an object named `sc`), we can initialize a `HiveContext` using the `sparkRHive.init()` function. We are doing this because we will be using the Hive database as our data source (normally, for data stored locally, the `sparkRSQL.init()` command is used):

```
> hiveContext <- sparkRHive.init(sc)
...#output truncated
```

Another rather long output will follow. It will contain status messages and warnings related to Hive initialization and other information messages about inspecting Hadoop versions, loading required Hadoop and Hive libraries, creating temporary HDFS, local directories for the logged-in user, and other details. The reference to the created `HiveContext` is available in the object named `hiveContext`:

```
> hiveContext
Java ref type org.apache.spark.sql.hive.HiveContext id 2
```

The `hiveContext` object is essential to retrieve any information about our data stored in Hive and every time we query or extract this data. For example, we may want to check what tables are contained within the `bikes` database. This can be achieved with the `tableNames()` function by referring to the `hiveContext` object and the `"bikes"` database name as follows:

```
> tableNames(hiveContext, "bikes")
[[1]]
[1] "station_data"
[[2]]
[1] "status_data"
[[3]]
[1] "trip_data"
[[4]]
[1] "weather_data"
```

In a similar way, we can extract all data from the `status_data` table using `table()`:

```
> status.data <- table(hiveContext, "bikes.status_data")
```

Note that that the resulting `status.data` R object is very small; it doesn't physically contain any values from our data stored in Hive. As the `SparkR` queries are evaluated lazily, the created R objects work simply as maps and they pull the requested data into the R workspace only when explicitly ordered by the user. This allows heavy data crunching on massive datasets without overloading the available memory resources in R. It has to be explained here that such objects are called `DataFrames`, which are more optimized and distributed collections of data equivalent to `data.frames` in R or tables in RDBMSs.

Secondly, note that when referring to the specific table its name has to be prefixed with a database name, for example, in order to query or retrieve data from the `status_data` table, we need to add the `bikes` prefix and separate both names with a full stop.

One of the great features of `SparkR` is that we can run SQL queries straight from the R console, just like with the R packages that support connectivity with other relational and non-relational databases presented in Chapters 6, *R with Relational Database Management Systems (RDBMSs)*, and Chapters 7, *R with Non-Relational (NoSQL) Databases*. For example, we may retrieve all data from the `status_data` table as we did in the preceding step using `table()`, but this time by passing a query within the `sql()` function:

```
> status.data <- sql(hiveContext, "FROM bikes.status_data SELECT *")
```

The resulting `DataFrame` is the same as the one produced with the `table()` function before. We can now explore the `status.data` object a little bit more closely, for example by printing the names of all columns:

```
> columns(status.data)
[1] "station_id"      "bikes_available" "docks_available" "time"
```

The `SparkR` package contains a number of its own implementations of popular core R functions and methods that can be applied on `DataFrames`. For example, we may calculate the overall number of rows in the data using the `count()` command:

```
> count(status.data)
...#output truncated
[1] 36647623
...#output truncated
```

The `count()` function initialized the Hadoop and Spark frameworks to perform the operation. You may also notice that the resulting value is increased by 1 compared with the result of the count query performed directly on the Hive table. As we mentioned earlier, the fact that we skipped the header line when importing the data from HDFS to Hive did not apply to querying the data from other tools, for example the RStudio Server session. We can confirm this by returning the first few rows of the data using the `head()` function:

```
> head(status.data)
...#output truncated
  station_id bikes_available docks_available                time
1         NA              NA              NA                time
2          2              15              12 "2014-09-01 00:00:03"
3          2              15              12 "2014-09-01 00:01:02"
4          2              15              12 "2014-09-01 00:02:02"
5          2              15              12 "2014-09-01 00:03:03"
6          2              15              12 "2014-09-01 00:04:02"
```

As you can see, the first row includes the variable names (if the data type is set to string) or missing values (if the data types of the corresponding fields are other than string). You may also notice that the execution of the `head()` function was followed by an output with information and warning messages—this is yet more evidence of lazy evaluation which triggers the initialization of required processes and returns only the requested amount of data without putting too much pressure on available resources.

We can delete the first row by simply getting rid of all rows with missing values or querying data without the first line. We will use the first option here, as we don't want any missing values anyway:

```
> status.data <- dropna(status.data)
> head(status.data)
  station_id bikes_available docks_available
1          2              15              12
2          2              15              12
3          2              15              12
4          2              15              12
5          2              15              12
6          2              15              12
                 time
1 "2014-09-01 00:00:03"
2 "2014-09-01 00:01:02"
3 "2014-09-01 00:02:02"
4 "2014-09-01 00:03:03"
5 "2014-09-01 00:04:02"
6 "2014-09-01 00:05:02"
```

The `SparkR` package also includes some useful data management functions such as `dtypes()` and `printSchema()`. The former provides the information on data types of each variable, whereas the latter displays the full schema of the table:

```
> dtypes(status.data)
[[1]]
[1] "station_id" "int"

[[2]]
[1] "bikes_available" "int"

[[3]]
[1] "docks_available" "int"

[[4]]
[1] "time"    "string"

> printSchema(status.data)
root
 |-- station_id: integer (nullable = true)
 |-- bikes_available: integer (nullable = true)
 |-- docks_available: integer (nullable = true)
 |-- time: string (nullable = true)
```

As the data type of our `time` variable has been set to string because it couldn't be loaded as a timestamp during the import from HDFS to Hive, we may now attempt to transform it to the actual timestamp variable again. In order to do so, we first need to remove the quote marks around the values. This can be achieved through the `regexp_replace()` function:

```
> status.data$time <- regexp_replace(status.data$time, "\"", "")
```

Following this transformation, we can now cast the `time` variable to a timestamp data type using the `cast()` command and store it as a new variable called `datetime`:

```
> status.data$datetime <- cast(status.data$time, "timestamp")
```

The following statements confirm that the transformations have been successful:

```
> printSchema(status.data)
root
 |-- station_id: integer (nullable = true)
 |-- bikes_available: integer (nullable = true)
 |-- docks_available: integer (nullable = true)
 |-- time: string (nullable = true)
 |-- datetime: timestamp (nullable = true)

> head(status.data)
  station_id bikes_available docks_available
1          2              15              12
2          2              15              12
3          2              15              12
4          2              15              12
5          2              15              12
6          2              15              12
                 time            datetime
1 2014-09-01 00:00:03 2014-09-01 00:00:03
2 2014-09-01 00:01:02 2014-09-01 00:01:02
3 2014-09-01 00:02:02 2014-09-01 00:02:02
4 2014-09-01 00:03:03 2014-09-01 00:03:03
5 2014-09-01 00:04:02 2014-09-01 00:04:02
6 2014-09-01 00:05:02 2014-09-01 00:05:02
```

The timestamps are extremely flexible data types in `SparkR`. They can be easily accessed to extract specific date and time information such as year, month, day of the month, hour, minute, seconds, and much more. Using our example, we will retrieve the hour for each row of the `status.data` and will assign the resulting values to a new variable called `hour`:

```
> status.data$hour <- hour(status.data$datetime)
> head(status.data)
  station_id bikes_available docks_available
1          2              15              12
```

```
2          2                      15                  12
3          2                      15                  12
4          2                      15                  12
5          2                      15                  12
6          2                      15                  12
                 time                    datetime  hour
1 2014-09-01 00:00:03 2014-09-01 00:00:03     0
2 2014-09-01 00:01:02 2014-09-01 00:01:02     0
3 2014-09-01 00:02:02 2014-09-01 00:02:02     0
4 2014-09-01 00:03:03 2014-09-01 00:03:03     0
5 2014-09-01 00:04:02 2014-09-01 00:04:02     0
6 2014-09-01 00:05:02 2014-09-01 00:05:02     0
```

We may wish to explore the processed `status.data DataFrame` a little bit further by obtaining basic descriptive statistics on the selected numeric variables in our data. This can be done using the `describe()` function:

```
> output1 <- describe(status.data, "bikes_available", "hour")
...#output truncated
> output1
DataFrame[summary:string, bikes_available:string, hour:string]
```

However, calling the name of the object that stores the output of the `describe()` command does not produce any informative results apart from displaying the format of the contents. As the `output1` object is another product of a lazily evaluated `SparkR` function, its contents can only be retrieved if the user explicitly requests to display it with the `showDF()` command:

```
> showDF(output1)
+-------+------------------+------------------+
|summary|   bikes_available|              hour|
+-------+------------------+------------------+
|  count|          36647622|          36647622|
|   mean| 8.214080684416578| 11.50165645126988|
| stddev| 4.195785032041624| 6.921465440622335|
|    min|                 0|                 0|
|    max|                27|                23|
+-------+------------------+------------------+
```

The preceding summary table presents some basic descriptive statistics such as count, arithmetic mean, standard deviation, and minimum and maximum values for two numeric variables of `bikes_available` and `hour`.

Knowing the basic descriptive statistics for selected variables, we may attempt to perform some more complex calculations and queries. The `SparkR` package adopts a method of chaining query operations similar to the one that was previously described when discussing the `dplyr` package. In the following example we will aggregate the average number of available bikes for each station per hour. In the first step, we will group the `DataFrame` by two factors we want to aggregate the statistics by: `station_id` and `hour`:

```
> status.data.grouped <- group_by(status.data, "station_id", "hour")
```

Secondly, we will use the `summarize()` function to calculate the arithmetic mean of available bikes and aggregate the results by previously identified grouping variables:

```
> output2 <- summarize(status.data.grouped,
+                     meanBikesAvail = mean(status.data$bikes_available))
```

The top 20 rows of the resulting `output2` object can be displayed using the `showDF()` command as used before:

```
> showDF(output2)
...#output truncated
+----------+----+------------------+
|station_id|hour|    meanBikesAvail|
+----------+----+------------------+
|        47|   8| 7.171029668411867|
|        58|   1| 8.236814321781042|
|        63|  16|10.043907844314141|
|        74|   9|10.604784382017323|
|         9|  14| 7.780260025636331|
|        25|  22| 5.190905760600788|
|        31|   0| 6.886633913402819|
|        36|  15|  8.51516542008902|
|        47|   9| 8.563218917556481|
|        58|   2| 8.233333333333333|
|        63|  17| 7.799658954742372|
|        74|  10|  10.92126742479824|
|         4|   0| 5.009871894944672|
|         9|  15| 7.807690542834855|
|        25|  23| 5.198130498533724|
|        31|   1|  6.91030525591003|
|        36|  16| 8.469488957252398|
|         4|   1| 5.046040853798485|
|         9|  16| 7.871849028881032|
|        31|   2| 6.935674931129476|
+----------+----+------------------+
only showing top 20 rows
```

Faster than Hadoop - Spark with R

If you want to print fewer or more rows, you can set the desired number of rows to display by altering the value of the `numRows` argument in the `showDF()` function (by default it's set to 20).

The output doesn't arrange the values in any user-friendly order. We can for instance sort the average number of bikes available by the values of `station_id` and `hour`-both in ascending order:

```
> showDF(arrange(output2, "station_id", "hour",
+                decreasing = c(FALSE, FALSE)))
...#output truncated
+----------+----+------------------+
|station_id|hour|     meanBikesAvail|
+----------+----+------------------+
|         2|   0| 13.45029615684834|
|         2|   1|13.425659857700252|
|         2|   2|13.335858585858587|
|         2|   3|13.365201465201466|
|         2|   4|13.359023675413289|
|         2|   5|13.356825571278106|
|         2|   6|13.849221611721612|
...#output truncated
```

Once we are happy with the contents and order of the `DataFrame`, we can pull the output from a `SparkRDataFrame` object into the native R `data.frame`. Make sure your memory allowance is sufficient to hold the R object in RAM:

```
> df.output2 <- as.data.frame(output2)
...#output truncated
```

It took 84 seconds to import the data from the server and put it in the `df.output2data.frame` object. It is generally recommended that you do this only in the final stages of your data processing activities or when you want to apply certain data transformations and calculations that are not possible from within the `SparkR` package or using SQL queries. Otherwise it is almost always much better and more convenient to use the `DataFrame` instead and perform all operations on Hive tables using `SparkR`.

In the next task we will merge the `output2DataFrame` with the station information data. Firstly, we need to use the `hiveContext` to *extract* the data to the `station.data` object and remove all rows with missing values:

```
> station.data <- sql(hiveContext, "FROM bikes.station_data SELECT *")
...#output truncated

> columns(station.data)
[1] "station_id"   "name"         "lat"          "long"
```

```
[5] "dockcount"      "landmark"       "installation"
> head(station.data)
...#output truncated
  station_id                                    name     lat
1         NA                                    name      NA
2          2    San Jose Diridon Caltrain Station 37.32973
3          3              San Jose Civic Center    37.33070
4          4             Santa Clara at Almaden   37.33399
5          5                  Adobe on Almaden    37.33141
6          6                  San Pedro Square    37.33672
       long dockcount  landmark installation
1        NA        NA  landmark installation
2 -121.9018        27  San Jose    8/6/2013
3 -121.8890        15  San Jose    8/5/2013
4 -121.8949        11  San Jose    8/6/2013
5 -121.8932        19  San Jose    8/5/2013
6 -121.8941        15  San Jose    8/7/2013

> station.data <- dropna(station.data)
```

Then we are able to perform the inner join by the station_id variable using the merge() function:

```
> output3 <- merge(output2, station.data, by = "station_id")
> printSchema(output3)
root
 |-- station_id_x: integer (nullable = true)
 |-- hour: integer (nullable = true)
 |-- meanBikesAvail: double (nullable = true)
 |-- station_id_y: integer (nullable = true)
 |-- name: string (nullable = true)
 |-- lat: double (nullable = true)
 |-- long: double (nullable = true)
 |-- dockcount: integer (nullable = true)
 |-- landmark: string (nullable = true)
 |-- installation: string (nullable = true)
```

We may use the subset() function to select only the variables of interest:

```
> output3 <- subset(output3, select = c("station_id_x", "name",
                                         "hour", "meanBikesAvail",
                                         "dockcount"))
> head(output3)
  station_id_x                              name hour
1            2 San Jose Diridon Caltrain Station    4
2            2 San Jose Diridon Caltrain Station    3
3            2 San Jose Diridon Caltrain Station    2
4            2 San Jose Diridon Caltrain Station    0
```

```
5              2 San Jose Diridon Caltrain Station       1
6              2 San Jose Diridon Caltrain Station       6
  meanBikesAvail dockcount
1       13.35902        27
2       13.36520        27
3       13.33586        27
4       13.45030        27
5       13.42566        27
6       13.84922        27
```

As you can see, the `SparkR` implementation of the `subset()` function is the same as the generic `subset()`, but it is also applicable to `DataFrame` objects. Having our data appropriately prepared, we may now calculate the percentage of docks available for each station per hour. In this case we want to aggregate a statistic by two variables: `name` of the station and `hour`:

```
> output3.grouped <- group_by(output3, "name", "hour")
```

We will call the newly estimated variable `percDocksAvail` and we will define a formula to calculate it for both grouping variables:

```
> output4 <- summarize(output3.grouped,
+                percDocksAvail = sum((output3$dockcount-
output3$meanBikesAvail)/output3$dockcount*100))
```

We will also clean up the results a little bit by sorting the names and the hours in ascending order:

```
> output4 <- arrange(output4, "name", "hour",
+              decreasing = c(FALSE, FALSE))
> showDF(output4, numRows = 60)
+-----------------+----+------------------+
|             name|hour|    percDocksAvail|
+-----------------+----+------------------+
|     2nd at Folsom|   0| 71.07595196724996|
|     2nd at Folsom|   1| 71.14528345191647|
|     2nd at Folsom|   2| 71.17756512493355|
|     2nd at Folsom|   3| 71.25747059957587|
|     2nd at Folsom|   4| 71.33354061069696|
|     2nd at Folsom|   5|  73.2508562242259|
|     2nd at Folsom|   6| 72.58145363408521|
|     2nd at Folsom|   7| 69.80820042043548|
|     2nd at Folsom|   8| 65.86005520828802|
|     2nd at Folsom|   9| 67.98038586682553|
|     2nd at Folsom|  10| 66.59820635594856|
...#output truncated
```

As the final activity in this task, we may show only the stations with the average percentage of available docks equal to or above 70% for a particular hour. This can be achieved by filtering the `DataFrame` using the `filter()` command and specifying the condition to filter the data on:

```
> output4.subset <- filter(output4, "percDocksAvail >= 70")
> count(output4.subset)
...#output truncated
[1] 30

> showDF(output4.subset, numRows = 30)
...#output truncated
+--------------------+----+-----------------+
|                name|hour|    percDocksAvail|
+--------------------+----+-----------------+
|       2nd at Folsom|   0|71.07595196724996|
|       2nd at Folsom|   1|71.14528345191647|
|       2nd at Folsom|   2|71.17756512493355|
|       2nd at Folsom|   3|71.25747059957587|
|       2nd at Folsom|   4|71.33354061069696|
|       2nd at Folsom|   5| 73.2508562242259|
|       2nd at Folsom|   6|72.58145363408521|
...#output truncated
```

We can perform many more analyses of the Bay Area Bike Share data using the `SparkR` package. In fact, the datasets give us plenty of information to establish which bikes may need some maintenance or servicing. The bike share scheme operators often run such tests to identify bikes which have been used more than others and to move them to slightly less busy stations or check whether they are in good working order. We will carry out a similar analysis in the following task. Our estimation of the most used bikes will be based on the value of trip `duration` (in seconds) available in the `trip_data` table. It has to be stressed, however, that this analysis may not produce desirable results. The duration of the trip does not necessarily mean that the bike was in constant use throughout the whole trip. Someone might have simply rented a bike, met a friend on the way, and had a long chat while standing next to the bike.

Another alternative is that a bike could have been stolen or vandalized and never returned to any of the docking stations. One way or another, it is a good idea to identify bikes with high values of trip duration and investigate their current condition. As before, we need to use the previously created `hiveContext` to link and explore the contents of the `trip_data` table:

```
> trip.data <- sql(hiveContext, "FROM bikes.trip_data SELECT *")
...#output truncated

> columns(trip.data)
 [1] "trip_id"     "duration"    "start_date" "start_st"
 [5] "start_term"  "end_date"    "end_st"     "end_term"
 [9] "bike_no"     "cust_type"   "zip"
> trip.data <- dropna(trip.data)
> head(trip.data)
  trip_id duration       start_date
1  913460      765 8/31/2015 23:26
2  913459     1036 8/31/2015 23:11
3  913455      307 8/31/2015 23:13
4  913454      409 8/31/2015 23:10
5  913453      789 8/31/2015 23:09
6  913452      293 8/31/2015 23:07
                                      start_st start_term
1           Harry Bridges Plaza (Ferry Building)         50
2                    San Antonio Shopping Center         31
3                                 Post at Kearny         47
4                              San Jose City Hall         10
5                           Embarcadero at Folsom         51
6 Yerba Buena Center of the Arts (3rd @ Howard)         68
         end_date                                  end_st
1 8/31/2015 23:39 San Francisco Caltrain (Townsend at 4th)
2 8/31/2015 23:28                 Mountain View City Hall
3 8/31/2015 23:18                       2nd at South Park
4 8/31/2015 23:17                   San Salvador at 1st
5 8/31/2015 23:22                   Embarcadero at Sansome
6 8/31/2015 23:12 San Francisco Caltrain (Townsend at 4th)
  end_term bike_no   cust_type    zip
1       70     288  Subscriber   2139
2       27      35  Subscriber  95032
3       64     468  Subscriber  94107
4        8      68  Subscriber  95113
5       60     487    Customer   9069
6       70     538  Subscriber  94118
```

Secondly, we create a subset of the dataset that will contain only the variables of interest: `duration`, `start_date`, and `bike_no`. We decided to select the starting date and time of the trip as we would like to measure the average usage of bikes in each month:

```
> bikes.used <- subset(trip.data,
+                     select = c("duration", "start_date",
+                                "bike_no"))
> printSchema(bikes.used)
root
 |-- duration: integer (nullable = true)
 |-- start_date: string (nullable = true)
 |-- bike_no: integer (nullable = true)
```

As we are interested in calculating the bike usage per month, we firstly need to convert the `start_date` to a timestamp and then extract the month of the year information from the date. The formats of dates coded as string variables (for example the `start_date` variable) can be converted to `UNIX` timestamps using the `unix_timestamp()` function and by specifying the `format` of the original date. Then, the `UNIX` timestamp can be transformed to any date with a required format through the `from_unixtime()` command. These procedures are as follows:

```
> bikes.used$datetime <- unix_timestamp(bikes.used$start_date,
+                       format = "MM/dd/yyyy HH:mm")
> bikes.used$datetime2 <- from_unixtime(bikes.used$datetime, "yyyy-MM-dd")
```

In the same way we had earlier extracted the hour information from a timestamp, we can retrieve the month values from the newly created `datetime2` variable using `month()`:

```
> bikes.used$month <- month(bikes.used$datetime2)

> head(bikes.used)
  duration       start_date bike_no   datetime   datetime2 month
1      765 8/31/2015 23:26     288 1441063560  2015-08-31     8
2     1036 8/31/2015 23:11      35 1441062660  2015-08-31     8
3      307 8/31/2015 23:13     468 1441062780  2015-08-31     8
4      409 8/31/2015 23:10      68 1441062600  2015-08-31     8
5      789 8/31/2015 23:09     487 1441062540  2015-08-31     8
6      293 8/31/2015 23:07     538 1441062420  2015-08-31     8
```

We may now proceed to the calculation of the total duration of all trips for each bike per month. As we want to perform this aggregation for each bike and month, we need to set up grouping variables for our `DataFrame` first:

```
> bikes.grouped <- group_by(bikes.used, "month", "bike_no")
```

With grouping variables indicated, we can perform the calculation of monthly bike usage. We will also sort the results by `month` and total duration of all trips for each bike (`sumDuration`). The `month` variable will be sorted in ascending order, whereas the `sumDuration` will be arranged in descending order as we want to show the bikes with the highest cumulative duration of all trips first:

```
> bikes.mostused <- summarize(bikes.grouped,
+                             sumDuration = sum(bikes.used$duration))
> bikes.mostused <- arrange(bikes.mostused, "month",
+                           "sumDuration", decreasing = c(FALSE, TRUE))
```

Of course, there is an obvious problem with such an arrangement of results. As we calculated the statistics for all bikes, the output will include a long list of all used bikes in a particular month; this is not the most user-friendly way of identifying bikes with the highest overall trip duration. We will therefore pull our aggregation output to an R object and will run a set of commands which will eventually result in displaying only five bikes with the highest cumulative duration of all trips for each month:

```
> df.bikes <- as.data.frame(bikes.mostused)
...#output truncated
> str(df.bikes)
'data.frame':   7373 obs. of  3 variables:
 $ month       : int  1 1 1 1 1 1 1 1 1 1 ...
 $ bike_no     : int  117 309 306 230 241 650 511 541 413 632 ...
 $ sumDuration: num  410665 342638 266102 250430 196195 ...
> df.bikes.split <- split(df.bikes, df.bikes$month)
> topUsage <- do.call(rbind,
+                     sapply(df.bikes.split,
+                     simplify = FALSE,
+                     function(x)x[order(x$sumDuration,
+                                 decreasing = TRUE), ][1:5,]))
```

In the preceding first statement we only imported the data from the processed `SparkRDataFrame` to a `data.frame` hold in memory by R. Then we printed the structure of this newly created object and we split it into a list with 12 components holding data for each of the 12 months of the `df.bikes data.frame`. Finally, we applied a function to sort the individual elements of all components of the list in decreasing order and to print just the top five rows of data for each component. We then combined all the resulting components into one new `data.frame` object named `topUsage`:

```
> topUsage
       month bike_no sumDuration
1.1        1     117      410665
1.2        1     309      342638
1.3        1     306      266102
```

```
1.4        1     230    250430
1.5        1     241    196195
2.616      2     132    663204
2.617      2       9    607711
2.618      2     662    558139
2.619      2      22    510893
2.620      2     490    352680
3.1216     3     440    411017
3.1217     3     654    294075
3.1218     3     310    263660
3.1219     3     589    224536
3.1220     3     419    221302
4.1836     4     196    620418
4.1837     4     687    304193
4.1838     4     574    261867
4.1839     4      32    235494
4.1840     4     374    161137
...#output truncated
```

The preceding output may allow bike share scheme operators to identify bikes with larger than average overall duration for all trips they made in a particular month. However, they may also be interested in the cumulative duration of all trips across the whole year. This calculation is simpler, as we don't have to group the data by the month variable:

```
> bikes.grouped2 <- group_by(bikes.used, "bike_no")
> bikes.mostused2 <- summarize(bikes.grouped2,
+                       sumDuration = sum(bikes.used$duration))
```

We can now restrict the output of our calculation to only the 20 bikes with the highest values of duration using the limit() command:

```
> bikes.mostused2 <- limit(arrange(bikes.mostused2, "sumDuration",
                      decreasing = TRUE), 20)

> showDF(bikes.mostused2)
...#output truncated
+--------+------------+
|bike_no|sumDuration|
+--------+------------+
|    535|   17634977|
|    466|    2611616|
|    680|    1955369|
|    415|    1276705|
|    262|    1231761|
|    376|    1205689|
|    613|    1154441|
|    440|    1145327|
|    374|    1089341|
```

```
|    484|   1081343|
|    542|   1071426|
|    589|   1054142|
|     85|   1029330|
|    599|   1022603|
|    306|   1014901|
|    618|   1011358|
|    419|   1009977|
|    312|   1005573|
|    328|   1002772|
|    437|    999758|
+-------+----------+
```

Although one bike seems to be in use for 17,634,977 seconds, which equates to 204 days over the period of 1 year, it is very unlikely that it was actually in constant use during that time. The duration merely indicates that the bike wasn't docked for a specific number of seconds. It is more probable that in this case the bike was stolen, vandalized, or maybe incorrectly docked to a bike station. However, the calculation is still meaningful as it can identify suspicious events and provides a trigger for further investigation.

At this point we will end the `SparkR` tutorial, but hopefully you are already motivated enough to continue exploring the Bay Area Bike Share open data using Spark, RStudio Server, and Hive. The datasets offer a large array of possible analyses to perform. For example, you may want to test whether there is any correlation between the weather indicators such as the wind speed or precipitation level on a specific date for a particular geographical area of Bay Area and the bike use on that day.

Most of these data transformations and analyses can only be carried out using the `SparkR` package. It provides users with a large amount of methods that support data aggregations, cross-tabulations, and data management activities such as recoding, renaming, casting variables into new data types, and many other simple calculations and descriptive statistics. It also provides more complex data analysis functions, such as performing simple Pearson correlations and fitting Generalized Linear Models as well as implementing machine learning algorithms. All `SparkR` functions and commands for its current version (we have used the most recent 1.6.1 release in this chapter) are documented online at: https://spark.apache.org/docs/1.6.1/api/R/00Index.html.

When you complete all `SparkR` data processing, don't forget to close the Spark context for the R session:

```
> sparkR.stop(sc)
...#output truncated
```

Also, if you used an HDInsight cluster remember to always delete the cluster and any of its associated storage resources in the Microsoft Azure Portal to avoid any further charges.

Summary

In this chapter we introduced you to the Apache Spark engine for fast Big Data processing. We explained how to launch a multi-node HDInsight cluster with Hadoop, Spark, and the Hive database installed and how to connect all these resources to RStudio Server.

We then used Bay Area Bike Share open data to guide you through the numerous functions of the `SparkR` package for data management, transformations, and analysis on data stored in Hive tables directly from the R console.

In Chapter 8, *Machine Learning Methods for Big Data in R*, we will explore another powerful dimension of Big Data analytics using R language: we will apply a variety of predictive analytics algorithms to large-scale data sets using the **H2O** platform for distributed machine learning of Big Data.

8
Machine Learning Methods for Big Data in R

So far in this book we have explored a variety of descriptive and diagnostic statistical methods that can easily be applied to out-of-memory data sources. But the true potential of modern data science resides in its predictive and prescriptive abilities. In order to harness them, versatile data scientists should understand the logic and implementations of techniques and methods commonly known as **machine learning algorithms**, that allow making robust predictions and foreseeing patterns of events. In this chapter we will introduce you to machine learning methods that are applicable to Big Data classification and clustering problems through the syntax of the R language. Moreover, the contents of this chapter will provide you with the following skills:

- You will understand the concept of machine learning and be able to distinguish between supervised/unsupervised methods and clustering/classification models
- You will carry out a high-performance **Generalized Linear Model** using the Spark MLlib module through the `SparkR` package for R on a multi-node Spark HDInsight cluster
- You will perform a **Naive Bayes** classification algorithm and design a **multi-layered Neural Network** (Deep Learning) in **H2O**, an open-source, Big Data distributed machine learning platform, connected with R through the `h2o` package to predict the classes of events on real-world data
- You will learn how to evaluate the performance metrics and accuracy measures of selected machine learning algorithms

What is machine learning?

We will begin this chapter with a brief introduction to the concept of machine learning by presenting an overview of the most frequently used predictive algorithms, their classification, and typical characteristics. We will also list a number of resources where you can find more information about the specifics of chosen algorithms and we will guide you through the growing number of Big Data machine learning tools available to data scientists.

Machine learning algorithms

Machine learning methods encapsulate data mining and statistical techniques allowing researchers to make sense of data, model the relationships between variables or features, and extend these models to predict the values or classes of events in the future. So how does this field differ from the already well-known statistical testing? In general, we can say that machine learning methods are less stringent about the required format and characteristics of the data; that is, many machine learning algorithms when predicting the outcome of numerical continuous response variables do not require the residuals of this variable to be normally distributed. Most statistical tests are more focused on inference and hypothesis testing, especially where one general statistic is calculated (for example, F statistic in ANOVA or regression), whereas machine learning models attempt to use the observed patterns to explain and predict future data. In fact, there is a large overlap between both concepts and many techniques can be classified as both machine learning and statistical tests. As we will see shortly they also use similar diagnostic tests to assess the generalizability of identified patterns and models, for example Mean Squared Error or R^2.

Due to the growing contributions from researchers coming from various scientific backgrounds, machine learning has evolved as a field that incorporates a large plethora of techniques and algorithms from computer science, statistics, data mining, but also cognitive and computational neuroscience, engineering, biomedical science, robotics, and artificial intelligence. In the era of rapid development of the Internet of Things, the practical applications of machine learning algorithms are widespread and easily noticeable in various areas of life. For example they are common building blocks in:

- Search engines such as Google, Bing, or Yahoo!
- Online recommendation systems for example Amazon, eBay, Netflix, or Spotify
- Personalized trending information and news feeds, for example Twitter and Facebook

- Highly sensitive spam filters and email virus scans for example Gmail and Outlook
- Predicting virus evolution and cancer/disease diagnosis
- Computer vision including facial, behavioral, and object recognition in dynamic real-life scenarios, for example self-driving cars, image searches, augmented reality systems, smart cameras for security, public safety, and emergency response
- Financial fraud detection, loan applications and credit scoring, identifying low and high-risk customers, spending habits, and cyber-crime prevention
- Fund management, macro-research, investment opportunities, and real-time stock market trading
- Insurance premium calculations, risk assessments, and claim predictions
- Political election predictions, sales analytics, and many others

The preceding list is not fully inclusive, but this is evidence of the widespread usage of machine learning in various industries and practical settings. In the following sections, we will attempt to categorize machine learning methods based on the structure of input data and the purpose of the algorithms.

Supervised and unsupervised machine learning methods

We will start our journey through machine learning techniques from the general division of all predictive algorithms into two major groups of supervised and unsupervised methods. The **supervised learning methods** include machine learning algorithms in which the training data contains labeled examples and the model learns the characteristics of specific classes during the training phase. A simple example of such a method will be an image recognition algorithm that classifies the recognized objects as either people, cars, houses, or other features of the surrounding landscape. In fact, such algorithms are currently being used in self-driving vehicles, which while moving around our roads constantly scan the environment and classify the encountered local and global landmarks based on their unique characteristics and match them with the labeled examples provided by the training data.

On the other hand, **unsupervised learningmethods** attempt to cluster similar objects without being aware of their labels. They accomplish this task by inferring the structure and extracting the features of the data. One example of an unsupervised learning method is the concept of dimensionality reduction. Similar highly-correlated variables are grouped together into principal components and used as a single factor in analysis. The goal of unsupervised learning is not to label similar examples, but merely extract them into a separate grouping. The labeling is almost always left to the researcher.

In reality, some machine learning algorithms incorporate characteristics of both supervised and unsupervised methods and hence they are called **semi-supervised**. They use small quantities of labeled and large amounts of unlabeled examples as their training sets, but their applications are generally not very common.

Classification and clustering algorithms

Another major categorization of machine learning algorithms is dependent on the desired goal of the learning. In classification algorithms a researcher wants to assign new examples to two or more classes based on the features of usually labeled training data. As the classification models make predictions about the classes based on the labelled training sets, these methods almost exclusively belong to the category of supervised learning. The classification methods include quite an impressive selection of algorithms, starting from rather simple binomial logistic regressions, through the Naive Bayes algorithms and extending to support **Vector Machines** and **Neural Networks**. During the practical tutorials you will have a chance to perform a logistic regression, Naive Bayes, and even a Neural Network on real-world Big Data classification problems using the Spark and H2O platforms for distributed machine learning.

Clustering methods are designed to group similar examples based on the patterns of values found in the features of training data. Most of the clustering algorithms favor continuous numeric predictor variables-somewhat contrary to classification algorithms, that generally prefer to predict classes from categorical input. The clustering methods are largely unsupervised-they simply categorize the observations without inferring their labels. Some of the most frequently used clustering algorithms include **Principal Components Analysis** and **k-means**.

Machine learning methods with R

Most, if not all, of the known machine learning methods can be easily implemented in R workflows by the means of a rich collection of base R and third-party packages, that support predictive analytics in the R language. The *CRAN Task View: Machine Learning & Statistical Learning*, available at `https://cran.r-project.org/web/views/MachineLearning.html`, provides an ample list of R libraries that address specific needs of researchers and analysts wanting to utilize the R language for machine learning purposes. However, as many predictive and data modeling tasks are, computationally, extremely expensive, most of the packages presented in this CRAN task view do not apply to Big Data problems-they require different, more optimized treatment and distributed architecture, that we are going to describe in the next part of this chapter.

If your datasets do not fall in the Big Data category, then you are strongly encouraged to test some of the listed packages and explore the predictive power of R. As we are not going to explain them in this chapter, the following books can help you in your first steps with machine learning algorithms in R applied to small and normal-sized datasets (by normal-sized we generally mean smaller than one-eighth of your RAM resources at most):

- Lantz, B. (2015). *Machine Learning with R*, 2nd edition. Packt Publishing
- James, G., Witten, D., Hastie, T., and Tibshirani, R. (2013). *An Introduction to Statistical Learning: with Applications in R.* Springer Texts in Statistics
- Kuhn, M., and Johnson, K. (2013). *Applied Predictive Modelling*. Springer
- Makhabel, B. (2015). *Learning Data Mining with R*. Packt Publishing
- Yu-Wei, C. (2015). *Machine Learning with R Cookbook*. Packt Publishing

Apart from these books, there is quite a lot of informative online resources elaborating on different aspects of machine learning applications or providing data scientists with detailed predictive know-how. Some of them are listed as follows:

- `http://www.dataschool.io/15-hours-of-expert-machine-learning-videos/`-15 hours of statistical learning video lectures and tutorials
- `http://www.kdnuggets.com/2015/06/top-20-r-machine-learning-packages.html`-links to the top 20 most popular machine learning R packages
- `http://www.analyticsvidhya.com/`-a good machine learning blog with a large number of R examples

- `https://lagunita.stanford.edu/courses/HumanitiesSciences/StatLearning/Winter2016/info`-Stanford University's statistical learning online course
- `https://www.coursera.org/learn/machine-learning`-Stanford University's machine learning online course on Coursera
- `https://www.coursera.org/learn/practical-machine-learning`-John Hopkins University's machine learning online course on Coursera
- `https://www.edx.org/course/data-science-machine-learning-essentials-microsoft-dat203x-0#`-data science and machine learning online course by Microsoft

Big Data machine learning tools

As mentioned before, most of the existing machine learning R packages do not scale well to address Big Data predictive tasks. Fortunately, a number of separate tools and open-source projects allow users to deploy complex machine learning algorithms on massive datasets and some of those listed below even support the R language:

- **H2O** (`http://www.h2o.ai/`): A machine learning platform for Big Data; allows the implementation of large number of classification and clustering algorithms and works well with the Hadoop ecosystem and cloud computing solutions; it supports R through the `h2o` package. We will explore the H2O platform in detail during two practical tutorials later on in this chapter.
- **Spark MLlib** (`http://spark.apache.org/mllib/`): Spark's library for machine learning that includes a selection of clustering, regression, and classification algorithms. It's scalable, fast, and flexible allowing compatibility with multiple languages, for example Scala, Java, Python, and R (through the `SparkR` package) and it can be even connected with H2O to create **Sparkling Water**-a fast, distributed platform for high-performance machine learning. We will guide you through the functionalities of the `SparkR` package and the `SparkMLlib` module in the next section.
- **Microsoft Azure ML** (`https://azure.microsoft.com/en-gb/services/machine-learning/` and `https://studio.azureml.net/`): A cloud-based machine learning platform with a GUI. It contains many in-built, ready-to-go, machine learning algorithms. Users can perform experiments with custom-made algorithms that can be designed and deployed using R or Python. Azure ML with R allows the creation of machine learning web services.

Other less R-oriented machine learning tools that support predictive analytics of relatively large datasets include **Weka** (http://www.cs.waikato.ac.nz/ml/weka/-it offers R connectivity), **scikit-learn** for Python (http://scikit-learn.org/stable/), and the **Machine Learning Toolbox** for Matlab (http://uk.mathworks.com/solutions/machine-learning/index.html?s_tid=gn_loc_drop).

Following this brief introduction to machine learning, in the next section you will have a chance to perform a logistic regression through the Spark MLlib module on a multi-node Hadoop cluster.

GLM example with Spark and R on the HDInsight cluster

In the first practical example of this chapter we will use the HDInsight cluster with Spark and Hadoop, and run a **Generalized Linear Model** (**GLM**) on the flight data available for you to download from the Packt Publishing website created for this book.

Preparing the Spark cluster and reading the data from HDFS

Before carrying out any analytics on the data, let's firstly double-check whether you have all of the required resources in place. In this tutorial, we will be using the same multi-node HDInsight cluster that you previously deployed following the instructions in Chapter 7, *Faster than Hadoop: Spark with R* and specifically the section on *Launching HDInsight with Spark and R/RStudio*. If you don't remember how to launch the HDInsight cluster on Microsoft Azure, detailed step-by-step guidelines have been provided earlier in the *HDInsight – A multi-node Hadoop cluster on Azure* section in Chapter 4, *Hadoop and MapReduce Framework for R*. As a result of the instructions in these two chapters, we have created a fully-operational and highly-scalable HDInsight cluster named swalkospark with the standard distribution of Spark 1.6.0 (HDI 3.4) installed on the Linux Ubuntu operating system. The cluster contains four worker nodes (D12 type), each with 4 cores and 28GB of RAM, and two head nodes of the same specification. During the cluster optional configuration, we have also installed R by providing an external link at https://hdiconfigactions.blob.core.windows.net/linuxrconfigactionv01/r-installer-v01.sh pointing to the R installation script.

Upon deployment of the cluster, we connected through `ssh` to its head node and installed RStudio Server using the following commands in the terminal window:

```
$ sudo apt-get install gdebi-core
...#output truncated

$ wget https://download2.rstudio.org/rstudio-server-0.99.902-amd64.deb
...#output truncated

$ sudo gdebi rstudio-server-0.99.902-amd64.deb
...#output truncated
```

We then issued the head node with the public IP, that allowed us to access RStudio from a browser. If you followed all these previously explained guidelines precisely, you shouldn't encounter any problems with starting the cluster.

As for the data, in this chapter we will make use of the well-known flight data that we briefly introduced earlier in Chapter 3, *Unleashing the Power of R from Within*, where we performed a number of data transformations and aggregations with the `ff` and `ffbase` packages. Previously, the dataset contained all flights to/from all American airports in September and October 2015, and it included a selection of 28 variables characterizing specific features of each flight. The data that we will be using in this chapter includes two separate files. The first file, named `flights_2014.csv` (169.7 MB), consists of 5,819,811 flights to/from American airports and covers the period between January 1, 2014 until December 31, 2014. This dataset will be used as training data for our models later on. The second file, `flights_jan_2015.csv` (13.7 MB), includes 469,968 flights that occurred only in January 2015 and will act as a test/validation dataset. Both files contain only nine variables each: day of week (`DAY_OF_WEEK`), departure time of a flight (`DEP_TIME`), departure delay (`DEP_DELAY`), arrival time (`ARR_TIME`), arrival delay (`ARR_DELAY`), whether a flight was canceled (`CANCELLED`) or diverted (`DIVERTED`), the air time (`AIR_TIME`), and the distance traveled (`DISTANCE`). All variables are integers; however as we are going to show you soon, they can also be converted into categorical variables to suit specific machine learning algorithms. As previously explained, the original data comes from the website of the Bureau of Transportation Statistics at http://www.transtats.bts.gov/DL_SelectFields.asp?Table_ID=236&DB_Short_Name=On-Time, but both processed files that we are going to use in this chapter are available for you to download as a ZIP package (57.1 MB) from Packt Publishing's website prepared for this book. Of course, the sizes of both datasets are quite small so you can easily download the files and implement the described techniques on relatively small instances and even locally on single-node clusters.

If you are up for a more true Big Data thrill you can visit the preceding website to obtain similar, much larger data across many years. As always in such cases, remember that larger datasets require much greater computational resources available on your personal or virtual machines; therefore make sure to scale up your architecture accordingly as we have already explained in the preceding chapters.

Assuming that you have downloaded the zipped data to your desktop, you can now copy the file across to the cluster. In a new terminal window type the following line of commands (don't forget to adjust the user and cluster names appropriately):

```
$ scp -r ~/Desktop/data.zip swalko@swalkospark-
ssh.azurehdinsight.net:~/data.zip
```

In the terminal window that you are currently using to `ssh` to the cluster, unzip the `data.zip` package and check the contents of your current directory:

```
$ unzip data.zip
Archive:  data.zip
   creating: data/
  inflating: data/flights_2014.csv
  inflating: data/flights_jan_2015.csv
$ ls
data  data.zip  rstudio-server-0.99.902-amd64.deb
```

Stay in the current `/home/<user>` directory and copy the data files to HDFS of the cluster. Firstly, create a new `data` folder on HDFS:

```
$ hadoop fs -mkdir /user/swalko/data
```

And then copy the files from the `data` folder on the local area to the `/user/<user>/data` directory on HDFS, for example:

```
$ hadoop fs -copyFromLocal data/ /user/swalko/data
```

You can check the contents of the `data` folder on HDFS using the standard `hadoop fs -ls` command:

```
$ hadoop fs -ls data
Found 2 items
-rw-r--r--   1 swalko supergroup   169677488 2016-05-23 14:52 data/flights_2014.csv
-rw-r--r--   1 swalko supergroup    13740996 2016-05-23 14:52 data/flights_jan_2015.csv
```

At this point, we are ready to initialize the `SparkR` package and prepare the data for logistic regression. As you may remember from the preceding chapter, in order to use Spark with R we firstly have to set the `SPARK_HOME` environmental variable in RStudio Server. To achieve this we need to extract its value from the R shell with `SparkR` already installed. Start `SparkR` directly from the command line:

```
$ sparkR
...#output truncated
```

Once in the R shell with `SparkR` initialized, retrieve the value of `SPARK_HOME`:

```
> Sys.getenv("SPARK_HOME")
[1] "/usr/hdp/2.4.2.0-258/spark"
```

After obtaining the `SPARK_HOME` value, we may leave the R shell using the `quit()` function and log in to the cluster's RStudio Server through your chosen browser. In RStudio, set the `SPARK_HOME` variable to the value that we just retrieved and check whether the assignment was successful:

```
> Sys.setenv(SPARK_HOME = "/usr/hdp/2.4.2.0-258/spark")
> Sys.getenv("SPARK_HOME")
[1] "/usr/hdp/2.4.2.0-258/spark"
```

Before we can start `SparkR` through RStudio we have to do one more very important thing. As Spark does not support CSV files as data sources natively, we need to install an external Spark connector called `spark-csv` that will provide this facility for us. This has to be done before starting `SparkR` and initializing `Spark Context` through the `sparkR.init()` function. The implementation of `spark-csv` library and its development can be reviewed at the package's GitHub at https://github.com/databricks/spark-csv. The section on features provides detailed a description of options and arguments that may come useful when reading data from CSV as a Spark `DataFrame`. The repository also explains various methods of `spark-csv` integration depending on the language and API used. In our case, we need to specify the `SPARKR_SUBMIT_ARGS` to include the reference to the current version of `spark-csv` as follows:

```
> Sys.setenv('SPARKR_SUBMIT_ARGS'='"--packages" "com.databricks:spark-csv_2.11:1.4.0" "sparkr-shell"')
> Sys.getenv("SPARKR_SUBMIT_ARGS")
[1] ""--packages" "com.databricks:spark-csv_2.11:1.4.0" "sparkr-shell""
```

Apart from the official GitHub page for the `spark-csv` package, its current stable version can be reviewed and obtained from http://spark-packages.org/package/databricks/spark-csv. The http://spark-packages.org/ website includes third-party packages (exactly 230 at the time of writing this chapter) created for Spark by its growing community of users; make sure to browse through its contents if you are considering integrating Spark with your current R and Big Data workflows.

After adding `spark-csv` to `SPARKR_SUBMIT_ARGS`, we may now load the required `rJava` package and then the `SparkR` library:

```
> library(rJava)
> library(SparkR, lib.loc = c(file.path(Sys.getenv("SPARK_HOME"), "R", "lib")))
...#output truncated
```

Following the confirmation output, both packages are ready to use. Create a new `Spark Context` in a similar way to how we did it in Chapter 7, *Faster than Hadoop: Spark with R*. The only difference is that we will now add the `sparkPackages` argument with the reference to `spark-csv` as its character value:

```
> sc <- sparkR.init(master="yarn-client",
+          appName="SparkRStudio",
+          sparkJars = c("/usr/hdp/2.4.2.0-258/hadoop/hadoop-nfs.jar",
+          "/usr/hdp/2.4.2.0-258/hadoop/hadoop-azure.jar",
+          "/usr/hdp/2.4.2.0-258/hadoop/lib/azure-storage-2.2.0.jar"),
+          sparkPackages="com.databricks:spark-csv_2.11:1.4.0")
...#output truncated
```

Once we have `Spark Context` in place, we can create `SQL Context` and define the custom schema for our data. The schema can be created through the `structType()` function and specific fields and their types are determined with the `structField()` command:

```
> sqlContext <- sparkRSQL.init(sc)
> schema <- structType(structField("DAY_OF_WEEK", "string"),
+                     structField("DEP_TIME", "integer"),
+                     structField("DEP_DELAY", "integer"),
+                     structField("ARR_TIME", "integer"),
+                     structField("ARR_DELAY", "integer"),
+                     structField("CANCELLED", "integer"),
+                     structField("DIVERTED", "integer"),
+                     structField("AIR_TIME", "integer"),
+                     structField("DISTANCE", "integer"))
```

As you can see in the preceding code, we decided to arbitrarily set the type of the `DAY_OF_WEEK` variable to `string` despite its native type of `integer` for two reasons: firstly, we want this field to act as a categorical variable, and secondly, we are not going to recode this variable during processing, so this transformation into `string` when defining the schema will just save us some time and one extra line of code.

Once the schema is formulated and stored in the `schema` object, we should be in a good position to import the `flights_2014.csv` file directly from HDFS into a Spark session as a `DataFrame`. Note that in the following code, we explicitly set the `source` the name of the external data source-to `spark-csv`. Also, we specify the representation of missing values in our data by defining the `nullValue` option and as our file includes column names in its header we set the `header` value to `true`:

```
> flights <- read.df(sqlContext,
+                    path = "/user/swalko/data/flights_2014.csv",
+                    source = "com.databricks.spark.csv",
+                    header = "true",
+                    schema = schema, nullValue = "NA")
```

The `flights` data is now available in the RStudio workspace as a `DataFrame` and we can explore it using a number of `base` and `SparkR` methods applicable to `DataFrame` objects, which we have already introduced in preceding chapters, for example `head()` and `str()` functions:

```
> head(flights)
...#output truncated
  DAY_OF_WEEK DEP_TIME DEP_DELAY ARR_TIME ARR_DELAY
1           2      854        -6     1217         2
2           3      853        -7     1225        10
3           4      856        -4     1203       -12
4           5      857        -3     1229        14
5           6      854        -6     1220         5
6           7      855        -5     1157       -18
  CANCELLED DIVERTED AIR_TIME DISTANCE
1         0        0      355     2475
2         0        0      357     2475
3         0        0      336     2475
4         0        0      344     2475
5         0        0      338     2475
6         0        0      334     2475
> str(flights)
...#output truncated
'DataFrame':  9 variables:
 $ DAY_OF_WEEK: chr "2" "3" "4" "5" "6" "7"
 $ DEP_TIME   : int 854 853 856 857 854 855
 $ DEP_DELAY  : int -6 -7 -4 -3 -6 -5
```

```
$ ARR_TIME    : int  1217 1225 1203 1229 1220 1157
$ ARR_DELAY   : int  2 10 -12 14 5 -18
$ CANCELLED   : int  0 0 0 0 0 0
$ DIVERTED    : int  0 0 0 0 0 0
$ AIR_TIME    : int  355 357 336 344 338 334
$ DISTANCE    : int  2475 2475 2475 2475 2475 2475
```

So far we have been able to re-use the multi-node Spark (on Hadoop) HDInsight cluster with RStudio Server from Chapter 7, *Faster than Hadoop: Spark with R* to successfully read in and briefly explore the CSV file with a custom schema directly from HDFS.

In the next section, we will briefly remind you of the most important features of logistic regression as one of the Generalized Linear Models and we will also clean and prepare the data for modeling.

Logistic regression in Spark with R

Generalized Linear Models (**GLMs**) extend the general linear model framework by addressing two issues: when the range of dependent variables is restricted (for example binary or count variables) and in cases when the variance of dependent variables relies on the mean. In GLMs we relax the assumption that the response variable and its residuals are normally distributed; in fact the dependent variable may follow a distribution that is a member of the exponential family, for example Poisson, binomial, and others. A typical GLM formula is made up of a standard **linear predictor** (known from general linear models) and two functions: a **link function** of the conditional mean, that defines how the mean depends on the linear predictor, and a **variance function** describing how the variance depends on the mean.

In a **logistic regression**, one example of GLMs, we are interested in determining the outcome of a categorical variable. In most cases, we deal with **binomial logistic regression** with the binary response variable, for example yes/no, lived/died, passed/failed, and others. Sometimes however the dependent variable may contain more than two categories; in those circumstances we are talking about **multinomial logistic regression** when we are trying to predict the value of the polytomous outcome variable, for example poor/good/excellent or conservative/liberal/independent. Simply speaking, logistic regression can be applied to classification problems when we want to determine a class of an event based on the values of its features-or in other words, the likelihood that the predicted binary value equals a specific class given a set of continuous and nominal independent variables.

Translating the preceding assumptions to our flight data, we want to predict whether an individual flight is delayed or not delayed. We will firstly train our model on the data that contains all flights in 2014 and then will fit it to the subset of flights in January 2015 to estimate its predictive power and accuracy. Before we can do it though, we ought to pre-process and clean the data by eliminating all cancelled and diverted flights. If a flight was cancelled or diverted, it was assigned the value 1 to either the CANCELLED or DIVERTED field respectively. Based on this, we may use the `filter()` function to extract only these flights that were not cancelled nor diverted:

```
> flights <- filter(flights, flights$CANCELLED == 0)
> flights <- filter(flights, flights$DIVERTED == 0)
```

After filtering the data, both the CANCELLED and DIVERTED variables can be deleted from the `flightsDataFrame`:

```
> flights <- flights[, -6:-7]
```

As you can probably notice, our original `flights_2014.csv` file does not include any binary variable with information as to whether a flight was delayed or not. It does however contain the ARR_DELAY variable, that indicates the arrival delay (in minutes) for each flight. The negative integer means that the flight arrived before the schedule, whereas the positive value confirms that the flight was indeed delayed. Relying on this information, we can simply transform the existing numeric ARR_DELAY variable into a nominal response variable with two levels: not delayed (for ARR_DELAY values smaller than or equal to zero) and delayed (for ARR_DELAY values greater than zero). Unfortunately, SparkR has not yet been equipped with functions resembling `recode()` from the car package and the generic `with()` and `within()` functions cannot be applied to DataFrames. Therefore, the only problem that we are facing right now is how we can achieve this transformation in SparkR without converting the flights Spark DataFrame into the R data.frame. As we know by now such a conversion is not particularly memory efficient, especially in cases when we deal with Big Data much greater than our flight datasets. Luckily for us, there is a practical workaround solution that can be operationalized in SparkR by registering a temporary table containing the flights data in the SQL Context and applying SQL queries to recode the ARR_DELAY values into a new binary categorical variable. We may achieve this transformation with the following two lines of R code:

```
> registerTempTable(flights, "flights")
> flights <- sql(sqlContext, "SELECT *, IF(ARR_DELAY > 0, 1, 0)
+               AS ARR_DEL from flights")
```

The first of the preceding lines registers a temporary table called `flights` (the second argument of the function) and the latter runs an SQL query that includes a simple `IF` statement to assign value `1` (delayed) to all observations where `ARR_DELAY` was greater than zero and value `0` (not delayed) to all other flights. In the process we have also created a new variable called `ARR_DEL` that will hold the newly computed classes.

Similarly, we may recode the `DEP_TIME` variable, that indicates the actual departure time of flights into a new `DEP_PART` categorical variable that will contain the departure part of the day with four levels: morning, afternoon, evening, and night. Just as in the earlier transformation, we have to register the `flightsDataFrame` as a temporary table to keep the changes from the previous recoding:

```
> registerTempTable(flights, "flights")
> flights <- sql(sqlContext, "SELECT *,
+                   CASE WHEN(DEP_TIME >= 500 AND DEP_TIME < 1200)
+                   THEN ('morning') WHEN(DEP_TIME >= 1200 AND DEP_TIME < 1700)
+                   THEN ('afternoon') WHEN(DEP_TIME >= 1700 AND DEP_TIME < 2100)
+                   THEN ('evening') ELSE('night') END AS DEP_PART from flights")
```

Finally, we can print the resulting structure of the processed `DataFrame` and one last time register it as a temporary table to retain the alterations:

```
> str(flights)
..#output truncated
'DataFrame':   9 variables:
 $ DAY_OF_WEEK: chr "2" "3" "4" "5" "6" "7"
 $ DEP_TIME   : int 854 853 856 857 854 855
 $ DEP_DELAY  : int -6 -7 -4 -3 -6 -5
 $ ARR_TIME   : int 1217 1225 1203 1229 1220 1157
 $ ARR_DELAY  : int 2 10 -12 14 5 -18
 $ AIR_TIME   : int 355 357 336 344 338 334
 $ DISTANCE   : int 2475 2475 2475 2475 2475 2475
 $ ARR_DEL    : int 1 1 0 1 1 0
 $ DEP_PART   : chr "morning" "morning" "morning" "morning" "morning" "morning"
> registerTempTable(flights, "flights")
```

At this stage we are now ready to input the variables into our logistic regression model. The `SparkR` framework allows R users to perform Generalized Linear Models through its implementation of the `glm()` function (originally known from the `stats` package)-now applicable to `SparkR DataFrame` objects. In the following line, we specify the model's formula with `ARR_DEL` as a response variable and five other variables as terms of the predictor. We also set the `family` of the error distribution to `binomial` in order to apply the `logit` link function to our logistic regression model:

```
> logit1 <- glm(ARR_DEL ~ AIR_TIME + DISTANCE + DAY_OF_WEEK + DEP_PART +
DEP_DELAY, data = flights, family = "binomial")
...#output truncated
```

It has to be said here that the current implementation of the `SparkR` package's `glm()` function is still a work in progress and it doesn't include many parameters known from the standard `glm()` method of the `stats` package. But as these are still early days for Spark and its integration with R, we believe that this and many other `SparkR` commands will increase in their functionalities very soon. However, it is very useful that the `glm()` function includes the `standardize` option, that by default is set to `TRUE` and it automatically standardizes the values of continuous independent variables before performing the GLM estimation.

The `glm()` command initializes the Spark MLlib module, that is responsible for implementation and evaluation of machine learning algorithms within the Spark framework. The coefficients of the performed logistic regression can be obtained through the `summary()` function applied on the name of the object that stores the output of the `glm()` method:

```
> summary(logit1)
$coefficients
                       Estimate
(Intercept)           -1.618186980
AIR_TIME               0.063242891
DISTANCE              -0.007757951
DAY_OF_WEEK_1          0.129620966
DAY_OF_WEEK_5          0.190648077
DAY_OF_WEEK_3          0.170850312
DAY_OF_WEEK_4          0.220783062
DAY_OF_WEEK_2          0.132066275
DAY_OF_WEEK_7          0.101398697
DEP_PART_morning      -0.232715802
DEP_PART_afternoon    -0.237307268
DEP_PART_evening      -0.316990963
DEP_DELAY              0.181978355
```

The mere coefficients are not the most informative estimates of the model. In practice, where machine learning or predictive analytics methods are concerned, researchers and scientists want to see how the model fits to real data and whether its accuracy allows us to generalize to new unlabelled observations or events. For that purpose, we will apply the model to a number of observations that will constitute a new test `DataFrame` with three continuous variables of `AIR_TIME`, `DISTANCE`, and `DEP_DELAY` centered at their mean values and all combinations of levels of two conditional variables: `DAY_OF_WEEK` and `DEP_PART`. In order to calculate and extract the mean estimates across all flights we are going to use the `mean()` function and then retrieve its results using the `select()` and `head()` methods. Remember that in `SparkR` functions are lazily evaluated and need to be explicitly called to output the results. In our case you can use the following code to calculate the arithmetic means of the `AIR_TIME`, `DISTANCE`, and `DEP_DELAY` variables:

```
> head(select(flights, mean(flights$AIR_TIME)))
...#output truncated
  avg(AIR_TIME)
1    111.3743
> head(select(flights, mean(flights$DISTANCE)))
...#output truncated
  avg(DISTANCE)
1    802.5352
> head(select(flights, mean(flights$DEP_DELAY)))
...#output truncated
  avg(DEP_DELAY)
1    10.57354
```

Once we have the means, we may now create a test `DataFrame` with the estimated averages of the continuous variables and all combinations of levels of the remaining categorical variables:

```
> test1 <- createDataFrame(sqlContext,
+         data = data.frame(AIR_TIME = 111.37,
+            DISTANCE = 802.54,
+            DEP_DELAY = 10.57,
+            DAY_OF_WEEK = factor(rep(c("1", "2",
+                                       "3", "4",
+                                       "5", "6", "7"), each=4)),
+            DEP_PART = factor(rep(c("morning", "afternoon",
+                                    "evening", "night"), times=7))))
...#output truncated
```

We can display the top 20 rows of our newly created Spark `DataFrame` using `showDF()`:

```
> showDF(test1)
...#output truncated
+--------+--------+---------+-----------+---------+
|AIR_TIME|DISTANCE|DEP_DELAY|DAY_OF_WEEK| DEP_PART|
+--------+--------+---------+-----------+---------+
|  111.37|  802.54|    10.57|          1|  morning|
|  111.37|  802.54|    10.57|          1|afternoon|
|  111.37|  802.54|    10.57|          1|  evening|
|  111.37|  802.54|    10.57|          1|    night|
|  111.37|  802.54|    10.57|          2|  morning|
|  111.37|  802.54|    10.57|          2|afternoon|
|  111.37|  802.54|    10.57|          2|  evening|
|  111.37|  802.54|    10.57|          2|    night|
|  111.37|  802.54|    10.57|          3|  morning|
|  111.37|  802.54|    10.57|          3|afternoon|
|  111.37|  802.54|    10.57|          3|  evening|
|  111.37|  802.54|    10.57|          3|    night|
|  111.37|  802.54|    10.57|          4|  morning|
|  111.37|  802.54|    10.57|          4|afternoon|
|  111.37|  802.54|    10.57|          4|  evening|
|  111.37|  802.54|    10.57|          4|    night|
|  111.37|  802.54|    10.57|          5|  morning|
|  111.37|  802.54|    10.57|          5|afternoon|
|  111.37|  802.54|    10.57|          5|  evening|
|  111.37|  802.54|    10.57|          5|    night|
+--------+--------+---------+-----------+---------+
only showing top 20 rows
```

We can now easily apply the model on the `test1DataFrame` to predict the values of the response variable. The `predict()` function used for this purpose will provide us with the probabilities of each observation being either the class of the outcome variable or the overall prediction based on the characteristics of its features:

```
> predicted <- predict(logit1, test1)
```

As the created object is in fact yet another Spark `DataFrame` we may retrieve its values using standard `showDF()` (note that we've truncated the output to the prediction estimates for the first observation only, but when we run the code, it will display predictions for all 28 rows of the `test1 DataFrame`):

```
> showDF(predicted, numRows = 28, truncate = FALSE)
...#output truncated
+--------+--------+---------+-----------+---------+-----------------------+---------+
------------------------------+-----------------------------------+---------+
-----------------------------------+----------+
```

```
|AIR_TIME|DISTANCE|DEP_DELAY|DAY_OF_WEEK|DEP_PART |features
|rawPrediction                                    |probability
|prediction|
+--------+--------+---------+-----------+---------+------------------------
-------------------------+------------------------------------------+---------
--------------------------+----------+
|111.37  |802.54  |10.57    |1          |morning
|(12,[0,1,2,8,11],[111.37,802.54,1.0,1.0,10.57])
|[-1.0195239529038107,1.0195239529038107]|[0.26512013910874027,0.7348798608
912597]|1.0         |
...#output truncated
```

The preceding output looks a bit messy and not very user-friendly, but we can see that the predicted class of the first observation has been found to equal 1.0, which simply means that, based on our logit1 model, the flight with mean air time, distance, and departure delay is predicted to arrive delayed at its destination airport if it departed on Monday morning. We can also see that the probability of this flight arriving delayed is 0.73, much higher than the probability of getting to its destination on time (0.27). Of course, you can play with the values of the terms to investigate how probabilities and the overall predictions change when certain flight metrics are increased or decreased. In the following example, we want to classify another Monday morning flight with arbitrary values of the remaining continuous variables:

```
> test2 <- createDataFrame(sqlContext,
+            data = data.frame(AIR_TIME=450,
+                              DISTANCE=3400,
+                              DEP_DELAY = -10,
+                              DAY_OF_WEEK = "1",
+                              DEP_PART = "morning"))
...#output truncated
> showDF(predict(logit1, test2), truncate = FALSE)
...#output truncated
|AIR_TIME|DISTANCE|DEP_DELAY|DAY_OF_WEEK|DEP_PART|features
|rawPrediction                                    |probability
|prediction|
+--------+--------+---------+-----------+--------+-------------------
|450.0   |3400.0  |-10.0    |1          |morning
|(12,[0,1,2,8,11],[450.0,3400.0,1.0,1.0,-10.0])|[1.4587987210916415,-1.4587
987210916415]|[0.8113488735461956,0.18865112645380439]|0.0          |
...#output truncated
```

In the preceding case, the flight was predicted to arrive on time (0.0) as the probability of not being delayed (0.81) was much higher than the probability of arriving delayed (0.19). This however, tells us nothing about the overall accuracy of the model. In order to estimate it, we can run the model over our training data to measure how well it picked up the classes of the response variable. For that reason, we will again use the `predict()` function, but this time we are going apply the model to our existing training dataset stored in `flightsDataFrame`:

```
> flightsPred <- predict(logit1, flights)
```

As we are interested in estimating the accuracy of the model, we will extract only the actual values of the response variable and their predictions computed with the logistic regression model:

```
> prediction <- select(flightsPred, "ARR_DEL", "prediction")
> showDF(prediction, numRows = 200)
...#output truncated
+--------+----------+
|ARR_DEL|prediction|
+--------+----------+
|      1|       1.0|
|      1|       1.0|
|      0|       0.0|
|      1|       1.0|
|      1|       0.0|
|      0|       0.0|
|      0|       0.0|
...#output truncated
+--------+----------+
only showing top 200 rows
```

There are two simple ways to quickly estimate the accuracy of our model. The first one is by calculating the overall accuracy rate: how well our model predicts both on-time and delayed flights. Secondly, we will measure how well our model identified delayed flights when in fact these flights arrived delayed. Following the first scenario, we will firstly assign value 1 to all observations where the actual observed value of the outcome variable agrees with its model prediction. In this case, we will use the `SparkR` implementation of the `ifelse()` statement:

```
> prediction$success <- ifelse(prediction$ARR_DEL == prediction$prediction, 1, 0)
```

The resulting `DataFrame` will be then registered as a temporary table in the `SQL Context`, so we can run a SQL query to calculate the sum of all rows that have been predicted correctly:

```
> registerTempTable(prediction, "prediction")
> correct <- sql(sqlContext, "SELECT count(success) FROM prediction WHERE
success = 1")
...#output truncated
```

Finally, we count the size of our dataset and we use this value as a denominator to estimate the overall success rate of the model. As the preceding step involved lazily evaluated SQL transformations, we also have to explicitly retrieve the value of the correct predictions using the `collect()` function and input it as a numerator in our overall accuracy rate formula:

```
> total <- count(prediction)
...#output truncated
> accuracy <- collect(correct) / total
...#output truncated

> accuracy
        _c0
1 0.8268416
```

The overall predictive accuracy of the model as applied on the training dataset was ~0.83. It means that almost 83% of the values have been correctly predicted for the response variable. But this accuracy rate only applies to both classes of the categorical outcome variable. As stated earlier we are more interested in being able to predict delayed rather than on-time flights. In this case, the steps are similar to before when calculating the overall accuracy, but this time we will create a subset of rows where the actual observed value of the outcome variable in the training set was 1 (delayed):

```
> prediction <- select(flightsPred, "ARR_DEL", "prediction")
> pred_del <- filter(prediction, prediction$ARR_DEL==1)
> registerTempTable(pred_del, "pred_del")
> showDF(pred_del, numRows = 200)
...#output truncated
+-------+----------+
|ARR_DEL|prediction|
+-------+----------+
|      1|       1.0|
|      1|       1.0|
|      1|       1.0|
|      1|       0.0|
|      1|       0.0|
|      1|       1.0|
|      1|       0.0|
```

```
|       1|       1.0|
|       1|       1.0|
|       1|       1.0|
...#output truncated
+--------+----------+
only showing top 200 rows
```

Using the newly registered temporary table, we may sum up the values in the `prediction` column and divide the result by the overall number of delayed flights in the training dataset:

```
> pred_cor <- sql(sqlContext, "SELECT count(prediction) FROM prediction
WHERE prediction = 1")
> total_delayed <- count(pred_del)
...#output truncated

> acc_pred <- collect(pred_cor) / total_delayed
...#output truncated

> acc_pred
        _c0
1 0.7997071
```

The resulting accuracy of correctly predicting delayed flights dropped to around 80%. It's generally not too bad considering the simplicity of the model-both the theoretical assumptions of the logistic regression and its easy practical implementation in `SparkR`. However, as mentioned earlier, the accuracy of the model is best assessed when one fits it to new, test data. This is the only way for us to investigate how well the model generalizes to unseen, future data. For this purpose, we will read in the second flight dataset that contains only flights that occurred in January 2015. As both datasets have the same structure, we will re-use the previously defined schema to upload the `flights_jan_2015.csv` file directly from HDFS with the help of the `spark-csv` library:

```
> jan15 <- read.df(sqlContext,
+                  path = "/user/swalko/data/flights_jan_2015.csv",
+                  source = "com.databricks.spark.csv",
+                  header = "true",
+                  schema = schema,
+                  nullValue = "NA")
```

We will also process the `jan15` object of `DataFrame` in exactly same way as we did with the training dataset earlier. Go a few pages back if you are looking for detailed explanations of each command:

```
> jan15 <- filter(jan15, jan15$CANCELLED == 0)
> jan15 <- filter(jan15, jan15$DIVERTED == 0)
```

```
> jan15 <- jan15[, -6:-7]
> registerTempTable(jan15, "jan15")
> jan15 <- sql(sqlContext, "SELECT *, IF(ARR_DELAY > 0, 1, 0)
+                AS ARR_DEL from jan15")

> registerTempTable(jan15, "jan15")
> jan15 <- sql(sqlContext, "SELECT *,
+             CASE WHEN(DEP_TIME >= 500 AND DEP_TIME < 1200)
+             THEN ('morning') WHEN(DEP_TIME >= 1200 AND DEP_TIME < 1700)
+             THEN ('afternoon') WHEN(DEP_TIME >= 1700 AND DEP_TIME <
2100)
+             THEN ('evening') ELSE('night') END AS DEP_PART from jan15")
```

In the final processing stage of the test data, we will select only the variables of interest-the ones that were used as the term and response variable in the logistic regression model:

```
> jan15 <- select(jan15, "DAY_OF_WEEK", "DEP_DELAY",
+                "AIR_TIME", "DISTANCE", "DEP_PART", "ARR_DEL")

> str(jan15)
...#output truncated
'DataFrame': 6 variables:
 $ DAY_OF_WEEK: chr "4" "5" "6" "7" "1" "2"
 $ DEP_DELAY  : int -5 -10 -7 -7 -7 -4
 $ AIR_TIME   : int 378 357 330 352 338 335
 $ DISTANCE   : int 2475 2475 2475 2475 2475 2475
 $ DEP_PART   : chr "morning" "morning" "morning" "morning" "morning"
"morning"
 $ ARR_DEL    : int 1 0 0 0 0 1
```

We are now ready to fit the `logit1` model to our new test dataset and evaluate its predictive performance. We may assume that the overall accuracy will drop further, but we can't really say by how much at this moment:

```
> janPred <- predict(logit1, jan15)
...#output truncated
> jan_eval <- select(janPred, "ARR_DEL", "prediction")
```

In order to calculate the overall model accuracy on the new data, we will simply follow the steps from when we estimated the same metric for the training set:

```
> jan_eval$success <- ifelse(jan_eval$ARR_DEL == jan_eval$prediction, 1, 0)
> registerTempTable(jan_eval, "jan_eval")
> correct <- sql(sqlContext, "SELECT count(success) FROM jan_eval WHERE
success = 1")
> total <- count(jan_eval)
> accuracy <- collect(correct) / total
...#output truncated
```

```
> accuracy
         _c0
1 0.8070536
```

Based on the preceding output, the overall accuracy rate dropped only to 81%; it is still a pretty good result. Being able to convincingly predict the class of four out of five events in the untrained data is quite reassuring. How about the model accuracy of predicting only the delayed flights then? Once again, for our new test data, we will apply the same calculation methods and techniques as earlier for the training set:

```
> jan_eval <- select(janPred, "ARR_DEL", "prediction")
> jan_del <- filter(jan_eval, jan_eval$ARR_DEL==1)
> registerTempTable(jan_del, "jan_del")
> jan_cor <- sql(sqlContext, "SELECT count(prediction) FROM jan_del WHERE prediction = 1")
> total_delayed <- count(jan_del)
...#output truncated

> acc_pred <- collect(jan_cor) / total_delayed
...#output truncated

> acc_pred
         _c0
1 0.6725848
```

Now, this must be a bit disappointing. A 67% accuracy rate means that two out of three delayed flights will be correctly identified by our model. It's not disastrous, but it's much lower than the overall model accuracy, which was probably inflated by the high predictive level of on-time flights. In this case the achieved model accuracy is still in fact quite acceptable, but it will need further investigation if we were to predict, for example whether a patient suffers from a medical condition or in other high-risk classification tasks.

In this section, we have guided you through the implementation of a logistic regression as an example of a Generalized Linear Model on a multi-node Spark and Hadoop HDInsight cluster using R's API for Spark with the `SparkR` package. In the next section, we will attempt to improve our predictive model by running other machine learning algorithms for example **Naive Bayes** and **Neural Networks** using the Big Data machine learning platform, which supports R and Hadoop, called **H2O**.

Naive Bayes with H2O on Hadoop with R

The growing number of machine learning applications in data science has led to the development of several Big Data predictive analytics tools as described in the first part of this chapter. It is even more exciting for R users that some of these tools connect well with the R language allowing data analysts to use R to deploy and evaluate machine learning algorithms on massive datasets. One such Big Data machine learning platform is **H2O**–open-source, hugely scalable, and fast data exploratory and machine learning software developed and maintained by California-based start-up **H2O.ai** (formerly known as **0xdata**). As H2O is designed to effortlessly integrate with cloud computing platforms such as Amazon EC2 or Microsoft Azure, it has become the obvious choice for large businesses and organisations wanting to implement powerful machine and statistical learning models on massively scalable in-house or cloud-based infrastructures.

Running an H2O instance on Hadoop with R

The H2O software can be installed either in standalone mode or on top of the existing architecture, for example Apache Hadoop and Spark. Its flexibility also means that it can be easily run on most popular operating systems and it connects well to other programming languages including R.

In this section we will show you how to install and run the H2O instance on the previously created multi-node Spark and Hadoop HDInsight cluster with R and RStudio Server installed. If you followed the preceding sections of this chapter your HDInsight cluster should be already fully-operational and ready to use. If you skipped the preceding parts of this chapter make sure to take a step back and read it carefully. For specific details on how to set up and configure the HDInsight cluster you may also wish to re-visit the section titled *HDInsight – A multi-node Hadoop cluster on Azure* in `Chapter 4`, *Hadoop and MapReduce Framework for R* and the section on *Launching HDInsight with Spark and R/RStudio* in `Chapter 7`, *Faster than Hadoop: Spark with R*.

The installation of H2O is relatively easy and it can be done directly from RStudio Server or R shell. The details of the procedures are explained on its website at `http://www.h2o.ai/download/h2o/r`. We will, however, begin the H2O integration with downloading and installing Linux Ubuntu dependencies for the H2O pre-requisites. We also assume that the cluster is already configured and it includes at least the Apache Hadoop, R, and RStudio Server. In the terminal window, `ssh` to the head node of your HDInsight cluster as shown on multiple occasions in this book, for example:

```
$ ssh swalko@swalkospark-ssh.azurehdinsight.net
```

Download and install `curl` and `openssl` dependencies for Ubuntu:

```
$ sudo apt-get install libssl-dev libcurl4-openssl-dev libssh2-1-dev
...#output truncated
```

Following the installation of the required Ubuntu libraries, download and install R pre-requisites for H2O:

```
$ sudo Rscript -e 'install.packages(c("statmod","RCurl","jsonlite"),
repos="https://cran.r-project.org/")'
...#output truncated
```

The H2O website also lists other R packages such as `methods`, `stats`, `graphics`, `tools`, and `utils`, which are required for H2O to run; however these packages are already part of the recent R and RStudio distributions and do not need to be re-installed. In the unlikely scenario that you are using older versions of RStudio, make sure to add the names of these packages to the previously run statement.

Once the required R packages are installed, we may now install H2O for R straight from the command line by invoking the R shell. Make sure to install the most recent stable version of H2O (at the time of writing this chapter the latest stable release was called Rel-Turchin 3.8.2.6); you can review all available H2O builds at `http://www.h2o.ai/download/h2o/r`. Type the following line of code to install H2O:

```
$ sudo Rscript -e 'install.packages("h2o", type="source",
repos=(c("http://h2o-release.s3.amazonaws.com/h2o/rel-turchin/3/R")))'
...#output truncated
```

The installation of H2O will only take a few seconds. It is important to say here that, by achieving this, we will only be able to use H2O on the node on that we installed it. Therefore, in order to benefit from combined computational resources of our HDInsight cluster and before we go any further and start implementing H2O machine learning algorithms directly from RStudio, let's run H2O on top of the existing Hadoop architecture in the cluster. Firstly, we need to check what version of Hadoop our cluster is equipped with:

```
$ hadoop version
Hadoop 2.7.1.2.4.2.0-258
Subversion git@github.com:hortonworks/hadoop.git -r
13debf893a605e8a88df18a7d8d214f571e05289
Compiled by jenkins on 2016-04-25T05:44Z
Compiled with protoc 2.5.0
From source with checksum 2a2d95f05ec6c3ac547ed58cab713ac
This command was run using /usr/hdp/2.4.2.0-258/hadoop/hadoop-
common-2.7.1.2.4.2.0-258.jar
```

Based on the version of Hadoop files installed in the `/usr/hdp/` directory, visit the `http://www.h2o.ai/download/h2o/hadoop` website and download the matching version of H2O. In our case we will download and unzip the `h2o-3.8.2.6-hdp2.4.zip` file:

```
$ wget
http://download.h2o.ai/download/h2o-3.8.2.6-hdp2.4?id=5e7ef63b-1eb6-dc1f-e8
cc-b568ba40f526 -O h2o-3.8.2.6-hdp2.4.zip
...#output truncated

$ unzip h2o-3.8.2.6-*.zip
...#output truncated
```

After unzipping the file, change the current directory (the `cd` command) to the newly created directory with H2O driver files for Hadoop:

```
$ cd h2o-3.8.2.6-*
```

At this stage, we are only one step away from initializing the H2O platform on Apache Hadoop. It's now a good time to remind ourselves of the actual configuration of the cluster's infrastructure that we have at our disposal. When setting up the cluster earlier in this chapter, we chose four worker nodes of `D12` type for the Azure virtual machines-each with 4 cores and 28GB of RAM, and two head nodes with the same `D12` type configuration on each node. It is recommended that the H2O cluster uses up to one-quarter of the available RAM for each node, so in our case it will be 7GB of RAM for each of the four nodes, therefore 28GB in total. Knowing this information, we can now start the H2O instance that will contain four Hadoop nodes-each with 7GB of RAM available for H2O.

The initialization of the distributed version of H2O is in fact yet another Hadoop job, hence the starting command of the H2O deployment statement begins with the already known `hadoop` method:

```
$ hadoop jar h2odriver.jar -nodes 4 -mapperXmx 7g -output
/user/swalko/output
...#output truncated
```

The `-nodes` option indicates the number of Hadoop nodes to be used by the H2O cluster, the `-mapperXmx` specifies the amount of memory designated for each node to run H2O jobs, and the `-output` option defines the output directory for H2O.

It may take up to two minutes to set up and start the H2O cluster. Quite a long output will follow with its final part resembling the following screenshot:

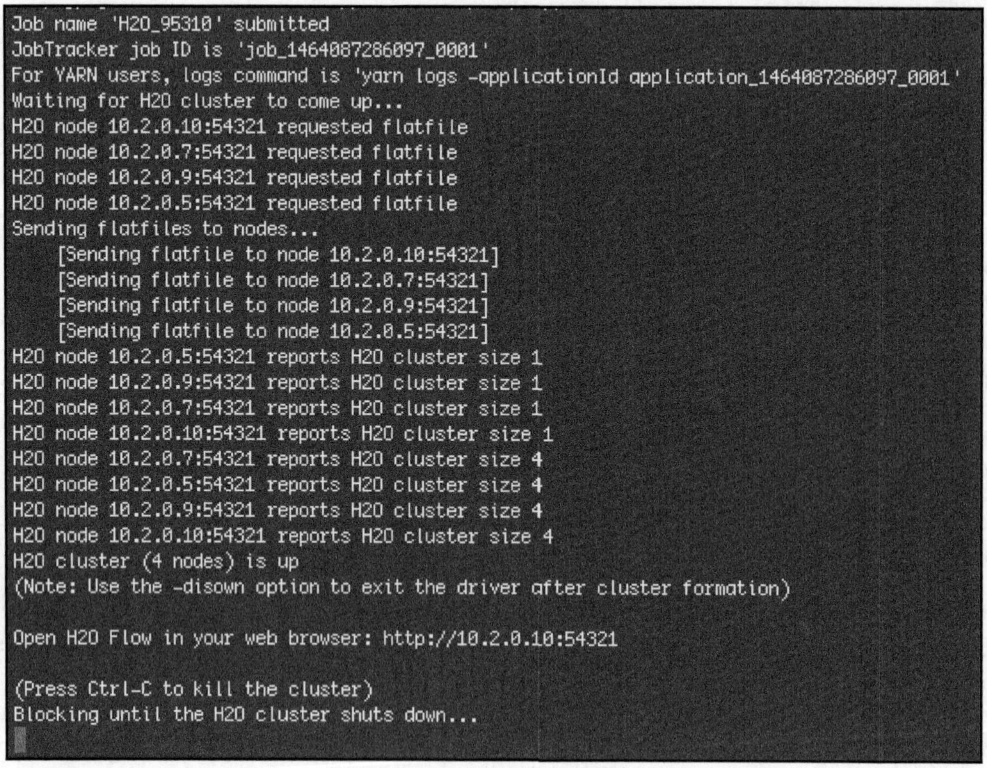

```
Job name 'H2O_95310' submitted
JobTracker job ID is 'job_1464087286097_0001'
For YARN users, logs command is 'yarn logs -applicationId application_1464087286097_0001'
Waiting for H2O cluster to come up...
H2O node 10.2.0.10:54321 requested flatfile
H2O node 10.2.0.7:54321 requested flatfile
H2O node 10.2.0.9:54321 requested flatfile
H2O node 10.2.0.5:54321 requested flatfile
Sending flatfiles to nodes...
    [Sending flatfile to node 10.2.0.10:54321]
    [Sending flatfile to node 10.2.0.7:54321]
    [Sending flatfile to node 10.2.0.9:54321]
    [Sending flatfile to node 10.2.0.5:54321]
H2O node 10.2.0.5:54321 reports H2O cluster size 1
H2O node 10.2.0.9:54321 reports H2O cluster size 1
H2O node 10.2.0.7:54321 reports H2O cluster size 1
H2O node 10.2.0.10:54321 reports H2O cluster size 1
H2O node 10.2.0.7:54321 reports H2O cluster size 4
H2O node 10.2.0.5:54321 reports H2O cluster size 4
H2O node 10.2.0.9:54321 reports H2O cluster size 4
H2O node 10.2.0.10:54321 reports H2O cluster size 4
H2O cluster (4 nodes) is up
(Note: Use the -disown option to exit the driver after cluster formation)

Open H2O Flow in your web browser: http://10.2.0.10:54321

(Press Ctrl-C to kill the cluster)
Blocking until the H2O cluster shuts down...
```

Your terminal window will now become busy and go into an idle state, but keep it open. In the event your machine learning algorithms cause problems, you can kill the cluster by pressing *Ctrl + C*.

In one of the last lines, the output displays the `IP` with a port number that can be used to connect to the H2O cluster. In our case the `IP` is `10.2.0.10` and the port number is the default value for H2O clusters: `54321`.

Having started the H2O instance, we may now log in to RStudio Server through the web browser. In the RStudio console, load the `h2o` package we installed earlier:

```
> library(h2o)
...#output truncated
```

We may now connect the H2O instance. As the connection is going to be made to the already running H2O cluster, make sure to set the `startH2O` argument to `FALSE` and provide the `IP` and `port` numbers retrieved before (these may differ; therefore adjust the values for your own cluster):

```
> h2o <- h2o.init(ip = "10.2.0.10", port = 54321, startH2O = F)
Connection successful!

R is connected to the H2O cluster:
    H2O cluster uptime:         3 minutes 53 seconds
    H2O cluster version:        3.8.2.6
    H2O cluster name:           H2O_95310
    H2O cluster total nodes:    4
    H2O cluster total memory:   27.61 GB
    H2O cluster total cores:    16
    H2O cluster allowed cores:  16
    H2O cluster healthy:        TRUE
    H2O Connection ip:          10.2.0.10
    H2O Connection port:        54321
    H2O Connection proxy:       NA
    R Version:                  R version 3.3.0 (2016-05-03)
```

The preceding output confirms the configuration of the H2O cluster: the number of nodes, memory allowance, enabled cores, and connection IP and port numbers. At any time during the data exploration or analysis in H2O you can retrieve this information using the `h2o.clusterInfo()` function. As our H2O cluster is alive and kicking, we may now upload the data and test H2O's analytical capabilities.

Reading and exploring the data in H2O

The H2O platform allows the use of a variety of file formats as its data sources, from individual locally stored CSV, JSON, and XLS files, through data stored in Amazon S3 buckets or HDFS to SQL tables and NoSQL collections. The list of supported formats is constantly growing. However the integration with Azure Storage Blobs and HDInsight HDFS is still problematic and therefore in this tutorial we will be using locally stored data on the cluster's hard drive. We will also continue fitting the models to the flight datasets that we introduced you to in the preceding section on logistic regression with `SparkR`. If you want to follow the instructions given in this part of the book, make sure to download the data from the Packt Publishing website created for this chapter.

We will begin with indicating the location of the `flights_2014.csv` file in the `swalko` user area on the cluster's hard drive:

```
> path1 <- "/home/swalko/data/flights_2014.csv"
```

We may then pass the path to the file in the `h2o.uploadFile()` function as a value of the `path` argument. The function also allows us to specify the name of the destination frame (`destination_frame` argument) in the H2O cloud. As we import a CSV file with variable names in the first row of the data, we should set the `header` to `TRUE` and we may also specify the column separator `sep`. The resulting object of the file import implemented with the code below is the `H2OFrame` named `flights14`:

```
> flights14 <- h2o.uploadFile(path = path1,
+                             destination_frame = "flights14",
+                             parse = TRUE, header = TRUE,
+                             sep = ",")
  |=============================================| 100%
```

Note that the `H2OFrame` objects do not use the memory resources of R, instead they are stored and processed in the H2O cloud's own RAM. By deploying the H2O cloud on a multi-node Hadoop cluster, we are able to perform heavy data transformations and complex predictive analytics by pushing the processing to the combined H2O memory allowance assigned from each available node of the cluster. The `H2OFrames` residing in R's workspace are therefore simply mappings to the data stored in H2O, but at any time they can be explicitly downloaded and converted into R `data.frame` objects through the `as.data.frame()` implementation in the `h2o` package.

Apart from being able to access the `H2OFrames` directly from the R environment, users can also view and list all `H2OFrames` generated during the session including subsets, models, and intermediary objects with the `h2o.ls()` function:

```
> h2o.ls()
        key
1 flights14
```

The data mapped by the `H2OFrames` can be easily explored through the standard `summary()` and `str()` methods, for example:

```
> summary(flights14)
 DAY_OF_WEEK     DEP_TIME          DEP_DELAY
 Min.   :1.000   Min.   :   1.0   Min.   :-251.000
 1st Qu.:2.000   1st Qu.: 927.4   1st Qu.:  -6.832
 Median :4.000   Median :1325.8   Median :  -1.524
 Mean   :3.924   Mean   :1334.8   Mean   :  10.642
 3rd Qu.:6.000   3rd Qu.:1731.4   3rd Qu.:   9.092
 Max.   :7.000   Max.   :2400.0   Max.   :2402.000
                 NA's   :122742   NA's   :122742
 ARR_TIME        ARR_DELAY         CANCELLED
 Min.   :   1    Min.   :-112.000  Min.   :0.00000
 1st Qu.:1110    1st Qu.: -12.277  1st Qu.:0.00000
 Median :1515    Median :  -4.606  Median :0.00000
 Mean   :1486    Mean   :   7.328  Mean   :0.02182
 3rd Qu.:1914    3rd Qu.:  10.736  3rd Qu.:0.00000
 Max.   :2400    Max.   :2444.000  Max.   :1.00000
 NA's   :129628  NA's   :141433
 DIVERTED          AIR_TIME         DISTANCE
 Min.   :0.000000  Min.   :  7.0    Min.   :  24.0
 1st Qu.:0.000000  1st Qu.: 59.0    1st Qu.: 361.3
 Median :0.000000  Median : 92.0    Median : 624.2
 Mean   :0.002483  Mean   :111.4    Mean   : 798.7
 3rd Qu.:0.000000  3rd Qu.:141.0    3rd Qu.:1021.0
 Max.   :1.000000  Max.   :706.0    Max.   :4983.0
                   NA's   :141433

> str(flights14)
Class 'H2OFrame' <environment: 0x509f610>
 - attr(*, "op")= chr "Parse"
 - attr(*, "id")= chr "flights14"
 - attr(*, "eval")= logi FALSE
 - attr(*, "nrow")= int 5819811
 - attr(*, "ncol")= int 9
 - attr(*, "types")=List of 9
  ..$ : chr "int"
  ..$ : chr "int"
```

```
 ..$ : chr "int"
 ..$ : chr "int"
 ..$ : chr "int"
 ..$ : chr "int"
 ..$ : chr "int"
 ..$ : chr "int"
 ..$ : chr "int"
 - attr(*, "data")='data.frame': 10 obs. of  9 variables:
 ..$ DAY_OF_WEEK: num  2 3 4 5 6 7 1 2 3 4
 ..$ DEP_TIME   : num  854 853 856 857 854 855 851 855 857 852
 ..$ DEP_DELAY  : num  -6 -7 -4 -3 -6 -5 -9 -5 -3 -8
 ..$ ARR_TIME   : num  1217 1225 1203 1229 1220 ...
 ..$ ARR_DELAY  : num  2 10 -12 14 5 -18 -14 30 -9 -17
 ..$ CANCELLED  : num  0 0 0 0 0 0 0 0 0 0
 ..$ DIVERTED   : num  0 0 0 0 0 0 0 0 0 0
 ..$ AIR_TIME   : num  355 357 336 344 338 334 330 332 330 338
 ..$ DISTANCE   : num  2475 2475 2475 2475 2475 ...
```

Note the different way that the `str()` function applied on the H2OFrame displays its output-it includes a set of attributes such as the number of rows and columns in the object and also the types of data stored by variables. The standard output of the generic `str()` function is included in the form of a small snapshot of the actual R `data.frame`, but only 10 observations are shown to save the memory resources allocated to R.

Before we move on to the implementation of specific machine learning algorithms, we will perform a few operations we previously carried out when processing the data for the purpose of logistic regression with SparkR. Firstly, we will subset the data by filtering out all flights that were either cancelled and diverted and deleting the CANCELLED and DIVERTED variables from the H2OFrame. We can accomplish these actions in H2O through the standard subscripting and common logical operators, for example:

```
> flights14 <- flights14[flights14$CANCELLED==0 & flights14$DIVERTED==0, ]
> flights14 <- flights14[, -6:-7]
```

The h2o package also allows us to quickly scan the data for the presence of missing values with the `h2o.nacnt()` function:

```
> h2o.nacnt(flights14)
[1] 0 0 0 0 0 0 0
```

We can also very easily transform the integer variables, for example DAY_OF_WEEK, into categorical factors:

```
> flights14$DAY_OF_WEEK <- as.factor(flights14$DAY_OF_WEEK)
```

One of the most powerful functionalities offered by the h2o package is the ability to define custom functions that can be executed on the H2OFrame objects. For example, we can quickly create a simple function which will calculate the arithmetic mean of departure delay for all flights:

```
avg_del <- function(x) { sum(x[,3])/nrow(x) }
```

This user-defined function can then be conveniently applied in the estimation of mean departure delays aggregated by each day of week. The h2o.ddply() method allows us to split the data by levels of the DAY_OF_WEEK categorical variable and calculate the mean departure delay for each subset:

```
> avg.del <- h2o.ddply(flights14, "DAY_OF_WEEK", FUN = avg_del)
```

Finally, we can display the entire contents of the resulting avg.delH2OFrame by downloading and converting it into the native R data.frame:

```
> as.data.frame(avg.del)
  DAY_OF_WEEK  ddply_C1
1           1 11.275567
2           2  9.935703
3           3 10.514751
4           4 12.141402
5           5 11.760211
6           6  8.504064
7           7  9.438214
```

From the above output we can clearly see that in general the largest departure delay occurs on Thursdays and Fridays.

In this section, we read the data into the H2O cloud and we run a number of simple transformations and data exploration activities available to users through the h2o package. The vignette of the package, which can be accessed at http://www.h2o.ai/product/integration/, provides further details on a number of other data manipulation techniques through h2o. In the next section we will present some of them while preparing our data for analytics using the Naive Bayes algorithm.

Naive Bayes on H2O with R

The h2o package for R supports a selection of popular machine learning and predictive models, for example **Generalized Linear Models**, **Gradient Boosted Regressions**, **Random Forests**, **k-Means**, and even **multi-layered Neural Networks**, commonly known as Deep Learning models. The list of available algorithms is expanding from version to version and it is quite likely that while you are reading this book there are already many other more specific models for you to use through the h2o package. In this section, we will turn our attention to one quite frequently implemented and relatively robust probabilistic supervised learning classifier called **Naive Bayes**. Although we are not going to dive into specifics and detailed theoretical assumptions of the model (feel free to visit some of the online and printed resources on supervised and unsupervised learning algorithms presented in the first part of this chapter), we will take a few minutes to explain some basic characteristics of this classification technique.

As the name suggests Naive Bayes is largely based on **the Bayes' theorem**, which simply assumes that a new parameter or even very weak information about an event can influence a probability of this event occurring if it is related to it. Without knowing anything about a specific event, we can only base our understanding of its probability on past experience and general knowledge; for example in assessing whether a flight arrival is delayed, the **prior probability** of all delayed flights is used. This probability however can then be influenced by other events with a certain **likelihood**, for example the number of passengers on a plane, the length of a flight, or adverse weather conditions. These likelihoods of additional features influencing the event will be used to calculate the **posterior probability** of an event, which in our case will measure how likely the flight is to be delayed when accounting for these other, related events. Similarly, to logistic regression, when the value of the updated posterior probability is above 0.5 we can assume that the flight is more likely to be delayed than not, and if it is larger than its prior probability we can claim that the related events are in fact **dependent events** and they were at least partially responsible for the arrival delay.

One of the characteristics of the Naive Bayes algorithm, which also puts certain constraints on the pre-processing of the data, is that it works very well with categorical/nominal features. Continuous, numeric variables can still be used as long as their values are *binned* into categories-it simply means that we need to collapse them into levels of factors. Incorrect or different binning of continuous values is a potential risk of the Naive Bayes algorithm as it may result in differing model predictions based on various transformation strategies and techniques.

Having said that, we may now attempt to convert our first numeric variable DEP_TIME into categorical departure parts of a day (DEP_PART). The h2o package provides us with a very useful function h2o.cut(), which allows users to efficiently recode the continuous variables into nominal features by specifying cut-off points for each level and assigning labels to the corresponding intervals of values. For example, from the preceding section on logistic regression, we know that we will like to assign the label "night" to all flight departure time (DEP_TIME) values between 1 and 459 and also between 2059 and 2400. The "night" labels and other parts of a day can be specified using the following code:

```
> flights14$DEP_PART <- h2o.cut(flights14$DEP_TIME,
+                               c(1, 459, 1159, 1659, 2059, 2400),
+                               labels = c("night", "morning",
+                                          "afternoon", "evening",
+                                          "night"))
```

Once executed, we may perform a frequency check of the labels assignment with the h2o.table() function:

```
> h2o.table(flights14$DEP_PART)
   DEP_PART   Count
1     night   30567
2   morning 2298597
3 afternoon 1723386
4   evening 1278974
5     night  346321

[5 rows x 2 columns]
...#output truncated
```

We can now remove both the DEP_TIME and ARR_TIME variables as we are not going to use them in the model:

```
> flights14$DEP_TIME <- flights14$ARR_TIME <- NULL
```

Just as in the `SparkR` package, the `h2o` library contains its own implementation of `IF ELSE` statements (`h2o.ifelse()`) which can be applied on `H2OFrame` objects. Using this method, we will create a binary categorical response variable named `ARR_DEL` which will assign value 1 (delayed) to all flights with a positive arrival delay (based on the value of the `ARR_DELAY` variable) and value 0 (not delayed) to all on-time flights. We will then remove the redundant variable `ARR_DELAY` and run a frequency check on the newly created `ARR_DEL` outcome variable:

```
> flights14$ARR_DEL <- as.factor(h2o.ifelse(flights14$ARR_DELAY > 0, 1, 0))
> flights14$ARR_DELAY <- NULL
> h2o.table(flights14$ARR_DEL)
  ARR_DEL    Count
1       0  3292647
2       1  2385731

[2 rows x 2 columns]
...#output truncated
```

We can also wrap the frequency table into a `prop.table()` function to display the ratio of both classes across all observations:

```
> prop.table(h2o.table(flights14$ARR_DEL))
  ARR_DEL     Count
1     NaN 0.5798569
2     NaN 0.4201430

[2 rows x 2 columns]
...#output truncated
```

From the output we learn that delayed flights make up 42% of all flights in the dataset.

At this stage we have three continuous variables left that should be transformed into sensible nominal features: `DEP_DELAY` (departure delay), `DISTANCE` (distance travelled between the airports of origin and destination) and `AIR_TIME` (the airtime of the flight). Starting from `DEP_DELAY`. It might be useful to firstly explore this variable a bit more to understand its distribution and make certain assumptions about the most meaningful arrangement of its cut-off points and categorical labels:

```
> summary(flights14$DEP_DELAY)
 DEP_DELAY
 Min.    :-112.00
 1st Qu.:  -6.37
 Median :  -1.34
 Mean   :  10.57
 3rd Qu.:   8.72
 Max.   :2402.00
```

Based on the summary statistics of the DEP_DELAY variable, it may make sense to divide the values into five bins with labels as follows: "very early" (from the minimum to -16), "somewhat early" (from -15 to -2), "on time" (from -1 to 1), "somewhat delayed" (from 2 to 15) and "very delayed" (from 16 up until the maximum value of 2402):

```
> flights14$DEP_DELAY <- h2o.cut(flights14$DEP_DELAY,
+                                c(-112, -15, -1, 1, 16, 2402),
+                                labels = c("very early",
+                                           "somewhat early",
+                                           "on time",
+                                           "somewhat delayed",
+                                           "very delayed"))
```

The frequency check executed on the resulting DEP_DELAY variable returned the following distribution of counts for each level:

```
> h2o.table(flights14$DEP_DELAY)
          DEP_DELAY   Count
1        very early   26020
2    somewhat early 3035002
3           on time  465419
4  somewhat delayed 1047625
5      very delayed 1104311

[5 rows x 2 columns]
```

We may now perform similar transformations of the remaining two variables of DISTANCE and AIR_TIME. In both cases we will bin the values into the "short", "medium", and "long" categories. The arbitrary decisions as to the selected cut-off points will be supported with user-defined quantile estimation and summary statistics:

```
> h2o.quantile(flights14$DISTANCE, prob = seq(0, 1, length = 4))
         0% 33.3333333333333% 66.6666666666667%
         31               442               896
       100%
       4983

> summary(flights14$DISTANCE)
 DISTANCE
 Min.    :  31.0
 1st Qu.: 362.9
 Median : 625.4
 Mean   : 802.5
 3rd Qu.:1031.5
 Max.   :4983.0

> flights14$DISTANCE <- h2o.cut(flights14$DISTANCE,
```

```
+                            c(31, 1000, 2000, 4983),
+                            labels = c("short", "medium", "long"))
> h2o.table(flights14$DISTANCE)
   DISTANCE   Count
1     short 4153865
2    medium 1170521
3      long  353285

[3 rows x 2 columns]

> h2o.quantile(flights14$AIR_TIME, prob = seq(0, 1, length = 4))
            0% 33.3333333333333% 66.6666666666667%
             7                68               123
          100%
           706
> summary(flights14$AIR_TIME)
 AIR_TIME
 Min.   :  7.0
 1st Qu.: 59.0
 Median : 92.0
 Mean   :111.4
 3rd Qu.:141.0
 Max.   :706.0
> flights14$AIR_TIME <- h2o.cut(flights14$AIR_TIME,
+                            c(7, 150, 300, 706),
+                            labels = c("short", "medium", "long"))
> h2o.table(flights14$AIR_TIME)
   AIR_TIME   Count
1     short 4454893
2    medium 1059431
3      long  164050
[3 rows x 2 columns]
```

Upon conversion of all continuous variables into categorical features we can finally print the resulting structure of the `flights14` H2OFrame:

```
> str(flights14)
Class 'H2OFrame' <environment: 0x41c2dd0>
 - attr(*, "op")= chr ":="
 - attr(*, "eval")= logi TRUE
 - attr(*, "id")= chr "RTMP_sid_ab22_56"
 - attr(*, "nrow")= int 5678378
 - attr(*, "ncol")= int 6
 - attr(*, "types")=List of 6
  ..$ : chr "enum"
  ..$ : chr "enum"
  ..$ : chr "enum"
```

Chapter 8

```
  ..$ : chr "enum"
  ..$ : chr "enum"
  ..$ : chr "enum"
 - attr(*, "data")='data.frame':   10 obs. of  6 variables:
  ..$ DAY_OF_WEEK: Factor w/ 7 levels "1","2","3","4",..: 2 3 4 5 6 7 1 2 3 4
  ..$ DEP_DELAY  : Factor w/ 5 levels "very early","somewhat early",..: 2 2 2 2 2 2 2 2 2 2
  ..$ AIR_TIME   : Factor w/ 3 levels "short","medium",..: 3 3 3 3 3 3 3 3 3 3
  ..$ DISTANCE   : Factor w/ 3 levels "short","medium",..: 3 3 3 3 3 3 3 3 3 3
  ..$ DEP_PART   : Factor w/ 5 levels "night","morning",..: 2 2 2 2 2 2 2 2 2 2
  ..$ ARR_DEL    : Factor w/ 2 levels "0","1": 2 2 1 2 2 1 1 2 1 1
```

At this point, we are now ready to implement the Naive Bayes algorithm to predict the classes of the response variable: ARR_DEL. For this purpose, the h2o package provides us with the h2o.naiveBayes() function, which in its minimal form requires us to specify the indices or names of the predictor variables and the name/index of the response variable as well as the H2OFrame object that contains all the variables that are used in the model. Additionally, we decided to fill the laplace argument with value 1:

```
> model1 <- h2o.naiveBayes(x = 1:5, y = 6, training_frame = flights14, laplace = 1)
  |=============================================| 100%
```

Laplace estimator

A few words are needed here to explain why we didn't leave the value of the **Laplace estimator** set to 0. In the Naive Bayes, the likelihoods of all features are multiplied by one another, but sometimes events and features may never occur for any of the levels of the class, so their conditional probability with the class will equal 0. In such cases, the null conditional probabilities may largely influence the final posterior probability and the overall accuracy of a model. In order to avoid such *zero* probabilities, a Laplace estimator can be simply set to 1 to add a small, non-zero, quantity to each of the features.

As we have run the Naive Bayes algorithm on our training set, we can extract the values of the model metrics and prediction accuracy by invoking the name of the model:

```
> model1
Model Details:
==============
```

```
H2OBinomialModel: naivebayes
Model ID:  NaiveBayes_model_R_1464255351942_5
Model Summary:
  number_of_response_levels min_apriori_probability max_apriori_probability
1                         2                 0.42014                 0.57986
H2OBinomialMetrics: naivebayes
** Reported on training data. **
MSE:  0.1472508
R^2:  0.395579
LogLoss:  0.4545171
AUC:  0.8301207
Gini:  0.6602414

Confusion Matrix for F1-optimal threshold:
              0       1    Error              Rate
0       2825098  467549 0.141998     =467549/3292647
1        678530 1707201 0.284412     =678530/2385731
Totals  3503628 2174750 0.201832    =1146079/5678378

Maximum Metrics: Maximum metrics at their respective thresholds
                        metric threshold    value idx
1                       max f1  0.429318 0.748693 183
2                       max f2  0.117998 0.784923 364
3                  max f0point5 0.764603 0.794782  49
4                  max accuracy 0.464678 0.798741 166
5                 max precision 0.989894 0.988227   1
6                    max recall 0.035772 1.000000 399
7               max specificity 0.990392 0.999956   0
8              max absolute_MCC 0.464678 0.583260 166
9    max min_per_class_accuracy 0.260206 0.760567 262
Gains/Lift Table: Extract with `h2o.gainsLift(<model>, <data>)` or
`h2o.gainsLift(<model>, valid=<T/F>, xval=<T/F>)`
```

From the model output we can see that the h2o package automatically recognizes the type of Naive Bayes model (H2OBinomialModel) and it presents us with initial, theoretical, minimum and maximum probabilities in the model summary table. Then the model metrics follow:

- The **Mean Squared Error** (**MSE**) assesses how well the estimators predict the values of the response variable. An MSE close to zero suggests that the used estimators predict the classes of the outcome variable with nearly perfect accuracy. It is useful to compare the values of MSE of two or more models, to choose the one that best predicts the outcome variable. As the MSE value in our model was ~0.15, in general its estimators are very good predictors of whether a flight will or will not be delayed.

- The next metric is a very frequently used coefficient of determination called **R squared** (**R2**), which indicates the proportion of the variance in the response variable that is explained by the model. The higher the value of R^2 the better, but usually researchers consider a level of R^2 equal to 0.6 as a good model (it ranges between 0 and 1). The observed value of R^2 in our model (0.4) is somewhat below this threshold, which simply suggests that only 40% of the variance in the ARR_DEL variables is predictable from the set of features. In such cases, we may ask ourselves whether the used features are in fact dependent events and can well predict the outcome variable.
- The third metric is the **Logarithmic Loss** (**LogLoss**), which measures the accuracy of our model by penalizing incorrect predictions based on their probability obtained from the model; therefore the lower the value of LogLoss the greater the accuracy and predictive power of our classifier. The LogLoss metric is often used in **Kaggle** (https://www.kaggle.com/) machine learning competitions to evaluate and compare the models submitted by contestants, so make sure to try to lower its value as much as possible if you have an ambition to win Kaggle data science contests.
- The penultimate model evaluation measure in our output is the **Area Under Curve** (**AUC**), which is frequently used as a metric to assess the accuracy of binary classifiers. The value of the AUC measure is related to the rate of true positives predicted by the model, or in other words, the observed ratio of correctly predicting the delayed flights in our example. If the classifier performs at the level of random guessing, the AUC will approximately equal 0.5. The increased performance of the model will make the AUC much closer to the ideal value of 1. In our case, the AUC of 0.83 means that the model is generally pretty good in accurately predicting the delayed flights.
- Finally, the Gini metric is the evaluation criterion similar in its interpretation to the AUC, but it is only sensitive to the rank of predictions. Using our data as an example, the predicted classes of specific flights are ordered from the largest to the smallest predictions. If the amount of correctly predicted classes is large in the top proportion of the ordered predictions, the overall Gini ratio will be high, meaning that the model correctly assigns strong predictions for specific observations. Our achieved Gini index 0.66 is quite high, once again confirming that overall the model is good in predicting the classes of the response variable.

Below the metrics table, the output displays the confusion matrix of predictions with prediction errors for each class. From the matrix, we can see that the overall accuracy of our model for both classes is close to 80%, because the combined error for all events equals 0.2. However, this overall error is lowered by the good performance of the model in predicting the true negatives (error of only 0.14), when the predicted class 0 (not delayed) agrees with the actual training data. For the true positives the accuracy is then almost 72%, because the error equals 0.28. Remember that these accuracy values have only been obtained on the training set and it is likely that their levels will drop when the model is applied on the new data.

Just below the confusion matrix of the predictions, the output shows a table (with maximum metrics at their respective thresholds) which provide users with cost-sensitive measures such as accuracy, precision, and recall. These estimates may additionally support us in choosing the best performing model. For example, the max F1 metric is often used in binary classifiers and it measures a weighted mean of the precision and recall. The closer its value to 1 the better the accuracy of the model.

The preceding individual sections of the combined output can be also accessed with separate functions; for example the h2o.auc() method will return the value of the AUC metric and h2o.performance() will present all performance-related estimates for the model:

```
> h2o.auc(model1)
[1] 0.8301207

> h2o.performance(model1)
H2OBinomialMetrics: naivebayes
** Reported on training data. **
MSE:      0.1472508
R^2:      0.395579
LogLoss:  0.4545171
AUC:      0.8301207
Gini:     0.6602414
Confusion Matrix for F1-optimal threshold:
              0        1    Error              Rate
0       2825098   467549 0.141998    =467549/3292647
1        678530  1707201 0.284412    =678530/2385731
Totals  3503628  2174750 0.201832   =1146079/5678378
...#output truncated
```

The achieved model performance has only been measured for the training set. We will now apply the same model on the new data with all flights that occurred in January 2015. Below, we will read the `flights_jan_2015.csv` file into the H2O cloud and process the resulting H2OFrame in the same way we did it with the training dataset; for this reason we will skip the explanations of the functions and we will only present the most important data manipulation activities. If you are not sure about the usage of specific methods, make sure to go back a few pages to read the whole explanation of actions when applied on the main, training data.

```
> path2 <- "/home/swalko/data/flights_jan_2015.csv"
> flightsJan15 <- h2o.uploadFile(path = path2,
+                                destination_frame = "flightsJan15",
+                                parse = TRUE, header = TRUE,
+                                sep = ",")
  |=========================================| 100%

> flightsJan15 <- flightsJan15[flightsJan15$CANCELLED==0 &
flightsJan15$DIVERTED==0, ]
> flightsJan15 <- flightsJan15[, -6:-7]

> flightsJan15$DAY_OF_WEEK <- as.factor(flightsJan15$DAY_OF_WEEK)
> flightsJan15$DEP_PART <- h2o.cut(flightsJan15$DEP_TIME,
+                         c(1, 459, 1159, 1659, 2059, 2400),
+                         labels = c("night", "morning",
+                                    "afternoon", "evening",
+                                    "night"))
> flightsJan15$DEP_TIME <- flightsJan15$ARR_TIME <- NULL
> flightsJan15$ARR_DEL <- as.factor(h2o.ifelse(flightsJan15$ARR_DELAY > 0,
1, 0))
> flightsJan15$ARR_DELAY <- NULL
> prop.table(h2o.table(flightsJan15$ARR_DEL)) #40% of delayed flights
  ARR_DEL     Count
1     NaN 0.5993317
2     NaN 0.4006661
[2 rows x 2 columns]
> flightsJan15$DEP_DELAY <- h2o.cut(flightsJan15$DEP_DELAY,
+                         c(-48, -15, -1, 1, 16, 1988),
+                         labels = c("very early",
+                                    "somewhat early",
+                                    "on time",
+                                    "somewhat delayed",
+                                    "very delayed"))
> h2o.table(flightsJan15$DEP_DELAY)
       DEP_DELAY  Count
1     very early   2898
2 somewhat early 253970
3        on time  36608
```

```
4   somewhat delayed  79365
5       very delayed  84171
[5 rows x 2 columns]

> flightsJan15$DISTANCE <- h2o.cut(flightsJan15$DISTANCE,
+                                   c(31, 1000, 2000, 4983),
+                                   labels = c("short",
+                                              "medium",
+                                              "long"))

> h2o.table(flightsJan15$DISTANCE)
  DISTANCE  Count
1    short 330417
2   medium  99576
3     long  26960

[3 rows x 2 columns]

> flightsJan15$AIR_TIME <- h2o.cut(flightsJan15$AIR_TIME,
+                                   c(8, 150, 300, 676),
+                                   labels = c("short",
+                                              "medium",
+                                              "long"))
> h2o.table(flightsJan15$AIR_TIME)
  AIR_TIME  Count
1    short 354738
2   medium  89290
3     long  12982

[3 rows x 2 columns]
> str(flightsJan15)
Class 'H2OFrame' <environment: 0x7e89228>
 - attr(*, "op")= chr ":="
 - attr(*, "eval")= logi TRUE
 - attr(*, "id")= chr "RTMP_sid_ab22_84"
 - attr(*, "nrow")= int 457013
 - attr(*, "ncol")= int 6
 - attr(*, "types")=List of 6
  ..$ : chr "enum"
  ..$ : chr "enum"
  ..$ : chr "enum"
  ..$ : chr "enum"
  ..$ : chr "enum"
  ..$ : chr "enum"
 - attr(*, "data")='data.frame': 10 obs. of  6 variables:
  ..$ DAY_OF_WEEK: Factor w/ 7 levels "1","2","3","4",..: 4 5 6 7 1 2 3 4 5 6
  ..$ DEP_DELAY  : Factor w/ 5 levels "very early","somewhat early",..: 2 2
```

```
 2 2 2 2 2 3 4
  ..$ AIR_TIME   : Factor w/ 3 levels "short","medium",..: 3 3 3 3 3 3 3
3 3
  ..$ DISTANCE   : Factor w/ 3 levels "short","medium",..: 3 3 3 3 3 3 3
3 3
  ..$ DEP_PART   : Factor w/ 5 levels "night","morning",..: 2 2 2 2 2 2 2
2 2
  ..$ ARR_DEL    : Factor w/ 2 levels "0","1": 2 1 1 1 2 1 1 2 1
```

Once the new data has been properly processed and all its continuous variables converted into appropriately binned categorical variables, we are now ready to apply the previously calculated Naive Bayes model to test its accuracy on January 2015 flights. This can be achieved with the h2o.predict() function, in which we will specify the model name as the object argument and the name of the H2OFrame that holds the test data as the newdata parameter:

```
> fit1 <- h2o.predict(object = model1, newdata = flightsJan15)
  |=========================================| 100%
```

After several seconds we are able to extract the probabilities for each class of the response variable and the predicted values of classes for each observation in the test dataset:

```
> fit1
  predict         p0         p1
1       0  0.8541997  0.1458003
2       0  0.8586930  0.1413070
3       0  0.8912964  0.1087036
4       0  0.8775638  0.1224362
5       0  0.8670818  0.1329182
6       0  0.8738735  0.1261265

[457013 rows x 3 columns]
```

We can then measure the performance of the model on the test set by running the h2o.performance() function on the flightsJan15 H2OFrame:

```
> h2o.performance(model1, newdata = flightsJan15)
H2OBinomialMetrics: naivebayes

MSE:      0.1510212
R^2:      0.3710932
LogLoss:  0.4675854
AUC:      0.8137276
Gini:     0.6274553

Confusion Matrix for F1-optimal threshold:
              0       1     Error           Rate
```

```
0           235400  38503 0.140572   =38503/273903
1            56263 126847 0.307263   =56263/183110
Totals 291663 165350 0.207360   =94766/457013
Maximum Metrics: Maximum metrics at their respective thresholds
                      metric threshold     value idx
1                     max f1  0.429318  0.728043 180
2                     max f2  0.101568  0.770639 373
3                max f0point5 0.958268  0.785523  47
4               max accuracy  0.431442  0.792717 179
5              max precision  0.988750  0.982967   3
6                 max recall  0.035771  1.000000 399
7            max specificity  0.990409  0.999916   0
8           max absolute_MCC  0.431442  0.563203 179
9 max min_per_class_accuracy  0.253972  0.745006 265

Gains/Lift Table: Extract with `h2o.gainsLift(<model>, <data>)` or
`h2o.gainsLift(<model>, valid=<T/F>, xval=<T/F>)`
```

As expected, the accuracy metrics of the model achieved on the new data are slightly lower than when the model was performed on the training set. Note that the overall accuracy in predicting the classes is still close to 80%, but the accuracy of predicting true positives decreased to the level of 69% (because the error for true positives equals 0.31). However, by running the Naive Bayes algorithm we were able to improve the predictive power of the model when compared to the performance of the logistic regression carried out in the SparkR package. As you may remember, the accuracy of logistic regression correctly predicting delayed flights was 67%, so a 2-point improvement in the model performance of the Naive Bayes algorithm is very welcome.

In the next section we will try to improve the model even further by carrying out a new generation of extremely powerful machine learning methods called Neural Networks.

Neural Networks with H2O on Hadoop with R

In this part of the chapter we will only briefly introduce you to a very extensive topic of **Artificial Neural Networks (ANN)** and their implementation in the h2o package for distributed machine learning in R. As a precise explanation of concepts related to Neural Networks goes far beyond the scope of this book, we will limit the theory to the bare minimum and guide you through a practical application of a simple Neural Network model performed on our flight data.

How do Neural Networks work?

It is very likely that at some point on your data analytical journey, you have come across the concept of Neural Networks in one way or another. Even if you are unaware of their mathematical underpinnings and theoretical assumptions, you've probably heard and followed many news stories about the advances of artificial intelligence and systems built on Neural Networks and their multi-layered versions in the form of Deep Learning algorithms.

As the name suggests, the idea of ANN derives from our understanding of how animal brains of mammalian organisms work and therefore is related to other fields such as neuroscience, cognitive science, medical, and biological sciences, but also computer science, statistics, and data mining. In simple terms, Neural Networks imitate processes occurring in animal nerve cells, in which an electrical signal travels along the body of a **neuron** to its **axon terminals** and then across the **synapse** to the **dendrites** of other neighboring neurons. What makes the electrical signal pass along the cell is the **action potential** generated when the **stimulus intensity** of the net inward current achieves the critical firing **threshold** and excites the **axon hillock** to propel the signal down the axon. The propagation of the signal across the synapse, which is the junction of two adjoining neurons, occurs as a result of complex chemical reactions including **neurotransmitters** and **post-synaptic receptors** that are responsible for further relay of the signal to the body of another neuron. These reactions and processes still remain largely unknown and there are many conflicting theories on how exactly the neurons communicate with one another.

To translate this simplified overview of neuronal activities into our data analysis, we can say that ANN are similar in both structure and operations to animal nerve cells, however they are not as complex as biological nervous systems. At least not yet. Just like living neural networks, they are organized in **layers** with interconnected **nodes** representing neurons. The **connections** between the nodes may be weighted. The propagation and learning mechanism of the input data is dependent on the **activation function**-an artificial and mathematical equivalent of the action potential threshold described earlier. The architecture of ANN called **network topology** can be customized to control the number of neurons, layers, how they are connected, what type of activation function should be used, and also whether there is any **back-propagation** of the network which adjusts the weights between the nodes and hence strengthens connections in the model-just as learning and experience strengthen connections between biological neurons.

A diagram depicting a simple Neural Network topology is presented in the following figure:

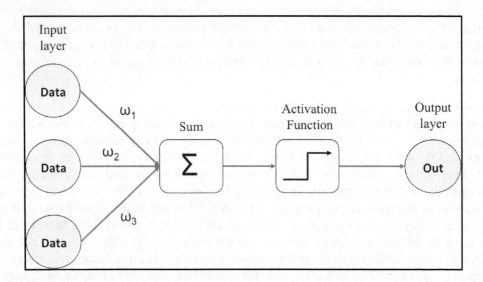

Some ANNs may contain many layers of large numbers of nodes that are interconnected in complex networks and designed to predict or learn abstract concepts and computationally expensive patterns. Such multi-layered algorithms are called **Multi-Layer Perceptrons (MLP)** and are used extensively in **Deep Learning** methods. The major issue, from the analytical point of view, is that some Deep Learning algorithms and their mechanisms have become so convoluted that it is extremely difficult for human analysts to understand the meaning of their outcomes. This leads us to another, more ethical problem; according to some scientists and artificial intelligence researchers, these complex self-learning structures may one day even excel human learning and analytical capabilities. This theoretical prediction has recently provoked a large increase of scientific and general public interest in powerful machine learning methods based on ANN by exciting many Hollywood film directors, innovators, and philosophers.

If you are interested in the details of how ANN and Deep Learning algorithms work, let us refer you to a very good introductory book on that subject titled *Neural Networks and Deep Learning* written by Michael Nielsen and available online at http://neuralnetworksanddeeplearning.com/.

Hopefully, after reading this brief introduction to Neural Networks you have now become motivated to explore this data modeling technique in practice. In the next section, we will apply two differently structured Neural Networks to our data to predict which flights will arrive delayed to their destination airports.

Running Deep Learning models on H2O

Following the short introduction on Neural Networks, we are going to continue using the same H2O cluster, we configured and started in the preceding tutorial on Naive Bayes with the `h2o` package for R on the HDInsight cluster with Hadoop. If you have already terminated the connection with the running H2O cluster, make sure to restart it as shown in the earlier tutorial. We advise those of you who have kept the R session and H2O cluster going, that you remove all old objects from your R workspace before continuing:

```
> rm(list=ls())
```

As the Neural Networks are less fussy about the types of variables used in the predictive model, we will limit the processing of data to the required minimum. Firstly, we shall upload the data again into a new `H2OFrame`, clean it to remove all canceled and diverted flights, and transform the continuous `ARR_DELAY` variable into the binary `ARR_DEL` response variable with the same levels as before: 0 (not delayed) and 1 (delayed):

```
> path1 <- "/home/swalko/data/flights_2014.csv"
> flights14 <- h2o.uploadFile(path = path1,
+                             destination_frame = "flights14",
+                             parse = TRUE, header = TRUE,
+                             sep = ",")
  |=============================================| 100%
> flights14 <- flights14[flights14$CANCELLED==0 & flights14$DIVERTED==0, ]
> flights14 <- flights14[, -6:-7]
> flights14$DAY_OF_WEEK <- as.factor(flights14$DAY_OF_WEEK)
> flights14$ARR_DEL <- as.factor(h2o.ifelse(flights14$ARR_DELAY > 0, 1, 0))
> flights14$ARR_DELAY <- flights14$ARR_TIME <- NULL

> str(flights14)
Class 'H2OFrame' <environment: 0x6740d68>
...#output truncated
 - attr(*, "data")='data.frame':  10 obs. of  6 variables:
  ..$ DAY_OF_WEEK: Factor w/ 7 levels "1","2","3","4",..: 2 3 4 5 6 7 1 2 3 4
  ..$ DEP_TIME   : num  854 853 856 857 854 855 851 855 857 852
  ..$ DEP_DELAY  : num  -6 -7 -4 -3 -6 -5 -9 -5 -3 -8
  ..$ AIR_TIME   : num  355 357 336 344 338 334 330 332 330 338
  ..$ DISTANCE   : num  2475 2475 2475 2475 2475 ...
  ..$ ARR_DEL    : Factor w/ 2 levels "0","1": 2 2 1 2 2 1 1 2 1 1
```

Once the training set is processed, we can go ahead and perform the same data manipulations for the test set. This is possible because the implementation of Neural Network algorithms in the h2o package allows users to specify both the training and validation frames in the same command:

```
> path2 <- "/home/swalko/data/flights_jan_2015.csv"
> flightsJan15 <- h2o.uploadFile(path = path2,
+                                 destination_frame = "flightsJan15",
+                                 parse = TRUE, header = TRUE,
+                                 sep = ",")
  |=============================================| 100%

> flightsJan15 <- flightsJan15[flightsJan15$CANCELLED==0 & flightsJan15$DIVERTED==0, ]
> flightsJan15 <- flightsJan15[, -6:-7]
> flightsJan15$DAY_OF_WEEK <- as.factor(flightsJan15$DAY_OF_WEEK)
> flightsJan15$ARR_DEL <- as.factor(h2o.ifelse(flightsJan15$ARR_DELAY > 0, 1, 0))
> flightsJan15$ARR_DELAY <- flightsJan15$ARR_TIME <- NULL

> str(flightsJan15)
Class 'H2OFrame' <environment: 0x62423c0>
...#output truncated
 - attr(*, "data")='data.frame': 10 obs. of  6 variables:
  ..$ DAY_OF_WEEK: Factor w/ 7 levels "1","2","3","4",..: 4 5 6 7 1 2 3 4 5 6
  ..$ DEP_TIME   : num  855 850 853 853 853 856 859 856 901 903
  ..$ DEP_DELAY  : num  -5 -10 -7 -7 -7 -4 -1 -4 1 3
  ..$ AIR_TIME   : num  378 357 330 352 338 335 341 333 353 345
  ..$ DISTANCE   : num  2475 2475 2475 2475 2475 ...
  ..$ ARR_DEL    : Factor w/ 2 levels "0","1": 2 1 1 1 1 2 1 1 2 1
```

We are now ready to apply our first Neural Network model on the data. In fact, it will even be a Deep Learning algorithm as we are going to create a multi-layered structure with three hidden layers which will contain 10, 5, and 3 neurons respectively. The `h2o.deeplearning()` function, which implements ANN, allows users to define an impressive number of arguments and parameters. We may not only customize the number and the size of network topology (`hidden`), but also specify activation functions (`activation`), how many times the dataset should be iterated (`epochs`), the size of training sample sizes for each iteration (`train_samples_per_iteration`), the adaptive learning rate (`adaptive_rate`), its time decay factor (`rho`), and epsilon (`epsilon`), the momentum and learning rate settings (`rate`, `rate_annealing`, `rate_decay`, `momentum_start`, `momentum_ramp`), l1 and l2 regularization coefficients (`l1` and `l2`), and many more options which control almost every aspect of the model to be generated.

The good news however is that most of the default settings and parameters of the `h2o.deeplearning()` method are quite well optimized for the majority of problems and require little customization for general daily use. In our model, we will additionally set the number of iterations (`epochs`) to 5 hoping that these extra 4 passes of the algorithm over the training dataset will result in an improved overall model. This is, however, likely to slow down its computation. The x and y arguments of the `h2o.deeplearning()` traditionally relate to the set of predictor variables and the response variable respectively. As mentioned earlier, we will also provide both the `training_frame` and the `validation_frame` within the same function:

```
> model2 <- h2o.deeplearning(x = 1:5, y = 6,
+                            training_frame = flights14,
+                            validation_frame = flightsJan15,
+                            hidden = c(10, 5, 3),
+                            epochs = 5)
|=================================================| 100%
```

The computation may take up to one minute to run. When it finishes, we may explore the performance of the achieved model with `summary()`. As the output is rather long, we will interpret it section by section starting from the basic information about the structure of the model:

```
> summary(model2)
Model Details:
==============

H2OBinomialModel: deeplearning
Model Key:  DeepLearning_model_R_1464255351942_18
Status of Neuron Layers: predicting ARR_DEL, 2-class classification,
bernoulli distribution, CrossEntropy loss, 211 weights/biases, 7.2 KB,
30,826,016 training samples, mini-batch size 1
  layer units       type dropout       l1       l2 mean_rate
1     1    12      Input  0.00 %
2     2    10  Rectifier  0.00 % 0.000000 0.000000  0.084909
3     3     5  Rectifier  0.00 % 0.000000 0.000000  0.004641
4     4     3  Rectifier  0.00 % 0.000000 0.000000  0.002806
5     5     2    Softmax         0.000000 0.000000  0.001516
  rate_RMS momentum mean_weight weight_RMS mean_bias bias_RMS
1
2 0.271416 0.000000   -0.039507   0.764668  0.061024 0.471607
3 0.004645 0.000000    0.033570   1.061414 -0.414059 1.171412
4 0.003273 0.000000   -0.057985   0.719442  0.066185 0.387324
5 0.000251 0.000000   -0.519681   2.445112  0.169129 0.227075
...#output truncated
```

The first part of the following output presented displays the main characteristics of the model. It confirms that we were interested in predicting the classes of the binary `ARR_DEL` variable with **Bernoulli distribution** and estimates the final quantity of the training samples. Then the output presents diagnostic stats about each layer with the information on the activation functions used in the model. The `layer` column identifies the layer index (we have five layers in total: one input, three hidden, and one output layer), whereas the `units` column gives the total number of neurons in each layer. The `type` field provides the name of the activation function used at each layer and the `dropout` defines the proportion of dropped features for each row of the training data (for the input layer) and the proportion of incoming weights that were dropped from training at each specific layer (for hidden layers). Other columns such as `L1` and `L2` show the values of regularization penalties for each layer and the remaining fields provide layer-specific learning statistics.

Moving on to the next part, we are presented with model performance metrics, which we previously described in detail when discussing the output of the Naive Bayes algorithm, and the equally well-known confusion matrix with the model errors for predicting each class of the outcome variable:

```
H2OBinomialMetrics: deeplearning
** Reported on training data. **
Description: Metrics reported on temporary training frame with 9970 samples
MSE:  0.1284521
R^2:  0.4714688
LogLoss:  0.4014959
AUC:  0.8838756
Gini:  0.7677511

Confusion Matrix for F1-optimal threshold:
          0     1   Error     Rate
0      4937   880  0.151281  =880/5817
1       970  3183  0.233566  =970/4153
Totals 5907  4063  0.185557  =1850/9970

Maximum Metrics: Maximum metrics at their respective thresholds
                     metric threshold     value idx
1                    max f1  0.278908  0.774830 230
2                    max f2  0.121966  0.823053 317
3              max f0point5  0.571107  0.841214 132
4              max accuracy  0.436460  0.825075 173
5             max precision  0.999970  1.000000   0
6                max recall  0.006239  1.000000 398
7           max specificity  0.999970  1.000000   0
8          max absolute_MCC  0.487465  0.643084 157
9 max min_per_class_accuracy  0.225812  0.797318 254
Gains/Lift Table: Extract with `h2o.gainsLift(<model>, <data>)` or
```

`h2o.gainsLift(<model>, valid=<T/F>, xval=<T/F>)`

From the following output, we can clearly see that the metrics of the model performed on the training set are generally much better than their estimates for the same training data in Naive Bayes. The values of mean squared error and logarithmic loss are closer to 0 and the R squared, AUC, and Gini estimates are much higher. Also, the overall accuracy of the model is increased; it now correctly predicts the classes of almost 81.5% of all flights and the accuracy of predicting the true positives (only the flights which were in fact delayed) reached the level of 77%. This is all very reassuring, but let's not forget that these values have been obtained solely on the training set, and to be precise, the model was performed on a sample of only 9,970 observations from the original training frame.

The next part of the preceding output provides the same model performance metrics, but this time when the model was run on the test data:

```
H2OBinomialMetrics: deeplearning
** Reported on validation data. **
Description: Metrics reported on full validation frame

MSE:      0.1357101
R^2:      0.4348545
LogLoss:  0.4255887
AUC:      0.861461
Gini:     0.7229219

Confusion Matrix for F1-optimal threshold:
              0        1     Error              Rate
0        236311    37592  0.137246     =37592/273903
1         52450   130660  0.286440     =52450/183110
Totals   288761   168252  0.197023     =90042/457013

Maximum Metrics: Maximum metrics at their respective thresholds
                       metric threshold     value idx
1                      max f1  0.379067  0.743734 195
2                      max f2  0.098955  0.796159 333
3                 max f0point5 0.693367  0.823799  96
4                max accuracy  0.515402  0.814732 148
5               max precision  0.999963  0.999231   0
6                  max recall  0.003078  1.000000 399
7             max specificity  0.999963  0.999843   0
8            max absolute_MCC  0.575156  0.617500 130
9      max min_per_class_accuracy 0.247664 0.773427 247
Gains/Lift Table: Extract with `h2o.gainsLift(<model>, <data>)` or
`h2o.gainsLift(<model>, valid=<T/F>, xval=<T/F>)`
```

The presented results look quite optimistic. Although the metrics for the full validation frame are slightly lower than when reported for a sample of the training frame, they are still better than for the test data in the Naive Bayes model in the preceding tutorial. In fact, we have improved the accuracy of correctly predicting delayed flights by another two points up to 71% (as compared to 67% achieved in the logistic regression on Spark and 69% returned by Naive Bayes in H2O).

The last section of the output includes the scoring history table with statistics on the progress of the algorithm, its duration, speed, number of iterations over the data, sample sizes, and the gradual optimization of performance metrics. We will skip this part of the output in this tutorial as it largely sums up the findings presented earlier.

Having observed improvements in the model accuracy by running just one Deep Learning algorithm, why don't we try again and perform another simple Neural Network? This one will iterate through the data twice (`epochs = 2`) and because we are going to use all other parameters of the `h2o.deeplearning()` function set to their default values, it will also include two hidden layers each with 200 neurons:

```
> model3 <- h2o.deeplearning(x = 1:5, y = 6,
+                            training_frame = flights14,
+                            validation_frame = flightsJan15,
+                            epochs = 2)
  |==================================================| 100%
```

The final output of the `model3` model is presented as follows:

```
> summary(model3)
Model Details:
==============
H2OBinomialModel: deeplearning
Model Key:  DeepLearning_model_R_1464255351942_23
Status of Neuron Layers: predicting ARR_DEL, 2-class classification,
bernoulli distribution, CrossEntropy loss, 43,202 weights/biases, 514.9 KB,
11,599,632 training samples, mini-batch size 1
  layer units      type dropout        l1        l2 mean_rate
1     1    12     Input  0.00 %
2     2   200 Rectifier  0.00 % 0.000000  0.000000  0.128657
3     3   200 Rectifier  0.00 % 0.000000  0.000000  0.444878
4     4     2   Softmax         0.000000  0.000000  0.021306
  rate_RMS momentum mean_weight weight_RMS mean_bias  bias_RMS
1
2 0.326028 0.000000   -0.009169   0.223656 -0.152386  0.184606
3 0.310810 0.000000   -0.026128   0.144947 -0.402533  0.690801
4 0.051171 0.000000   -0.012173   0.301910  0.302851  0.121955
H2OBinomialMetrics: deeplearning
** Reported on training data. **
```

Description: Metrics reported on temporary training frame with 9970 samples
MSE: 0.1261093
R^2: 0.4811087
LogLoss: 0.3944227
AUC: 0.8848212
Gini: 0.7696425

Confusion Matrix for F1-optimal threshold:
```
           0     1    Error         Rate
0       4871   946 0.162627     =946/5817
1        927  3226 0.223212     =927/4153
Totals  5798  4172 0.187864    =1873/9970
```
Maximum Metrics: Maximum metrics at their respective thresholds
```
                      metric threshold     value idx
1                     max f1  0.293545  0.775015 230
2                     max f2  0.146018  0.823183 308
3                max f0point5  0.642361  0.843293 109
4               max accuracy  0.509799  0.825677 151
5              max precision  0.999979  1.000000   0
6                 max recall  0.004105  1.000000 398
7            max specificity  0.999979  1.000000   0
8          max absolute_MCC  0.538712  0.644073 143
9    max min_per_class_accuracy  0.262271  0.800626 245
```
Gains/Lift Table: Extract with `h2o.gainsLift(<model>, <data>)` or
`h2o.gainsLift(<model>, valid=<T/F>, xval=<T/F>)`
H2OBinomialMetrics: deeplearning
** Reported on validation data. **
Description: Metrics reported on full validation frame
MSE: 0.1334476
R^2: 0.4442761
LogLoss: 0.4178745
AUC: 0.866789
Gini: 0.733578

Confusion Matrix for F1-optimal threshold:
```
            0      1    Error           Rate
0      240116  33787 0.123354   =33787/273903
1       53189 129921 0.290476   =53189/183110
Totals 293305 163708 0.190314   =86976/457013
```
aximum Metrics: Maximum metrics at their respective thresholds
```
                   metric threshold     value idx
1                  max f1  0.406905  0.749217 191
2                  max f2  0.114253  0.801369 328
3             max f0point5  0.744956  0.823677  83
4            max accuracy  0.553407  0.817200 143
5           max precision  0.999959  0.998330   0
6              max recall  0.002322  1.000000 399
7         max specificity  0.999959  0.999631   0
```

```
8              max absolute_MCC  0.604300 0.618785 127
9 max min_per_class_accuracy    0.280785 0.778559 239
...#output truncated
```

The values of performance metrics for `model3` that run on a sample of the training set are slightly better than in `model2` achieved in the previous attempt, with the R squared, AUC, and Gini coefficients achieving slightly higher values and the LogLoss and MSE metrics falling slightly below the values observed in `model2`. In terms of the overall predictive accuracy of the model, `model3` was found to correctly predict the classes of the response variable at just over 81% for both the training and validation frames. As for the accuracy of correctly predicting the delayed flights only, `model3` achieved 78% accuracy for the training frame and again 71% on the test dataset. Both models achieved very similar accuracy estimates with `model3` being just marginally better in the overall model performance metrics.

Following all the activities you may close the connection and shut down the running H2O cluster with the following line of code. Type Y to confirm you want to disable the H2O cloud:

```
> h2o.shutdown()
Are you sure you want to shutdown the H2O instance running at
http://10.2.0.10:54321/ (Y/N)? Y
```

We hope that the preceding examples inspired you to explore the exciting field of machine learning and Big Data predictive analytics, which you were introduced to in this chapter. As you have probably realized, the subject is extremely broad and includes a large spectrum of varying techniques. Many of them can be in fact applied to similar or even the same classification and clustering problems. Luckily, the R language is supported by several cutting-edge start-ups, and open-source projects such as Apache Spark or H2O, which allow R users to implement a vast array of high-performance, distributed machine learning algorithms which can be run on highly-scalable commodity architecture through the code written in their favorite programming language.

Summary

We began this chapter by introducing you gently to the rich and abundant world of machine learning algorithms and open-source tools which facilitate their application of large datasets.

We then moved on to practical tutorials during which we presented you with three different machine learning methods run on a multi-node Microsoft Azure HDInsight cluster with Hadoop, Spark, and RStudio Server installed. In the first example you learnt how to perform a logistic regression through the Spark MLlib module using the `SparkR` package for R with HDFS as a data source.

In two further tutorials, we explored the powerful capabilities of H2O-an open-source, highly-optimized platform for Big Data machine learning models run through the `h2o` package for R. We applied the Naive Bayes algorithm to predict the classes of the outcome variable and then we compared the achieved performance and accuracy metrics with two models generated by the Neural Networks and Deep Learning techniques.

In the final chapter of this book, we will summarize the material presented throughout the previous chapters and will address the potential areas of development of the R language to include support for real-time, streaming fast data and further optimization of Big Data processing workflows.

9
The Future of R - Big, Fast, and Smart Data

Congratulations on reaching the final chapter. In the last part of this book we will review the Big Data approaches presented earlier and will discuss the future of Big Data analytics using R . Whenever possible you will be provided with links and references to online and printed resources which you may use to expand your skills further in selected topics on Big Data with R. After reading this chapter you will be able to:

- Summarize major Big Data technologies available on the market and explain how they can be integrated with the R language
- Indicate the current position of R and its distributions in the landscape of statistical tools for Big Data analytics
- Identify potential opportunities for future development of the R language and how it can become an integral part of Big Data workflows

The current state of Big Data analytics with R

This section will serve as a critical evaluation and summary of the R language's ability to process very large, out-of memory data and its connectivity with a variety of existing Big Data platforms and tools.

Out-of-memory data on a single machine

We began the book with a brief revision of the most common techniques used to analyze data with the R language (`Chapter 2`, *Introduction to R Programming Language and Statistical Environment*). We guided you from importing the data into R, through data management and processing methods, cross-tabulations, aggregations, hypothesis testing, and visualizations. We then explained major limitations of the R language in terms of its requirement of memory resources for data storage and its speed of processing. We said that the data must fit within the available RAM installed on your computer if you were to use only a single machine for data processing in the R language. However, as a system runs other processes and applications simultaneously with the R environment, realistically the size of data should not exceed 50% of RAM. In fact, this upper limit depends not only on the size of raw data, but also on the planned data processing strategy and the type of transformations and data analysis techniques the researcher is going to carry out.

There are at least two common cases, where even a small dataset of ~100 MB processed in R may cause a significant slowdown of the entire operating system on a machine with 4 GB of RAM. Firstly, some users like to store all the results of intermediary processing steps into separate objects. This approach allows them to go back to the previous stages or transformations without repeating all intermediary steps in case the last transformation fails to return the requested output. Conceptually, there may be nothing wrong with this methodology, but it is not efficient in terms of memory resources. Imagine 15-20 such transformations that may return 15-20 additional objects of a similar size, ~100 MB, stored by R in the machine's RAM. Moreover, as each transformation consumes some memory resources itself, the analysts implementing such a methodology would quickly experience RAM issues caused by this approach. The better method would be to test the R code on small samples of larger datasets and remove the redundant objects from the R environment as soon as they are not used. Secondly, even small datasets of around 100MB can be problematic if the analysis involves iterative, computationally expensive algorithms, for example Neural Networks with many hidden layers. Such processes can cause large memory overheads and may consume many times more RAM than the raw data itself.

In `Chapter 3`, *Unleashing the Power of R From Within*, we presented you with a number of R packages that allow users to process and analyze data that may not fit the available RAM resources on a single machine. A large section of the chapter was dedicated to the `ff` package, which chunks a dataset into partitions stored on a hard drive and creates a small `ffdf` object in the R environment which merely contains mapping information about these partitions.

The `ffbase` package extends the practical applications of this approach by making it possible to carry out essential data management tasks and more complex cross-tabulations or aggregations on out-of-memory data without any need for cluster computing. As the `ffbase` package works well with other R libraries which allow analytics and statistical modeling of `ffdf` objects, for example `biglm`, `biglars`, or `bigrf`-the `ff` approach may solve many issues usually encountered by analysts who process large datasets with R on a single machine. However, the trade-off of this approach is the speed of processing. Due to chunking of data, the partitions are stored on a hard drive rather than in memory, which affects the execution time of R functions.

A similar methodology of dealing with out-of-memory datasets is applied by `bigmemory`-another package that we presented in Chapter 3, *Unleashing the Power of R from Within*. It creates a copy of a data file on a hard drive as a `backingfile` and it stores mapping information and metadata in a separate `descriptorfile`. As the `descriptorfile` is very small it can be easily imported into R with only minimal consumption of RAM. Running descriptive statistics, cross-tabulations, aggregations, and more complex analytics to include Generalized Linear Models and clustering algorithms on `bigmemory`-cached data is possible through a number of third-party R packages, for example `biglm` or `bigpca` and other libraries which have been created as part of the `bigmemory` project: `biganalytics`, `bigtabulate`, `bigalgebra`, and `synchronicity`. The versatility of aggregations and statistical techniques available for objects created in `bigmemory` allows users a very wide selection of methods and analytical approaches. This flexibility and the speed of processing are the main reasons for the growing popularity of `bigmemory` and its supporting libraries among R users who attempt to analyze large, often out-of-memory, datasets on a single machine with limited RAM resources.

Faster data processing with R

The second major limitation of R is that it is slower than the C family of languages and also Python. At the beginning of Chapter 3, *Unleashing the Power of R from Within*, we gave three major reasons why R generally lags behind:

- R is an interpreted language and, even though almost 40% of its code is written in C language, it is still slower mostly due to inefficient memory management
- Base R functions are executed as single-threaded processes, that is they are evaluated line-by-line with only one CPU being active
- The R language is not formally defined, which leaves its optimization to specific implementations of R

In the same chapter, we presented a number of packages and approaches which partly alleviate commonly experienced problems with the processing speed. Firstly, the users can explicitly optimize their code by benefiting from multi-threading by the means of the `parallel` and `foreach` packages. We also briefly mentioned R's support for GPU processing, however this approach is only available to users and analysts who can access specific infrastructure that implements GPUs, for example through cloud computing solutions, like Amazon EC2, which allow users to deploy computing clusters equipped with GPUs. Then we introduced you to a new distribution of R, originally developed by Revolution R Analytics, but acquired and rebranded last year by Microsoft as Microsoft R Open. By default, it includes support for multi-threading and offers quite impressive speed of R code execution on a single machine equipped with a multi-core processor.

The speed of processing is especially important when carrying out computationally expensive statistical modeling and predictive analytics. When working on a single machine, many iterative algorithms and more complex machine learning methods, for example Neural Networks, may run for a long time until they converge and return a satisfactory output. In `Chapter 3`, *Unleashing the Power of R from Within*, we introduced you to a new parallel machine learning tool called **H2O**, which utilizes the multi-core architecture of a computer and, as it can be run in a single-node mode, it can significantly speed-up algorithm execution on a single machine. We explored this approach further in `Chapter 8`, *Machine Learning Methods for Big Data in R*, when we performed a number of classification algorithms on a relatively large multi-node cluster.

As long as the data fits within RAM resources available on your computer, the considerable increase in speed can be also achieved by using a very versatile and powerful `data.table` package. It provides users with lightning fast data import, subsetting, transformations, and aggregation methods. Its functions usually return the outputs 3-20 times faster than corresponding methods from base R. Even if your data exceeds the available RAM, `data.table` workflows can be taken to cloud-based virtual machines to lower overall costs of data processing and management activities. In the section titled *Boosting R performance with data.table package and other tools* in `Chapter 3`, *Unleashing the Power of R from Within*, we provided you with a practical tutorial explaining the most essential features and operations available through the `data.table` package.

Hadoop with R

In *Online Chapter, Pushing R Further* (https://www.packtpub.com/sites/default/files/downloads/5396_6457OS_PushingRFurther.pdf), we showed you how you can set up, configure, and deploy cheap cloud-based virtual machines that can be easily scaled up by increasing their storage, memory, and processing resources according to your specific needs and requirements. We also explained how to can install and run R and its implementation-**RStudio Server** in the cloud and extend the techniques described in the preceding sections to larger datasets and more demanding processing tasks.

In `Chapter 4`, *Hadoop and MapReduce Framework for R*, we took another step forward and scaled the resources up by combining multiple cloud-operated virtual machines into multi-node clusters, which allowed us to perform heavy data crunching using the **Apache Hadoop ecosystem** directly from the RStudio Server console. Owing to high scalability of Hadoop and its **HDFS** and **MapReduce frameworks**, R users can easily manipulate and analyze massive amounts of data by directing the workflows to Hadoop through the `rhadoop` family of packages for R, for example `rhdfs`, `rmr2`, and `plyrmr`. The size of data which can be processed using this approach is only limited by the specification of the infrastructure in place and may be flexibly increased (or decreased) to reflect current data processing demands and budget restrictions. The raw data can be imported from a variety of data sources (for example HBase database with `rhbase` package) and stored as HDFS chunks for later processing by **Mapper** and **Reducer** functions.

The **Hadoop ecosystem** has revolutionized the way we process and analyze huge amounts of information. Its connectivity with the R language and widely accessible cloud computing solutions have enabled lone data analysts and small teams of data scientists with very limited budgets and processing resources to perform memory efficient heavy-load data tasks from the comfort of their desks without any need for investing hefty sums into local server rooms or data centers. Although this approach seems very comprehensive, there are still several issues that one should consider before implementing it at production level.

One of the criticisms is the uncertainty about data security and privacy when Hadoop analytics is used in the cloud-based model. Who owns the data? Who can have access to the data and how do we ensure that the data stays secure, safe, and unchanged in the process? These are only a few examples of questions most data analysts ask themselves before deploying Hadoop-operated workflows. Secondly, Hadoop is not the fastest tool and is not optimized for iterative algorithms. Very often mapper functions will produce huge amounts of data which have to be sent across the network for sorting before the **reducers** take place. Although the application of **combiners** may reduce the amount of information transfer across the network, this approach may not work for large number of iterations.

Finally, Hadoop is a complex system and it requires a wide array of skills including the knowledge of networking, the Java language, and other technical or engineering abilities to successfully manage the flow of data between the nodes and configure the hardware of clusters. These qualities are very often beyond the traditional skill set of statistically-minded data analysts or R users and hence require co-operation with other experts in order to carry out highly-optimized data processing in Hadoop.

Spark with R

At least one of the Hadoop limitations-the speed of processing, can be partially solved by another Big Data tool, **Apache Spark**, which can be built on top of the existing Hadoop infrastructure and uses HDFS as one of its data sources. Spark is a relatively new framework optimized for fast data processing of massive datasets and it is slowly becoming the preferred Big Data platform in the industry. As explained in Chapter 7, *Faster than Hadoop: Spark with R*, Spark connects well with the R language through its SparkR package. Analysts can create Spark RDDs directly from R using a number of data sources, from individual data files in CSV or TXT format to data stored in databases or HDFS.

As the SparkR package comes pre-installed with Spark distributions, R users can quickly transfer their data processing tasks to Spark without any additional configuration stages. The package itself offers a very large number of functionalities and data manipulation techniques: descriptive statistics, recoding of variables, easy timestamp formatting and date extraction, data merging, subsetting, filtering, cross-tabulations, aggregations, support for SQL queries, and custom-made functions. As all the transformations are lazily evaluated, the intermediary outputs are rapidly executed and returned and they can also be explicitly imported as native R data frames. This flexibility and its speed of processing makes Spark an ideal platform for (near) real-time data processing and analysis. Unfortunately, as the SparkR package is a work in progress, it still doesn't support the collection and analytics of streaming data in R. According to the authors and maintainers of the SparkR package, this issue will be addressed in the foreseeable future to allow deployment of real-time data applications directly from R connected to Spark.

R with databases

One of the strongest *selling* points of R is, that unlike other statistical packages, it can import data from numerous sources and almost unlimited data formats. As the Big Data is often stored, not as separate files, but in the form of tables in RDBMSs, R can easily connect to a variety of traditional databases and perform basic data processing operations remotely on the server through SQL queries without explicitly importing large amounts of data to the R environment.

In `Chapter 5`, *R with Relational Database Management Systems (RDBMSs)*, we presented three applications of such connectivity with a **SQLite database** run locally on a single machine, a **MariaDB database** deployed on a virtual machine, and finally **a PostgreSQL database** hosted through the **Amazon Relational Database Service** (RDS)-a highly-scalable Amazon Web Services solution for relational databases. These examples provide practical evidence of suitability of SQL databases for Big Data analytics using the R language. SQL databases can be easily implemented in data processing workflows with R as great data storage containers or for the purpose of essential data cleaning and manipulations at early stages of the data product cycle. This functionality is possible due to well-maintained and widely used third-party packages such as `dplyr`, `DBI`, `RPostgres`, `RMySQL`, and `RSQLite`, which support R's connectivity with a large number of open-source SQL databases.

Furthermore, more flexible, non-relational databases which store data in the form of documents contained within collections have recently found very fertile ground among R users. As many NoSQL databases, for example MongoDB, Cassandra, CouchDB, and many others, are open-source, community projects, they have rapidly stolen the hearts and minds of R programmers. In `Chapter 6`, *R with Non-Relational (NoSQL) Databases*, we provided you with two practical examples of Big Data applications where NoSQL databases, **MongoDB** and **HBase**, were used with R to process and analyze large datasets. Most NoSQL databases are highly-scalable and designed for near real-time analytics or data stream processing. They are also extremely flexibly in terms of the variety of types of data they hold and their ability to manage and analyze unstructured data. Moreover, the R packages that support integration of specific NoSQL databases with R are generally very well-maintained and user friendly. Three such libraries that allow connectivity with the popular MongoDB database were presented in `Chapter 6`, *R with Non-Relational (NoSQL) Databases*, namely `mongolite`, `RMongo`, and `rmongodb`. Additionally, we presented the functionalities and methods available in the `rhbase` packageâ©©one of the building blocks of the `rhadoop` family of packages, which can be used for manipulations and transformations of data stored in HBase database-a component of the Hadoop ecosystem.

Machine learning with R

As we have previously explained, R may struggle with computationally expensive iterative algorithms that are performed on a single machine due to its memory limitations and because of the fact that many R functions are single-threaded. Earlier we had said, however, that one particular platform called **H2O** allows R users to benefit from multi-threading and therefore may facilitate fast statistical modeling by utilizing the full computational power of a machine. If scaled out across multiple nodes in a cluster of commodity hardware, the H2O platform can easily apply powerful predictive analytics and machine learning techniques to massive datasets in a memory-efficient manner. The benefits of **high scalability** and **distributed processing** offered by H2O can now be experienced by R users first hand through the `h2o` package, which provides a user-friendly interface between R and H2O. As all the heavy data crunching is run in an H2O cluster, R does not consume significant amounts of memory. It also allows you to make use of all available cores across the cluster and hence improve the performance of algorithms considerably. In `Chapter 8`, *Machine Learning Methods for Big Data in R*, we guided you through a practical tutorial of H2O with R, which implemented a **Naive Bayes classification algorithm** and variants of multi-layered Neural Networks to predict the values of unlabeled examples of the real-world large scale dataset.

Also, in the same chapter, we presented you with an alternative method of carrying out Big Data machine learning tasks using **Spark MLlib**, one of Spark's native libraries specialized in performing clustering, regressions, and classification algorithms through the Spark platform. As before, its integration with R was possible due to the `SparkR` package, and although the package is still a work in progress and it offers only limited selection of built-in machine learning algorithms, we were able to easily perform a Generalized Linear Model on a large dataset. It is therefore possible to run similar algorithms on much bigger, out-of-memory data without the need to import data into R.

The future of R

In the following brief sections, we are going to try to imagine how R may develop within the next several years to facilitate Big, Fast, and Smart data processing.

Big Data

We hope that by reading this book you have gained an appreciation for the R language and what can potentially be achieved by integrating it with currently available Big Data tools. As the last few years have brought us many new Big Data technologies, it has to be said that the full connectivity of R with these new frameworks may take some time. The availability of approaches utilizing R to process large datasets on a single machine is still quite limited due to traditional limitations of the R language itself. The ultimate solution to this problem may only be achieved by defining the language from scratch, but this is obviously an extreme and largely impractical idea. There is a lot of hope associated with **Microsoft R Open**, but as these are still quite early days for this new distribution, we need to wait and test its functionalities in a large variety of Big Data scenarios before we can assess its usability in large scale analytics. It is however very easy to predict that various R distributions will soon be optimized to support multi-threading by default and allow users to run more demanding computations in parallel by utilizing all available cores without requiring analysts to explicitly adjust the code.

In terms of the memory requirements, the key is to efficiently reduce the reliance of the R language on RAM resources by optimizing methods of garbage collection and, if possible, to transfer some of the computations to a hard drive. Although this may slow down the processing, hopefully the implementation of multi-threading during the majority of base R functions can compensate for any potential trade-offs caused by the partial engagement of hard drives in order to execute the code. It may also be worth applying lazy evaluation to most of the data management and transformation functions, this would considerably speed up the execution of the code and will only import the data into R when users are satisfied with the final output. As there are already packages that support this approach, for example `ff`, `ffbase`, and `bigmemory` as explained before, the main task for the R community is to further explore opportunities offered by these packages and develop them to include a larger variety of scalable data transformation functions and statistical, as well as modeling, algorithms.

In the large and still growing landscape of many Big Data tools, it is vital to invest our energy into providing as many diverse channels for integrating R as possible. However, it is also important to focus on the quality of the packages that allow this connectivity. Currently, what happens very often very useful packages are too complex for the average R user or data analyst, either because of a lack of informative documentation explaining functionalities in detail, scarce examples and practical applications of a package, or obsolete and user-unfriendly syntax that implements functions that require users to know other programming languages which are uncommon for more traditionally educated data scientists and researchers. This is a very real problem that should be addressed promptly in order to ensure that R users interested in Big Data approaches can find and apply appropriate methods in an informed and well-documented fashion. Otherwise, the the R language and its specific applications for Big Data analytics will remain a domain dominated by a few experts or academics and will never reach every interested user who might benefit from these solutions, either in the industry or at the production level.

We therefore hope that the next several years will encourage Big Data leaders to integrate their products with the R language, which should result in comprehensive R packages with a variety of built-in, highly optimized functions for data manipulation and analysis, and well-maintained documentation which will explore and present their usage in easy-to-grasp and accessible language.

Fast data

Fast data analytics are the backbone of many data applications that consume and process information in real-time. Particularly in these times of the Internet of Things, fast data processing can often determine the future of a product or a service and may directly translate to its market success or its failure. Although the R language can process streaming or (near) real-time data, such examples are very rare and are largely dependent on a variety of factors: employed architecture, infrastructure, and telecommunication solutions in place, the amount of data collected by a specific unit of time, and the complexity of analysis and data processing required to produce a requested output. The topic of streaming or real-time data processing in R involves so many separate components and is so new that it would require another publication to describe all the processes and operations in detail. This is also one of the areas where R will probably develop within the next few years.

The current state of fast data analytics using the R language allows users to process small amounts of data imported either as individual files of different formats or scrapped of online sources through dynamically updated REST APIs. Some third-party packages, for example `twitteR`, enable R users to mine the contents of well-established web-based APIs in real-time, however very often their usability is limited by restrictions imposed by the data owners and their application servers. Even if the large data mining of real-time resources was allowed, the problem would then be the latency in data processing through R. One of the ways to alleviate the consequences of this issue is to use either one of the NoSQL fast databases optimized for real-time data crunching, for example MongoDB, and/or employ Spark functionalities to power analytics of streaming information. Unfortunately, Spark currently lacks the integration of stream analytics in the `SparkR` package, but according to the Spark developers this functionality will soon be available and R users will be able to consume and process Spark RDDs in real-time.

Smart data

Smart data encapsulates the predictive or even prescriptive power of statistical methods and machine learning techniques available to data analysts and researchers. Currently, R is positioned as one of the leading tools on the market in terms of the variety of algorithms and statistical models it contains. Its recent integration with Big Data machine learning platforms like H2O and Spark MLlib, as well as its connectivity with the Microsoft Azure ML service, puts the R language at the very forefront in the ecosystem of tools designed for Big Data predictive analytics. In particular, R's interface with H2O offered by the `h2o` package already provides a very powerful engine for distributed and highly-scalable classification, clustering, and Neural Networks algorithms that perform extremely well with a minimum configuration required from users. Most of the built-in `h2o` functions are fast, well-optimized, and produce satisfactory results without setting any additional parameters. It is very likely that H2O will soon implement a greater diversity of available algorithms and will provide further extension of functions and methods that may allow users to manipulate and transform the data within the H2O cluster, without the need for data pre-processing in other tools, databases, or large virtual machines.

Within the next several years we may expect many new machine learning start-ups to be created which will aim at strong connectivity with R and other open-source analytical and Big Data tools. This is an exciting area of research and hopefully the coming years will shape and strengthen the position of the R language in this field.

Where to go next

After reading this book and going through all its tutorials you should have enough skills to let you perform scalable and distributed analysis of very large datasets using the R language. The usefulness of the material contained in this book hugely depends on other tools your current Big Data processing stack includes. Although we have presented you with a wide array of applications and frameworks which are common ingredients of Big Data workflows, for example Hadoop, Spark, SQL, and NoSQL databases, we appreciate that your personal needs and business requirements may vary.

In order to address your particular data-related problems and accomplish Big Data tasks, which may include a myriad of data analytics platforms, other programming languages, and various statistical methods or machine learning algorithms, you may need to develop a specific skill set and make sure to constantly grow your expertise in this dynamically evolving field. Throughout this book, we have included a large number of additional online or printed resources which may help you in filling in any gaps in the Big Data skills you may have had and will keep you motivated to explore other avenues for your personal development as a data analyst and R user. Make sure to re-visit any chapters of interest for reference to external sources of additional knowledge, but, most importantly, remember that success comes with practice, so don't wait any more, fire up your preferred R distribution and get your hands dirty with real-world Big Data problems.

Summary

In the last chapter of this book, we have summarized the current position of the R language in the diverse landscape of Big Data tools and frameworks. We have also identified the potential opportunities of the R language to evolve into a leading Big Data statistical environment by tackling some of the most frequently encountered limitations and barriers. Finally, we have explored and elaborated on the requirements which R language will most likely meet within the next several years to provide even greater support for user-friendly, Big, Fast, and Smart data analytics.

Index

0
0xdata 437

A
Access Research Knowledge
  reference 49
activation function 459
ad hoc tests 58
Adjusted Correlation Coefficient (radj) 64
Akaike Information Criterion (AIC) 91
Amazon EC2
  MongoDB, installing on 322
Amazon RDS database instance
  launching 281, 282, 284, 285, 286, 287, 288
Amazon RDS
  about 245
  data, uploading to 290, 291
Amazon Relational Database Service (RDS) 477
Ambari
  about 136
  exploring 213, 214
Analysis of Variance (ANOVA)
  about 25, 59
  example 60, 61, 62
Apache Hadoop 130
Apache Hadoop ecosystem 475
Apache Nutch 12
Apache Spark 476
applied data science, with R 47
apply() example
  using on data.frame 109
  with big.matrix object 108
Area Under Curve (AUC) 453
arrays 37, 38
Artificial Neural Networks (ANN) 458
Attribute-Relation File Format (ARFF) 89
axon hillock 459
axon terminals 459
Azure
  Hortonworks Sandbox, deploying on 138, 139, 140, 141, 142, 143, 144, 145, 146, 147, 148, 149, 150, 151, 152, 153, 154, 155, 156, 157

B
Bay Area Bike Share
  about 383
  reference 383
Bernoulli distribution 464
Big Data analytics, with R
  current state 471
Big Data
  about 8, 10, 11
  processing, MongoDB used 325
Big Files 12
bigmemory functions 107
bigmemory package
  memory, expanding with 97, 98, 99, 100, 101, 102
bigpca package 105
binomial 88
binomial logistic regression 425
Bonferonni's test 62
bootstrapping 113
BSON format 317, 319

C
Cassandra Query Language (CQL) 318
Cassandra
  about 136, 318
  reference 318

chaining 123
classification algorithms 416
cloglog function 88
Cloudera 14
clustering methods 416
CodeSchool
  reference 27
collections 319
Combiner function 132
combiners 475
Comma-Separated Values (CSV) 10
Comprehensive R Archive Network (CRAN)
  reference 19, 26
contingency tables 53, 56
CouchDB
  about 318
  reference 318
Coursera
  reference 27
covariance 63
CRAN High-Performance Computing Task View
  reference 116
CULA tools
  reference 116

D

data aggregations
  about 53, 56
  with ff package 76, 77, 78, 79, 80, 81, 82, 83, 84, 85, 86, 87
  with ffbase package 76, 77, 78, 79, 80, 81, 82, 83, 84, 85, 86, 87
data frames 38, 40
data manipulations 56
data transformations
  with ff package 76, 77, 78, 79, 80, 81, 82, 83, 84, 85, 86, 87
  with ffbase package 76, 77, 78, 79, 80, 81, 82, 83, 84, 85, 86, 87
data visualizations packages
  about 70
  ggplot2 70
  ggvis 70
  googleVis 71
  htmlwidgets 71

lattice 71
plotly 71
rCharts 71
shiny 70
data.frame
  apply(), using on 109
  for() loop, using on 109
data.table package
  aggregations 120
  chaining 123
  complex aggregations 123
  lightning-fast subsets 120
  R performance, boosting with 118
  used, for fast data import and manipulation 118, 119, 120
data
  getting, into HDFS 385, 386
  importing, from different formats 48, 49, 50
  importing, from HDFS to Hive 386, 387, 388, 389, 391
  importing, to HBase 363, 364
  importing, to HDFS 363, 364
  preparing 290
  uploading, to Amazon RDS 290
databases 15, 16
DataCamp
  reference 27
DataNodes 130
db.t2.micro instance 285
Deep Learning methods 460
Deep Learning models
  running, on H2O 461, 462, 463, 464, 466
dendrites 459
dependent events 446
dependent variable 88
descriptive statistics
  obtaining, of data 51, 52
Distributed File System 13
distributed processing 478
documents 319

E

EC2 instance
  preparing, for use 255
edX

reference 27
Elastic Cloud Computing (EC2) 14
embedded documents 321
Energy Demand Reasearch Project 194
Exploratory Data Analysis (EDA) 28, 50

F

F statistic 61
Fast Fourier Transform (FFT) algorithm 50
fastICA algorithm 116
ff package 76
ffbase package 76
Fisher iterations 96
Fisher scoring iterations 95
for() loop
  using, on data.frame 109
  with ffdf object 108
foreach package
  example 113
FORK cluster 111
Fortran algorithms 17

G

Gaussian distribution 88
Generalized Linear Model (GLM)
  about 419
  with ff package 87, 88, 89
  with ffbase package 87, 88, 89
ggplot2 70
ggvis 70
Gini metric 453
GLM tutorials
  reference 88
Google File System model 13
googleVis 71
gputools package 115
Gradient Boosted Regressions 446
Graphics Processing Units (GPUs)
  utilizing, with R 115
GraphX component 374
Gray Sort Benchmark
  reference 14

H

H2O builds
  reference 438
H2O instance
  running, on Hadoop 437, 438, 439, 440
H2O platform 478
  about 118, 413
  data, reading 442, 444, 445
  Deep Learning models, running on 461, 462, 463, 464, 466
  reference 418
H2O.ai 118, 437
Hadoop 130
Hadoop 2.7.2 release
  reference 130
Hadoop commands
  reference 168
Hadoop Distributed File System (HDFS)
  about 16, 129, 130
  characteristics 130, 131
Hadoop ecosystem 475
Hadoop native tools 134
Hadoop, with R 475, 476
Hadoop
  about 12, 14, 103, 106
  H2O instance, running with R 437, 438, 439, 440
  learning 136, 137
  reference 137
  word count example 169
harmonic mean 59
HBase 477
HBase database 16
HBase
  about 135, 317
  reference 317
  with R 355
HDFS 475
HDFS commands
  reference 168
HDInsight cluster
  configuring 203, 204, 205, 206, 207, 208, 209, 210
  connecting to 215, 217

creating 194, 195
Network Security Group, creating 200, 201, 202
Resource Group, creating 195, 196
setting up 203, 204, 205, 206, 207, 208, 209, 210
smart energy meter readings analysis 229, 230, 231, 232, 233
starting 211, 212
Virtual Network, deploying 197, 198, 199
HDInsight
  about 184, 194
  launching, with R/RStudio 375
  launching, with Spark 375
Hive 135, 375
HiveQL 386
horizontally scalable 9
Hortonworks Sandbox
  deploying, on Azure 138, 139, 140, 141, 142, 143, 144, 145, 146, 147, 148, 149, 150, 151, 152, 153, 154, 155, 156, 157
Hortonworks
  about 14
  reference 135
htmlwidgets
  about 71
  reference 71
htop tool 158
hypothesis testing 56

I

identity link function 88
independent t-test example 57, 59

J

Java Runtime Environment (JRE) 152
Java
  word count example, in Hadoop 159, 160, 161, 162, 163, 164, 165, 166, 167, 168
Julia 24

K

k-means 416
k-Means 446

k-means clustering 103
Kaggle
  reference 453
KDnuggets
  reference 22

L

Land Registry Price Paid Data
  reference 327
Laplace estimator 451
lattice 71
Likert scale 57
linear predictor 425
link function 88, 425
lists 41, 42
Logarithmic Loss (LogLoss) 453
logistic regression 70, 425
logistic regression example
  with biglm 89, 90, 91, 92, 93, 94
  with ffbase 89, 90, 91, 92, 93, 94
logistic regression, in Spark
  with R 425, 426, 427, 428, 429, 430, 432
logistic regression
  about 89
logit function 88

M

machine learning 414
machine learning algorithms
  about 413
  classification 416
  clustering 416
  semi-supervised 416
  supervised 415
  unsupervised 416
Machine Learning Depository 89
machine learning methods, R 417
Machine Learning Toolbox for Matlab
  reference 419
machine learning, with R 478
Mahout 136
Mapper function 131
mapper functions 475
MapReduce 103, 106
MapReduce application

initializing 236
MapReduce commands
  reference 168
MapReduce framework 12, 131
MapReduce frameworks 475
MapReduce Job Tracker 135
MapReduce model
  Mapping procedure 13
  Reduce stage 13
  Shuffle stage 13
MapReduce word count example 132, 133, 134
MapReduce, in R
  word count example 180, 181, 182, 183, 184, 185, 186, 187, 189, 190, 192, 193
MariaDB database 477
MariaDB
  preparing, for use 257, 258, 260, 261, 262, 264, 265, 266
  reference 245
  working with 266, 267, 269, 270, 271, 272, 273
Math Kernel Library (KML) 117
matrices 35, 37
maximum-likelihood estimations 113
Mean Squared Error (MSE) 452
memory
  expanding, with bigmemory package 97, 98, 99, 100, 101, 102
Microsoft R Open 479
Microsoft R Open (MRO) 117
MLlib 17, 374
MongoDB 477
MongoDB data models 319
MongoDB
  about 317, 319
  data, importing into 326, 328, 329, 330, 331
  installing, on Amazon EC2 322, 323, 324, 325
  reference 317
  used, for processing Big Data 325, 326
mongolite package
  using 350, 354
MOT test
  reference 290
Multi-Layer Perceptrons (MLP) 460
multi-layered Neural Network 413

multi-layered Neural Networks 446
multi-threading, with Microsoft R Open distribution 117
multinomial 88
multinomial logistic regression 425
multiple R objects, versus R memory limitations 50
multiple regression
  example 65, 67, 68, 70
mutexes 104

N

Naive Bayes classification algorithm 413, 478
NameNode 130
NEED dataset
  reference 278
Neo4j
  about 318
  reference 318
network topology 459
Neural Networks 436
  about 416
  working 459, 460
neuron 459
neurotransmitters 459
non-continuous distribution 88
non-relational databases
  Cassandra 318
  CouchDB 318
  HBase 317
  MongoDB 317
  Neo4j 318
  reviewing 316
normalized data models 320
Northern Ireland Life and Times Survey (NILT) 49
NoSQL databases 315
  about 316
Not Only SQL (NoSQL) 15
NVIDIA CUDA Zone 117

O

Open Data Institute
  reference 388
OpenCL 116
operational Linux virtual machine

configuring 159
operational Microsoft Azure HDInsight cluster
  launching 355

P

p-value 58
package, data manipulations
  lubridate 56
packages, data manipulations
  stringi 56
  stringr 56
parallel machine learning, with R 118
parallel package
  example 110, 111, 112, 113
  reference 111
parallel processing, in R
  future 115
parallelism, in R 106
Pearson's r correlation coefficient 64
Pearson's r correlation
  example 64, 65
Pearson's r correlations
  example 63
Perl
  about 46
  reference 46
plotly
  about 71
  reference 71
Poisson regression 70
post-synaptic receptors 459
posterior probability 446
PostgreSQL 283
PostgreSQL database 477
  reference 245
PostgreSQL
  querying, remotely 304, 305, 306, 307, 312
predictor variables 88
Principal Components Analysis 416
Principle Components Analysis (PCA) 104
prior probability 446
probit function 88
Public Use File (PUF)
  download link 248
Python 24

R

R code
  writing 126
R community
  reference 26
R core 19
R data objects
  exporting 42, 43, 44, 45, 46, 47
R data structure
  lists 41
R data structures
  about 32
  arrays 37, 38
  data frames 38, 40
  lists 42
  matrices 35, 37
  scalars 35
  vectors 32, 33, 34, 35
R packages
  references 26
R performance
  boosting, with data.table package 118
R Project CRAN
  reference 19
R repositories
  URLs, setting to 30
R squared (R2) 453
R, future
  about 478
  Big Data 479
  fast data 480, 481
  smart data 481
R-bloggers
  reference 26
R/RStudio
  HDInsight, launching with 375
R
  about 17, 18, 19, 21
  applied data science 47
  Graphics Processing Units, utilizing with 115
  growth 22
  learning 26
  reference, for online sources 26
  traditional limitations 74

working 23
Random Access Memory (RAM) 9
Random Forests 446
RazorSQL
  about 289, 290
  reference 290
  setting up 291, 292, 294, 295
rCharts
  about 71
  reference 71
Reducer function 131
reducers 475
references 320
Relational Database Management System
    (RDBMS) 15, 135
Relational Database Management Systems
    (RDBMSs) 243
Relational Database Service (RDS) 280
RepoForge 157
Resource Manager User Interface 135
Revolution Analytics
  reference 363
Revolution R Open (RRO) 117
RHadoop packages
  configuring 177
  installing 177
rhbase package
  using 367, 368, 372
RMongo package
  using 346, 349
rmongodb package
  using 333
RStudio IDE
  about 19
  reference 19, 28
RStudio Server 475
  installing 215
  installing, on Linux RedHat/CentOS virtual
    machine 169, 170, 171, 172, 173, 174,
    175, 176
  new inbound security rule, adding for port 8787
    218, 219, 220
  preparing, for use 255
  virtual network's IP address, modifying for head
    node 221

RStudio
  about 20
  reference, for blog 26
  requisites, gathering 28, 29

S

S language 17, 18
S language, versus R language
  reference 18
S-Plus language 18
Sandbox
  about 138
  reference 138
scalars 35
Scheme programming language 18
semi-supervised learning methods 416
shiny
  about 70
  reference 70
simple regression 65
single-node Hadoop, in Cloud 137, 138
Singular Value Decompositions (SVD) 104
SOCK cluster 111
Spark 16
Spark cluster
  preparing 419, 420, 421, 422
Spark MLlib 478
  about 418
  reference 418
Spark SQL 17, 374
Spark Streaming 17, 374
Spark, with R 476
Spark
  about 136, 374
  benefits 374
  HDInsight, launching with 375
  on multi-node HDInsight cluster 375
Sparkling Water 418
SparkR
  using, for Bay Area Bike Share analysis 393,
    394, 395, 396, 397, 398, 399, 401, 402,
    406, 407, 408
split-apply-combine operation 103
SQLite database 477
  data, importing into 248, 249, 250

SQLite, with R 247
SQLite
  connecting to 250, 253, 254, 255
  reference 244, 248
StackExchange
  reference 27
StackOverflow
  reference 27
standard Eta Squared effect size 62
Stata 10
statistical inference 56
statistical learning online course
  reference 418
statistical learning video lectures
  reference 417
Statistical Package for the Social Sciences (SPSS) 10
stepwise regression 70
stimulus intensity 459
Storm 136
Structured Query Language (SQL) 15, 243, 245, 246
supervised learning methods 415
synapse 459

T

t statistic 64
t-statistic 58
t.test function 57
t2.micro RDS instance 281
Task Views 26
third-party packages, Spark
  reference 17
threshold 459
TIOBE Programming Community Index
  reference 22
traditional limitations, R
  about 74
  out-of-memory data 74
  processing speed 75
Trellis graphics 71

U

unsupervised learning methods 416
URLs
  setting, to R repositories 30
useR! conferences 19

V

variance function 425
Vector Machines 416
vectors 32, 33, 34, 35
vertically scalable 9

W

Weka
  reference 419

Y

Yet Another Resource Negotiator (YARN) 135

Z

Zookeper 136